Lecture Notes in Computer Science

Edited by G. Goos, J. Hartmanis, and J. van Leeuwen

Springer
Berlin
Heidelberg
New York
Barcelona
Hong Kong
London
Milan
Paris
Tokyo

Serge Vaudenay   Amr M. Youssef (Eds.)

Selected Areas in Cryptography

8th Annual International Workshop, SAC 2001
Toronto, Ontario, Canada, August 16-17, 2001
Revised Papers

Springer

Series Editors

Gerhard Goos, Karlsruhe University, Germany
Juris Hartmanis, Cornell University, NY, USA
Jan van Leeuwen, Utrecht University, The Netherlands

Volume Editors

Serge Vaudenay
EPFL, LASEC
1015 Lausanne, Switzerland
E-mail: serge.vaudenay@epfl.ch

Amr M. Youssef
University of Waterloo, CACR
Waterloo, Ontario N2L 3G1, Canada
E-mail: a2youssef@cacr.math.uwaterloo.ca

Cataloging-in-Publication Data applied for

Die Deutsche Bibliothek - CIP-Einheitsaufnahme

Selected areas in cryptography : 8th annual international workshop ; revised
papers / SAC 2001, Toronto, Ontario, Canada, August 16 - 17, 2001. Serge
Vaudenay ; Amr M. Youssef (ed.). - Berlin ; Heidelberg ; New York ;
Barcelona ; Hong Kong ; London ; Milan ; Paris ; Tokyo : Springer, 2001
    (Lecture notes in computer science ; Vol. 2259)
    ISBN 3-540-43066-0

CR Subject Classification (1998): E.3, D.4.6, K.6.5, F.2.1-2, C.2, H.4.3

ISSN 0302-9743
ISBN 3-540-43066-0 Springer-Verlag Berlin Heidelberg New York

Springer-Verlag Berlin Heidelberg New York
a member of BertelsmannSpringer Science+Business Media GmbH

http://www.springer.de

© Springer-Verlag Berlin Heidelberg 2001
Printed in Germany

Typesetting: Camera-ready by author, data conversion by Steingräber Satztechnik GmbH, Heidelberg
Printed on acid-free paper   SPIN: 10846068   06/3142   5 4 3 2 1 0

Preface

SAC 2001, the eighth annual workshop on selected areas in cryptography, was held at the Fields Institute in Toronto, Ontario, Canada. Previous SAC workshops were held at Queen's University in Kingston (1994, 1996, 1998, and 1999), at Carlton University in Ottawa (1995 and 1997) and at the University of Waterloo (2000). The conference was sponsored by the center for applied cryptographic research (CACR) at the University of Waterloo, Certicom Corporation, Communications and Information Technology Ontario (CITO), Ecole Polytechnique Fédérale de Lausanne, Entrust Technologies, and ZeroKnowledge. We are grateful to these organizations for their support of the conference.

The current SAC board includes Carlisle Adams, Doug Stinson, Ed Dawson, Henk Meijer, Howard Heys, Michael Wiener, Serge Vaudenay, Stafford Tavares, and Tom Cusick. We would like to thank all of them for giving us the mandate to organize SAC 2001.

The themes for SAC 2001 workshop were:

- Design and analysis of symmetric key cryptosystems.
- Primitives for private key cryptography, including block and stream ciphers, hash functions, and MACs.
- Efficient implementations of cryptographic systems in public and private key cryptography.
- Cryptographic solutions for web and internet security.

There were 57 technical papers submitted to the conference from an international authorship. Every paper was refereed by at least 3 reviewers and 25 papers were accepted for presentation at the conference. We would like to thank the authors of all the submitted papers, both those whose work is included in these proceedings, and those whose work could not be accommodated.

In addition to these 25 papers, two invited presentations were given at the conference: one by Moti Yung from CertCo, USA, entitled "Polynomial Reconstruction Based Cryptography " and the other by Phong Nguyen from the Ecole Normale Supérieure, France, entitled "The two faces of lattices in cryptology". Thanks to both Moti and Phong for their excellent talks and for kindly accepting our invitation.

The program committee for SAC 2001 consisted of the following members: Stefan Brands, Matt Franklin, Henri Gilbert, Howard Heys, Hideki Imai, Shiho Moriai, Kaisa Nyberg, Rich Schroeppel, Doug Stinson, Stafford Tavares, Serge Vaudenay, Michael Wiener, Amr Youssef , and Yuliang Zheng.

On behalf of the program committee we would like to thank the following sub-referees for their help in the reviewing process: Joonsang Baek, Guang Gong, Ian Goldberg, Darrel Hankerson, Keiichi Iwamura, Mike Just, Masayuki Kanda, Liam Keliher, Mira Kim, Kazukuni Kobara, Frédéric Légaré, Henk Meijer, Alfred John Menezes, Miodrag Mihaljevic, Ulf Möller, Dalit Naor, Daisuke Nojiri, Mohammad Ghulam Rahman, Palash Sarkar, Akashi Satoh, Junji Shikata, Takeshi Shimoyama, Ron Steinfeld, Anton Stiglic, Edlyn Teske, Yodai Watanabe, Huapeng Wu, Daichi Yamane, and Robert Zuccherato.

We would like to thank all the people involved in organizing the conference. In particular we would like to thank Pascal Junod for his effort in making the reviewing process run smoothly. Special thanks are due to Frances Hannigan for her help in the local arrangements and for making sure that everything ran smoothly during the workshop. Finally we would like to thank all the participants of SAC 2001.

August 2001   Serge Vaudenay and Amr Youssef

Organization

Program Committee

S. Brands	Zero Knowledge Systems (Canada)
M. Franklin	UC Davis (USA)
H. Gilbert	France Telecom (France)
H. Heys	Memorial University of Newfoundland (Canada)
H. Imai	University of Tokyo (Japan)
S. Moriai	NTT (Japan)
K. Nyberg	Nokia Research Center (Finland)
R. Schroeppel	Sandia National Lab (USA)
D. Stinson	University of Waterloo (Canada)
S. Tavares	Queen's University (Canada)
S. Vaudenay (co-chair)	EPFL (Switzerland)
M. Wiener	Entrust Technologies (Canada)
A. Youssef (co-chair)	University of Waterloo (Canada)
Y. Zheng	Monash University (Australia)

Local Organizing Committee

S. Vaudenay	EPFL (Switzerland)
A. Youssef	University of Waterloo (Canada)
F. Hannigan	University of Waterloo (Canada)
P. Junod	EPFL (Switzerland)

Sponsoring Institutions

EPFL
University of Waterloo
Entrust Technologies
Certicom
Zero Knowledge Systems
CITO

Table of Contents

Invited Talk I

Elliptic Curves and Efficient Implementation I

Cryptanalysis II

Elliptic Curves and Efficient Implementation II

Public Key Systems

Invited Talk II

Protocols and Mac

Author Index

Weaknesses in the Key Scheduling Algorithm of RC4

Scott Fluhrer[1], Itsik Mantin[2], and Adi Shamir[2]

[1] Cisco Systems, Inc.,
170 West Tasman Drive, San Jose, CA 95134, USA
sfluhrer@cisco.com
[2] Computer Science department, The Weizmann Institute,
Rehovot 76100, Israel
{itsik,shamir}@wisdom.weizmann.ac.il

Abstract. In this paper we present several weaknesses in the key scheduling algorithm of RC4, and describe their cryptanalytic significance. We identify a large number of weak keys, in which knowledge of a small number of key bits suffices to determine many state and output bits with non-negligible probability. We use these weak keys to construct new distinguishers for RC4, and to mount related key attacks with practical complexities. Finally, we show that RC4 is completely insecure in a common mode of operation which is used in the widely deployed Wired Equivalent Privacy protocol (WEP, which is part of the 802.11 standard), in which a fixed secret key is concatenated with known IV modifiers in order to encrypt different messages. Our new passive ciphertext-only attack on this mode can recover an arbitrarily long key in a negligible amount of time which grows only linearly with its size, both for 24 and 128 bit IV modifiers.

1 Introduction

RC4 is the most widely used stream cipher in software applications. It was designed by Ron Rivest in 1987 and kept as a trade secret until it leaked out in 1994. RC4 has a secret internal state which is a permutation of all the $N = 2^n$ possible n bits words, along with two indices in it. In practical applications $n = 8$, and thus RC4 has a huge state of $log_2(2^8! \times (2^8)^2) \approx 1700$ bits.

In this paper we analyze the Key Scheduling Algorithm (KSA) which derives the initial state from a variable size key, and describe two significant weaknesses of this process. The first weakness is the existence of large classes of weak keys, in which a small part of the secret key determines a large number of bits of the initial permutation (KSA output). In addition, the Pseudo Random Generation Algorithm (PRGA) translates these patterns in the initial permutation into patterns in the prefix of the output stream, and thus RC4 has the undesirable property that for these weak keys its initial outputs are disproportionally affected by a small number of key bits. These weak keys have length which is

S. Vaudenay and A. Youssef (Eds.): SAC 2001, LNCS 2259, pp. 1–24, 2001.
© Springer-Verlag Berlin Heidelberg 2001

divisible by some non-trivial power of two, i.e., $\ell = 2^q m$ for some $q > 0$[1]. When $RC4_n$ uses such a weak key of ℓ words, fixing $n + q(\ell - 1) + 1$ bits of K (as a particular pattern) determines $\Theta(qN)$ bits of the initial permutation with probability of one half and determines various prefixes of the output stream with various probabilities (depending on their length).

The second weakness is a related key vulnerability, which applies when part of the key presented to the KSA is exposed to the attacker. It consists of the observation that when the same secret part of the key is used with numerous different exposed values, an attacker can rederive the secret part by analyzing the initial word of the keystreams with relatively little work. This concatenation of a long term secret part with an attacker visible part is a commonly used mode of RC4, and in particular it is used in the WEP (Wired Equivalent Privacy) protocol, which protects many wireless networks. Our new attack on this mode is practical for any key size and for any modifier size, including the 24 bit recommended in the original WEP, and the 128 bit recommended in the revised version WEP2.

The paper is organized in the following way: In Section 2 we describe RC4 and previous results about its security. In Section 3 we consider a slightly modified variant of the Key Scheduling Algorithm, called KSA*, and prove that a particular pattern of a small number of key bits suffices to completely determine a large number of state bits. Afterwards, we show that this weakness of KSA*, which we denote as the *invariance weakness*, exists (in a weaker form) also in the original KSA. In Section 4 we show that with high probability, the patterns of initial states associated with these weak keys also propagate into the first few outputs, and thus a small number of weak key bits determine a large number of bits in the output stream. In Section 5 we describe several cryptanalytic applications of the invariance weakness, including a new type of distinguisher. In Sections 6 and 7 we describe the second weakness, which we denote as the *IV weakness*, and show that a common method of using RC4 is vulnerable to a practical attack due to this weakness. In Section 8, we show how both these weaknesses can separately be used in a related key attack. In the appendices, we examine how the IV weakness can be used to attack a real system (appendix A), how the invariance weakness can be used to construct a ciphertext-only distinguisher and to prove that RC4 has low sampling resistance (appendices B and C), and how to derive the secret key from an early permutation state (appendix D).

2   RC4 and Its Security

2.1   Description of RC4

RC4 consists of two parts (described in Figure 1): A key scheduling algorithm KSA which turns a random key (whose typical size is 40-256 bits) into an initial

---

[1]   Here and in the rest of the paper ℓ is the number of words of K, where each word contains n bits.

KSA(K)	PRGA(K)
Initialization:	Initialization:
For $i = 0 \ldots N - 1$	$i = 0$
$S[i] = i$	$j = 0$
$j = 0$	Generation loop:
Scrambling:	$i = i + 1$
For $i = 0 \ldots N - 1$	$j = j + S[i]$
$j = j + S[i] + K[i \bmod \ell]$	Swap$(S[i], S[j])$
Swap$(S[i], S[j])$	Output $z = S[S[i] + S[j]]$

Fig. 1. The Key Scheduling Algorithm and the Pseudo-Random Generation Algorithm

permutation S of $\{0, \ldots, N - 1\}$, and an output generation part PRGA which uses this permutation to generate a pseudo-random output sequence.

The PRGA initializes two indices i and j to 0, and then loops over four simple operations which increment i as a counter, increment j pseudo randomly, exchange the two values of S pointed to by i and j, and output the value of S pointed to by $S[i] + S[j]$[2]. Note that every entry of S is swapped at least once (possibly with itself) within any N consecutive rounds, and thus the permutation S evolves fairly rapidly during the output generation process.

The KSA consists of N loops that are similar to the PRGA round operation. It initializes S to be the identity permutation and i and j to 0, and applies the PRGA round operation N times, stepping i across S, and updating j by adding $S[i]$ and the next word of the key (in cyclic order).

2.2   Previous Attacks on RC4

Due to the huge effective key of RC4, attacking the PRGA seems to be infeasible (the best known attack on this part requires time that exceeds 2^{700}). The only practical results related to the PRGA deal with the construction of distinguishers. Fluhrer and McGrew described in [FM00] how to distinguish RC4 outputs from random strings with 2^{30} data. A better distinguisher which requires 2^8 data was described by Mantin and Shamir in [MS01]. However, this distinguisher could only be used to mount a partial attack on RC4 in broadcast applications.

The fact that the initialization of RC4 is very simple stimulated considerable research on this mechanism of RC4. In particular, Roos discovered in [Roo95] a class of weak keys that reduces their effective size by five bits, and Grosul and Wallach showed in [GW00] that for large keys whose size is close to N words, RC4 is vulnerable to a related key attack.

---

[2]   Here and in the rest of the paper all the additions are carried out modulo N

More analysis of the security of RC4 can be found in [KMP+98], [Gol97] and [MT98].

3   The Invariance Weakness

Due to space limitations we prove here the invariance weakness only for a simplified variant of the KSA, which we denote as KSA* and describe in Figure 2. The only difference between them is that KSA* updates i at the *beginning* of the loop, whereas KSA updates i at the *end* of the loop. After formulating and proving the existence of this weakness in KSA*, we describe the modifications required to apply this analysis to the real KSA.

3.1   Definitions

We start the round numbering from 0, which means that both KSA and KSA* have rounds $0, \ldots, N-1$. We denote the indices swapped in round r by i_r and j_r, and the permutation S after swapping these indices is denoted as S_r. Notice that by using this notation, $i_r = r$ in the real KSA. However, in KSA* this notation becomes somewhat confusing, when $i_r = r + 1$. For the sake of completeness, we can say that $j_{-1} = 0$, S_{-1} is the identity permutation and $i_{-1} = \begin{cases} -1 & KSA \\ 0 & KSA^* \end{cases}$.

Definition 1. *Let S be a permutation of $\{0, \ldots, N-1\}$, t be an index in S and b be some integer. Then if $S[t] \stackrel{\mathrm{mod}\ b}{\equiv} t$, the permutation S is said to b-conserve the index t. Otherwise, the permutation S is said to b-unconserve the index t.*

Definition 2. *A permutation S of $\{0, \ldots, N-1\}$ is b-conserving if $I_b(S) = N$, and is almost b-conserving if $I_b(S) \geq N - 2$.*

KSA(K)[a]	KSA*(K)
For $i = 0 \ldots N - 1$	For $i = 0 \ldots N - 1$
$\quad S[i] = i$	$\quad S[i] = i$
$i = 0$	$i = 0$
$j = 0$	$j = 0$
Repeat N times	Repeat N times
$\quad j = j + S[i] + K[i \bmod \ell]$	$\quad i = i + 1$
$\quad Swap(S[i], S[j])$	$\quad j = j + S[i] + K[i \bmod \ell]$
$\quad i = i + 1$	$\quad Swap(S[i], S[j])$

[a] KSA is rewritten in a way which clarifies its relation to KSA*

Fig. 2. KSA vs. KSA*

We denote the number of indices that a permutation b-conserves as $I_b(S)$. To simplify the notation, we often write I_r instead of $I_b(S_r)$.

Definition 3. *Let b, ℓ be integers, and let K be an ℓ word key. Then K is called a b-exact key if for any index r, $K[r \bmod \ell] \equiv (1-r) \pmod{b}$. In case $K[0] = 1$ and $MSB(K[1]) = 1$, K is called a* special b-exact *key.*

Notice that for this condition to hold, it is necessary (but not sufficient) that $b \mid \ell$.

3.2   The Weakness

Theorem 1. *Let $q \leq n$ and ℓ be integers and $b \stackrel{def}{=} 2^q$. Suppose that $b \mid \ell$ and let K be a b-exact key of ℓ words. Then the permutation $S = KSA^*(K)$ is b-conserving.*

Before getting to the proof itself, we will prove an auxiliary lemma

Lemma 1. *If $i_{r+1} \equiv j_{r+1} \pmod{b}$, then $I_{r+1} = I_r$.*

Proof. The only operation that might affect S (and maybe I) is the swapping operation. However, when i and j are equivalent ($\bmod\ b$) in round $r + 1$, S_{r+1} b-conserves position i_{r+1} (j_{r+1}) if and only if S_r b-conserved position j_r (i_r). Thus the number of indices S b-conserves remains the same.

Proof. (of Theorem 1) We will prove by induction on r that for any $-1 \leq r \leq N-1$, it turns out that $i_r \equiv j_r \pmod{b}$ and $I_b(S_r) = N$ and . This in particular implies that $I_{N-1} = N$, which makes the output permutation b-conserving.

For $r = -1$ (before the first round), the claim is trivial because $i_{-1} = j_{-1} = 0$ and S_{-1} is the identity permutation which is b-conserving for every b. Suppose that $j_r \equiv i_r$ and S_r is b-conserving. Then $i_{r+1} = i_r + 1$ and

$$j_{r+1} = j_r + S_r[i_{r+1}] + K[i_{r+1} \bmod \ell] \stackrel{\bmod\ b}{\equiv} i_r + i_{r+1} + (1 - i_{r+1}) = i_r + 1 = i_{r+1}$$

Thus, $i_{r+1} \equiv j_{r+1} \pmod{b}$ and by applying Lemma 1 we get $I_{r+1} = I_r = N$ and therefore S_{r+1} is b-conserving.

KSA* thus transforms special patterns in the key into corresponding patterns in the initial permutation. The fraction of determined permutation bits is proportional to the fraction of fixed key bits. For example, applying this result to $RC4_{n=8, \ell=6}$ and $q = 1$, 6 out of the 48 key bits completely determine 252 out of the 1684 permutation bits (this is the number of bits encapsulated in the LSBs).

3.3   Adjustments to KSA

The small difference between KSA* and KSA (see Figure 2) is essential in that KSA, applied to a b-exact key, does not preserve the equivalence ($\bmod\ b$) of i and j even after the first round. Analyzing its execution on a b-exact key gives

$$j_0 = j_{-1} + S_{-1}[i_0] + K[i_0] = 0 + S_{-1}[0] + K[0] = K[0] \stackrel{\bmod\ b}{\equiv} 1 \stackrel{\bmod\ b}{\not\equiv} 0 = i_0$$

and thus the structure described in Section 3.2 cannot be preserved by the cyclic use of the key words. However, it is possible to adjust the invariance weakness to the real KSA, and the proper modifications are formulated in the following theorem:

Theorem 2. *Let $q \leq n$ and ℓ be integers and $b \overset{def}{=} 2^q$. Suppose that $b \mid \ell$ and let K be a special b-exact key of ℓ words. Then*

$$Pr[KSA(K) \text{ is almost } b\text{-conserving}] \geq 2/5$$

where the probability is over the rest of the key bits.

Due to space limitations, the formal proof of this theorem (which is based on a detailed case analysis) will appear only in the full version of this paper. However, we can explain the intuition behind this theorem by concentrating on the differences between Theorems 1 and 2, which deal with KSA* and KSA respectively. During the first round, two deviations from KSA* execution occur. The first one is the non-equivalence of i and j which is expected to cause non-equivalent entries to be swapped during the next rounds, thus ruining the delicate structure that was preserved so well during KSA* execution. The second deviation is that S b-unconserves two of the indices, $i_0 = 0$ and $j_0 = K[0]$. However, we can cancel the ij discrepancy by forcing $K[0]$ (and j_0) to 1. In this case, the discrepancy in $S[j_0]$ ($S[1]$) causes an improper value to be added to j in round 1, thus repairing its non-equivalence to i during this round. At this point there are still two unconserved indices, and this aberration is dragged across the whole execution into the resulting permutation. Although these corrupted entries might interfere with j updates, the pseudo-random j might reach them *before* they are used to update j (i.e., before i reaches them), and send them into a region in S where they cannot affect the next values of j^3. The probability of this lucky event is amplified by the fact that the corrupted entries are $i_0 = 0$ which is not touched (by i) until the termination of the KSA due to its distance from the current location of i, and $j_1 = 1 + K[1] > N/2$ (recall that $MSB(K[1]) = 1$), that is far the position of i ($i_1 = 1$), which gives j many opportunities to reach it before i does. The probability of $N/2$ pseudo random j's to reach an arbitrary value can be bounded from below by 2/5, and extensive experimentation indicates that this probability is actually close to one half.

4   Key-Output Correlation

In this section we will analyze the propagation of the weak key patterns into the generated outputs. First we prove Claim 4 which deals with the highly biased behavior of a significantly weakened variant of the PRGA (where the swaps are avoided), applied to a b-conserving permutation. Next, we will argue that the

---

[3] if a value is pointed to by j before the swap, it will not be used as $S[i]$ (before the swap) for at least $N - 1$ rounds, and in particular it will not affect the values of j during these rounds.

prefix of the output of the original PRGA is highly correlated to the prefix of this swapless variant, when applied to the same initial permutation. These facts imply the existence of biases in the PRGA distribution for these weak keys.

Claim. Let RC4* be a weakened variant of RC4 with no swap operations. Let $q \leq n$, $b \overset{def}{=} 2^q$ and S_0^4 be a b-conserving permutation. Let $\{X_r\}_{r=1}^{\infty}$ be the output sequence generated by applying RC4* to S_0, and $x_r \overset{def}{=} X_r \bmod b$. Then the sequence $\{x_r\}_{r=1}^{\infty}$ is independent of the rest of the key bits.

Since there are no swap operations, the permutation does not change and remains b-conserving throughout the generation process. Notice that all the values of S are known $\bmod\ b$, as well as the initial indices $i = j = 0 \equiv 0 \pmod{b}$, and thus the round operation (and the output values) can be simulated $\bmod\ b$, independently of S. Consequently the output sequence $\bmod b$ can be predicted, and deeper analysis implies that it is periodic with period $2b$, as exemplified in Figure 3 for $q = 1$.

i	j	$S[i]$	$S[j]$	$S[i] + S[j]$	Out
0	0	0	0	0	/
1	1	1	1	0	0
0	1	0	1	1	1
1	0	1	0	1	1
0	0	0	0	0	0
1	1	1	1	0	0
⋮	⋮	⋮	⋮	⋮	⋮

Fig. 3. The rounds of RC4*, applied to a 2-conserving permutation

1st word | 1 | $\cdots$ | 1 | 1 | 1 |

2nd word | n | $\cdots$ | 3 | 2 | 1 |

3th word | n | $\cdots$ | 3 | 2 | 1 |

⋮

ℓ^{th} word | n | $\cdots$ | 3 | 2 | 1 |

Fig. 4. The stage in which each one of the bits is exposed during the related key attack

Recall that at each round of the PRGA, S changes in at most two locations, and thus we can expect the prefix of the output stream generated by RC4 from some permutation S_0, to be highly correlated with the stream generated from the same S_0 (or a slightly modified one) by RC4*. In particular the stream generated by RC4 from an almost b-conserving permutation is expected to be highly correlated with the (predictable) substream $\{x_r\}$ from Claim 4. This correlation is demonstrated in Figure 8, where the function $h \longrightarrow Pr[1 \leq \forall r \leq h\ Z_r \equiv x_r \bmod 2^q]$ (for special 2^q-exact keys) is empirically estimated for $n = 8$, $\ell = 16$ and different q's. For example, a special 2-exact key completely determines 20 output bits (the LSBs of the first 20 outputs) with probability $2^{-4.2}$ instead of 2^{-20}, and a special 16-exact key completely determines 40 output bits (4 LSBs from each of the first 10 outputs) with probability $2^{-2.3}$, instead of 2^{-40}.

---

[4] The term S_0 is used here for the common purpose of indicating the initial permutation of the PRGA.

We have thus demonstrated a strong probabilistic correlation between some bits of the secret key and some bits of the output stream for a large class of weak keys. In the next section we describe how to use this correlation to cryptanalyze RC4.

5   Cryptanalytic Applications of the Invariance Weakness

5.1   Distinguishing RC4 Streams from Randomness

In [MS01] Mantin and Shamir described a significant statistical bias in the second output word of RC4. They used this bias to construct an efficient algorithm which distinguishes between RC4 outputs and truly random sequences by analyzing only one word from $O(N)$ different outputs streams. This is an extremely efficient distinguisher, but it can be easily avoided by discarding the first two words from each output stream. If these two words are discarded, the best known distinguisher requires about 2^{30} output words (see [FM00]). Our new observation yields a significantly better distinguisher for most of the typical key sizes. The new distinguisher is based on the fact that for a significant fraction of keys, a significant number of initial output words contain an easily recognizable pattern. This bias is flattened when the keys are chosen from a uniform distribution, but it does not completely disappear and can be used to construct an efficient distinguisher even when the first two words of each output sequence are discarded.

Notice that the probability of a special 2^q-exact key to be transformed into a 2^q-conserving permutation, does not depend of the key length ℓ (see Theorem 2). However, the number of predetermined bits is linear in ℓ, and consequently the size of this bias (and thus the number of required outputs) also depends on ℓ. In Figure 5 we specify the quantity of data (or actually the number of different streams) required for a reliable distinguisher, for different key sizes. In particular, for 64 bit keys the new distinguisher requires only 2^{21} data instead of the previously best number of 2^{30} output words.

It is important to notice that the specified output patterns extend over several dozen output words, and thus the quality of the distinguisher is almost unaffected by discarding the first few words. For example, discarding the first two words causes the data required for the distinguisher to grow by a factor of between $2^{0.5}$ and 2^2 (depending on ℓ). Another important observation is that the biases in the LSBs distribution can be combined in a natural way with the biased distribution of the LSBs of English texts into an efficient distinguisher of RC4 streams from randomness in a ciphertext-only attack in which the attacker does not know the actual English plaintext which was encrypted by RC4. This type of distinguishers is discussed in Appendix B.

5.2   RC4 Has Low Sampling Resistance

Biryukov, Shamir and Wagner defined in [BSW00] a new security measure of stream ciphers, which they denoted as their *Sampling Resistance*. The strong

ℓ	q	b	$k_1{}^a$	$k_2{}^b$	p^c	$P_{RND}{}^d$	$P_{RC4}{}^e$	Data
4	1	2	12	15	2^{-3}	2^{-15}	$2 \cdot 2^{-15}$	2^{15}
6	1	2	14	18	2^{-4}	2^{-18}	$2 \cdot 2^{-18}$	2^{18}
8	1	2	16	21	2^{-5}	2^{-21}	$2 \cdot 2^{-21}$	2^{21}
10	1	2	18	24	2^{-6}	2^{-24}	$2 \cdot 2^{-24}$	2^{24}
12	1	2	20	27	2^{-7}	2^{-27}	$2 \cdot 2^{-27}$	2^{27}
14	1	2	22	30	2^{-8}	2^{-30}	$2 \cdot 2^{-30}$	2^{30}
16	1	2	24	34	2^{-10}	2^{-34}	$2 \cdot 2^{-34}$	2^{34}

[a] number of predetermined bits ($q(\ell - 1) + n + 1$)
[b] number of determined output bits
[c] probability of these k_1 key bits to determine these k_2 output bits (taken from Figure 8)
[d] $= 2^{-k_2}$
[e] $\approx P_{RND} + 2^{-k_1} p$

Fig. 5. Data required for a reliable distinguisher, for different key sizes

correlation between classes of RC4 keys and corresponding output patterns can be used to prove that RC4 has relatively low sampling resistance, which improves the efficiency of time/memory/data tradeoff attacks. Further details can be found in Appendix C.

6 RC4 Key Setup and the First Word Output

In this section, we consider related key attacks where the attacker has access to the values of all the bits of certain words of the key. In particular, we consider the case where the key presented to the KSA is made up of a *secret key* concatenated with an attacker visible value (which we will refer to as an Initialization Vector or *IV*). We will show that if the same secret key is used with numerous different initialization vectors, and the attacker can obtain the first word of RC4 output corresponding to each initialization vector, he can reconstruct the secret key with minimal effort. How often he can do this, the amount of effort and the number of initialization vectors required depends on the order of the concatenation, the size of the IV, and sometimes on the value of the secret key. This observation is especially interesting, as this mode of operation is used by several deployed encryption systems ([Rei01], [LMSon]) and the first word of plaintexts is often an easily guessed constant such as the date, the sender's identity, etc, and thus the attack is practical even in a ciphertext-only mode of attack. However, the weakness does not extend to the Secure Socket Layer (SSL) protocol that browsers use, as SSL uses a cryptographic hash function to combine the secret key with the IV.

In terms of keystream output, this attack is interested only in the first word of output from any given secret key and IV. Hence, we can simplify our model of the output. The first output word depends only on three specific permutation elements, as shown in the figure below showing the state of the permutation immediately after KSA. When those three words are as shown, the value labeled Z will be output as the first word.

	1			X			$X + D$	
	X			D			Z	

In addition, we will define the *resolved* condition as any time within the KSA where i is greater than or equal to 1, X and Y, where X is defined as $S_i[1]$ and Y is defined as $X + S_i[X]$ (that is, $X + D$). When this resolved condition occurs, with probability greater than $e^{-3} \approx 0.05$, none of the elements $S[1]$, $S[X]$, $S[Y]$ will participate in any further swaps[5]. In that case, the value will be determined by the values of $S_i[1]$, $S_i[X]$ and $S_i[Y]$[6]. With probability less than $1 - e^{-3} \approx 0.95$, at least one of the three values will participate in a swap, which will destroy the resolved condition and set that element to an effectively random value. This will make the output value effectively random. Our attack involves examining messages with specific IV values such that, at some point, the KSA is in a resolved condition, and where the value of $S[Y]$ gives us information on the secret key. When we observe sufficiently many IV values, the actual value of $S[Y]$ occurs detectably often.

7 Details of the Known IV Attack

Whenever we discuss a concatenation of an IV and a secret key, we denote the secret key as SK, the size of the IV by I, and the size of SK as $\ell - I$. The variable K still represents the RC4 key, which in this case is the concatenation of these two (e.g. in section 7.1 $K[1 \ldots \ell] = IV[0] \ldots IV[I-1]SK[0] \ldots SK[\ell - 1 - I]$). The numbering of the rounds, as well as the terms i_r, j_r and S_r are as defined in section 3.1.

7.1 IV Precedes the Secret Key

First consider the case where the IV is prepended to the secret key. In this circumstance, assuming we have a known I word IV, and a secret key ($SK[0] \ldots SK[\ell - 1 - I]$), we attempt to derive information on a particular word B of the secret key ($SK[B]$ or $K[I+B]$) by searching for IV values such that after round I (that is after $I + 1$ rounds), $S_I[1] < I$ and $S_I[1] + S_I[S_I[1]] = I + B$. Then, with high likelihood (probability $\approx e^{-\frac{2B}{N}}$ if we model the intermediate swaps as random),

---

[5] In our case we assume that $c \approx 1$ (since i is small), that the remaining swaps in the key setup touch words with random j's, and that the three events are independent.

[6] And, in particular, if 1, X, Y are mutually distinct, then $S_i[Y]$ will be output as the first word.

we will be in a resolved condition after round $I + B$, and so the most probable output value will be $S_{I+B}[I + B]$. We further note that, at round $I + B$, the following assignments will take place:

$$j_{I+B} = j_{I+B-1} + K[B] + S_{I+B-1}[I + B]$$

$$S_{I+B}[I + B] = S_{I+B-1}[j_{I+B}]$$

Using algebra, we see that if we know the value of j_{I+B-1} and S_{I+B-1}, then given the first output word (which we will designate Out), we can make the probabilistic assumption that $Out = S_{I+B}[I + B]$, and then predict the value based on the assumption:

$$K[B] = S_{I+B-1}^{-1}[Out] - j_{I+B-1} - S_{I+B-1}[I + B]$$

where $S_r^{-1}[V]$ denotes the location within the permutation S_r where the value V appears. Since $Out = S_{I+B}[I + B]$ more than 5% of the time, this prediction is accurate that often, and effectively random less than 95% of the time. By collecting sufficiently many values from different IVs, we can reconstruct $K[B]$.

In the simplest scenario (3 word chosen IVs), the attack works as follows[7]: suppose that we know the first A words of the secret key ($K[3], \ldots, K[A + 2]$, with $A = 0$ initially), and we want to know the next word $K[A+3]$. We examine a series of IVs of the form $(A+3, N-1, V)$ for approximately 60 different values for V. At the first round, j is advanced by $A + 3$, and then $S[i]$ and $S[j]$ are swapped, resulting in the key setup state which is shown schematically below, where the top array is the combined IV and secret key presented to the KSA, and the bottom array is a portion of the permutation, and where the positions of the i, j variables are indicated.

$A + 3$	$N - 1$	V	$K[3]$		$K[A + 3]$	
0	1	2			$A + 3$	
$A + 3$	1	2			0	
i_0					j_0	

Then, on the next round, i is advanced, and then the advance on j is computed, which happens to be 0. Then, $S[i]$ and $S[j]$ are swapped, resulting in the below structure:

$A + 3$	$N - 1$	V	$K[3]$		$K[A + 3]$	
0	1	2			$A + 3$	
$A + 3$	0	2			1	
	i_1				j_1	

Then, on the next round, j is advanced by $V + 2$, which implies that each distinct IV assigns a different value to j, and thus beyond this point, each IV

---

[7] This scenario was first published by Wagner in [Wag95].

acts differently, approximating the randomness assumption made above. Since the attacker knows the value of V and $K[3], \ldots K[A + 2]$, he can compute the exact behavior of the key setup until before round $A+3$. At this point, he knows the value of j_{A+2} and the exact values of the permutation S_{A+2}. If the value at $S_{A+2}[0]$ or $S_{A+2}[1]$ has been disturbed, the attacker discards this IV. Otherwise, j is advanced by $S_{A+2}[i_{A+3}] + K[A + 3]$, and then the swap is done, resulting in the below structure:

$A+3$	$N-1$	V	$K[3]$		$K[A+3]$	
0	1	2			$A+3$	
$A+3$	0	$S[2]$			$S_{A+3}[A+3]$	
					i_{A+3}	

The attacker knows the permutation S_{A+2} and the value of j_{A+2}. In addition, if he knows the value of $S_{A+3}[A + 3]$, he knows its location in S_{A+2}, which is the value of j_{A+3}, and hence he would be able to compute $K[A + 3]$. We also note that i_{A+3} has now swept past 1, $S_{A+3}[1]$ and $S_{A+3}[1] + S_{A+3}[S_{A+3}[1]]$, and thus the resolved condition exists, and hence with probability $p > 0.05$, by examining the value of the first word of RC4 output with this IV, the attacker will be able to compute the correct value of $K[A + 3]$. Hence, by examining approximately 60 IVs with the above configuration, the attacker can rederive $K[A + 3]$ with a probability of success greater than 0.5.

By iterating the above process across the secret key, the attacker can rederive ℓ words of secret key using 60ℓ chosen 3 word IVs.

The next thing to note is that the attack works for IVs other than those in the specific $(A + 3, N − 1, V)$ form. Any I word IV that, after I rounds, leaves $S_I[1] < I$ and $S_I[1] + S_I[S_I[1]] = I + B$ will suffice for the above attack. In addition, since the attacker is able to simulate the first I rounds of the key setup, he is able to determine which IVs have this property. By examining all IVs that have this property, we can extend this into a known IV attack, without using an excessive number of IVs[8]. The probabilities to find the next word, and the expected number of IVs needed to obtain 60 IVs of the proper form, are given in Figure 6.

7.2   IV Follows the Secret Key

In the case that the IV is appended to the secret key, we need to take a different approach. The previous analysis attacked individual key words. When the IV follows the secret key, what we do instead is select IVs that give us the state of

---

[8] Note that different IVs that lead to the same intermediate values of j, are not properly modeled by our random swap model. It is possible that specific values of j will suggest specific incorrect keyword values, independently of the actual IV words. One way to overcome this difficulty, is to take only IVs which induce distinct values of j. An alternative approach is to try all the high probability key words in parallel, instead of concentrating only on the most probable one.

IV Length	Probability	Expected IVs required
3	4.57×10^{-5}	1310000
4	4.50×10^{-5}	1330000
5	1.65×10^{-4}	364000
6	1.64×10^{-4}	366000
7	2.81×10^{-4}	213000
8	2.80×10^{-4}	214000
9	3.96×10^{-4}	152000
10	3.94×10^{-4}	152000
11	5.08×10^{-4}	118000
12	5.04×10^{-4}	119000
13	6.16×10^{-4}	97500
14	6.12×10^{-4}	98100
15	7.21×10^{-4}	83200
16	7.18×10^{-4}	83600

Fig. 6. For various prepended IV and known secret key prefix lengths, the probability that a random IV will give us information on the next secret key word, and the expected number of IVs required to derive the next secret key word.

the permutation at an early phase of the key setup, such as immediately after all the words of the secret key have been used for the first time. Given that only a few swaps have occurred up to that point, it is reasonably straight-forward to reconstruct those swaps from the permutation state, and hence obtain the secret key (see Appendix D for one such method).

To illustrate the attack in the simplest case, suppose we have an A word secret key, and a 2 word IV. Further suppose that the secret key was weak in the sense that, immediately after A rounds of KSA, $S_{A-1}[1] = X$, $X < A$, and $X + S_{A-1}[X] = A$. This is a low probability event ($p \approx 0.00062$ if $A = 13$)[9]. For such a weak secret key, the attacker can assume the value of $j_{A-1} + S_{A-1}[A]$, and then examine IVs with a first word of $W = V - (j_{A-1} + S_{A-1}[A])$ (this assumption does increase the amount of work by a factor of N, and forces us to verify the assumption, which we can do by observing a consistent predicted value of S_{A-1}). With such IVs, the value of j_A will be the preselected value V. Then, $S[A]$ and $S[V]$ are swapped, and so $S_A[A] = S_{A-1}[V]$. Here, assuming V was neither 1 nor $S_{A-1}[1]$, then the resolved condition has been established, and with probability > 0.05, $S_{A-1}[V]$ will be the first word output. Then, by

---

[9] A straightforward assumption that the permutation S_{A-1} is equidistributed gives a much lower probability $13/256 \times 1/256 \approx 0.00020$, however, S_{A-1} is not equidistributed; the first A bytes are biased towards small values.

examining such IVs with the second word being at least 60 different values, we can observe the output a number of times and derive the value of $S_{A-1}[V]$ with good probability. By selecting all possible values of V, we can directly observe the state of the S_{A-1} permutation, from which we can rederive the secret key. We will denote this result as *key recovery*.

If $X + S_{A-1}[X] = A + 1$, a similar analysis would appear to apply. By assuming $S_{A-1}[A]$, $S_{A-1}[A+1]$ and j_{A-1}, we can swap $S_{A-1}[V]$ into $S_{A+1}[A+1]$ for $N-2$ distinct IVs for any particular V. However, the value of j_{A+1} is always the same for any particular V, and so the probabilities that a particular IV outputs the value $S[V]$ are not independently distributed. This effect causes the reading of the permutation state to be 'noisy', that is, for some values of V, we see $S[V]$ as the first word far more often than our analysis expected, and for other values of V, we see it far less often. Because of this, some of the entries $S_{A-1}[V]$ cannot be reliably recovered. Simulations assuming a 13 word secret key and $n = 8$ have shown that an average of 171 words of the S_{A-1} permutation state can be successfully reconstructed, including an average of 8 words of $(S_{A-1}[0], \ldots, S_{A-1}[12])$, which immediately give you effectively 8 key words. With this information, the key is reduced enough that it can be brute forced. We will denote this result as *key reduction*.

If we have a 3 word IV, then there are more types of weak secret keys. For example, consider a secret key where $S_{A-1}[1] = 1$ and $S_{A-1}[A] = A$. Then, by assuming j_{A-1}, we can examine IV where the first word has a value W so that the new value of j_A is 1, and so $S_{A-1}[1]$ and $S_{A-1}[A]$ are swapped, leaving the state after round A to be:

$SK[0]$	$SK[1]$		$SK[A-1]$	W	$IV[1]$	$IV[2]$
0	1		$A-1$	A	$A+1$	$A+2$
$S_{A-1}[0]$	A		$S_{A-1}[A-1]$	1	$S_{A-1}[A+1]$	$S_{A-1}[A+2]$
	j_A			i_A		

Then, by assuming $S_{A-1}[A+1]$ (which with high probability is $A+1$, and will always be at most $A+1$), we can examine IVs with the second word $IV[1] = V - (1 + S_{A-1}[A+1])$, for an arbitrary V, which will cause $j_{a+1} = V$ and swap the value of $S_{A-1}[V]$ into $S_{A+1}[A+1]$. Assuming V isn't either 1 or A, then the resolved condition have been set up, and using a number of values for the third IV word Z, we can deduce the value of $S_{A-1}[V]$ for an arbitrary V, giving us the permutation after A rounds.

There are a number of other types of weak keys that the attacker can take advantage of, summarized in Figure 7.

The last weak secret key listed in Figure 7 is especially interesting, in that the technique that exposes the weakness is rather different than that of the other weak secret keys listed. Immediately after A rounds, the state is:

Condition	IV Settings			Probability	Result
	First	Second	Third		
$S_{A-1}[1]=1$ $S_{A-1}[A]=A$	Swap with 1	Swap with Y	Cycle	0.0037	Key recovery
$S_{A-1}[1]=2$ $S_{A-1}[A+1]=A+1$	Swap with 1	Cycle	Swap with Y	0.0070	Key reduction
$S_{A-1}[1]=X<A$ $S_{A-1}[X]+X=A$	Swap with Y	Cycle	Cycle	0.0007	Key recovery
$S_{A-1}[1]=X<A$ $S_{A-1}[X]+X=A+1$	Cycle	Swap with Y	Cycle	0.0009	Key recovery
$S_{A-1}[1]=X<A$ $S_{A-1}[X]+X=A+2$	Cycle	Cycle	Swap with Y	0.0007	Key reduction
$S_{A-1}[1]=A$	Swap with $S_{A-1}^{-1}[1]$	Swap with Y	Cycle	0.0037	Key recovery
$S_{A-1}[1]=A+1$	Swap with Y	Swap with $S_{A-1}^{-1}[N-1]$	Cycle	0.0036	Key recovery
$S_{A-1}[1]=A+2$	Cycle	Swap with Y	Swap with $S_{A-1}^{-1}[N-1]$	0.0038	Key reduction
$S_{A-1}[1]=N-2$ $S_{A-1}[A+2]=A+2$	Swap with Y	Cycle	Swap with 1	0.0034	Key reduction
$S_{A-1}[1]=N-1$ $S_{A-1}[A+1]=A+1$	Swap with Y	Swap with 1	Cycle	0.0036	Key recovery
$S_{A-1}[1]=X<A$ $S_{A-1}[A]=Z$ $X+Z>A+2$	Swap with X	Cycle	Cycle	0.1007	Key reduction

Fig. 7. Weak secret keys with 3 word postfix IVs. Listed are the conditions on the S_{A-1} permutation that distinguish them, the IV properties that the attacker searches for to reveal $S[Y]$, the probability that this class of weak key will occur with $n = 8$ and a 16 word secret key, and the result of the attack on the weak key.

$SK[0]$	$SK[1]$		$SK[V]$		W	Z	
0	1		V		A	$A+1$	
$S_{A-1}[0]$	V		$S_{A-1}[V]$		Z	$S_{A-1}[A+1]$	

The initial IV word causes $S_{A-1}[V]$ and $S_{A-1}[A]$ to be swapped, leaving the state as:

$SK[0]$	$SK[1]$		$SK[V]$		W	Z	
0	1		V		A	$A+1$	
$S_{A-1}[0]$	V		Z		$S_{A-1}[V]$	$S_{A-1}[A+1]$	
			j_A		i_A		

Now, to inquire about the value of $S_{V+Z}[W+Const]$, we examine numerous IVs with second and third words that all set the value of j_{A+2} to be W. The

KSA will continue for $V + Z - (A + 2)$ more rounds until i now points to the element $S_{V+Z}[V + Z]$. At this point, since we haven't gone through a great number of rounds since we knew the value of j (since $V + Z - (A+2) \leq A - 4$), then with high probability, $j_{V+Z+1} = W + Const$, where $Const$ is a constant term that depends only on the state of the permutation S_A. If this is true, then $S_{V+Z+1}[V + Z] = S_{V+Z}[W + const]$, and if the elements $S[1]$ and $S[V]$ have not been disturbed (again, this happens with high probability), the resolved condition has been achieved, and the first output word will be biased towards $S_{V+Z}[W + const]$. In addition, because the value of $const$ will be the same independent of W, its value can easily be determined, thus allowing the attacker to observe many of the values of S_{V+Z}. This class of weak keys requires far more known IVs to exploit, but also occurs relatively frequently.

If we have a 4 word[10] IV, then the same general approach as the previous analysis can be used to recover virtually all secret keys, given sufficient IVs. First, we assume j_{A-1}, $S_{A-1}[A]$, $S_{A-1}[A+1]$, $S_{A-1}[A+2]$, $S_{A-1}[A+3]$ [11]. Then, based on this assumption, we search for IVs that, after round $A + 3$, sets $S_{A+3}[1] = V$ and $S_{A+3}[V] = Z$ for $V, Z < A + 4, V + Z \geq A + 4$, and we note the value of $j_{A+3} = W$. Then, we save the value of $V + Z$, the value W and the value output as the first word for that particular IV. With nontrivial probability, the value of this word will be $S_{V+Z}[W + const_{V+Z}]$, where $const_{V+Z}$ is a constant term that depends on the secret key, and the value $V + Z$. Since that value is independent of the IV, we can collect numerous possible values of $S_{V+Z}[W + const_{V+Z}]$ for various values of $V + Z$, and use that to first reconstruct $const_{V+Z}$, and then reconstruct S_{V+Z}.

8   Related-Key Attacks on RC4

In this section, we discuss two related-key attacks based on weaknesses discussed previously in this paper. They work within the following model: the attacker is given a black box that has a randomly chosen RC4 key K inside it, an output button and an input tape of $|K|$ words. In each step the attacker can either press the output button to get the next output word, or write Δ on the tape, which causes the black-box to restart the output generation process with a new key defined as $K' = K \oplus \Delta$. The purpose of the attacker is to find the key K (or some information about it).

8.1   Related-Key Attack Based on the Invariance Weakness

This attack works when the number of key words, is a power of two. It consists of n stages where in stage q the qth bit of every key word is exposed[12]. The predicate *CheckKey* takes as input an RC4 blackbox and a parameter q (the stage number) and decides whether the key in the box is special 2^q-exact. This purpose can be

---

[10]   This approach generalizes in the obvious way to longer IVs.

[11]   Note that $S_{A-1}[x] \leq x$ for $x \geq A$. This limits the size of the search required.

[12]   In fact, $K[1]$ is fully revealed during the first stage (see Figure 4)

achieved by randomly sampling key bits that are irrelevant for the 2^q-exactness of the key and estimating the expected length of q-patterned output. For a special 2^q-exact key the expected length will be significantly longer than in a random output (where it is less than 2) and thus *CheckKey* works in time $O(1)$. The procedure *Expand* takes as input an RC4 blackbox and a parameter q (the stage number), assumes that the key in the box is special 2^{q-1}-exact, and makes it special 2^q-exact. The method for doing so is by enumerating all the possibilities for the qth bits ($2^{\ell-1}$ such possibilities) and invoking CheckKey to decide when the key in the box is special 2^q-exact. *Expand* works in a slightly different way for $q = 1$ and $q = n$. For $q = 1$, except for the LSBs, it determines the complete $K[0]$ (by forcing it to 1) and $MSB(K[1])$. For $q = n$, there is only one 2^n-exact key and consequently we can calculate the output produced from this key and replace *CheckKey* by simple comparison. The time complexity of this stage is $O(2^{n+\ell})$ for $q = 1$ and $O(2^{\ell-1})$ for any other q.

The total time required for the attack is thus $O(2^{n+\ell}) + (n - 1)O(2^{\ell}) = O(2^{n+\ell})$. For typical RC4$_{n=8}$ key with 32 bytes, the complexity of exhaustive search is completely impractical (2^{256}), whereas the complexity of the new attack is only $O(2^{n+\ell}) = O(2^{40})$.

8.2   Related-Key Attack Based on the Known IV Weakness

In this section we use the known IV weaknesses to develop an efficient related key attack on RC4.

The attack consists of 3 stages, where in the first two stages we gain information on the first three words of the secret key, and in the third stage we iterate down the key, and expose each word of the key successively. The stages of the attack are as follows:

Step 1. This step attempts to find values of $K[0]$, $K[1]$ such that $S_1[1] = 1$, and reveal the value of $K[2]$. The procedure is to select random values of (X, Y), and for each such random value, write onto the tape 240 vectors with the initial four words (X, Y, Z, W) for $Z \in \{0, N/4, N/2, 3N/4\}$ and with 60 distinct random values of W, and for each such vector, press the output button. If X and Y are such that $S_1[1] = 1$ (for the modified key), then the output of the first word will be biased towards $3 + (K[2] \oplus Z)$, unless that value happens to be 1. Hence, for at least 3 of the selected values of Z, the first word outputs will be biased towards one of *const*, *const* $+ N/4$, *const* $+ N/2$, *const* $+ 3N/4$. This is detectable, and also by examining the value of *const*, the attacker can reconstruct the value of $K[2]$. We expect to try N random values of (X, Y) before finding a pair that is appropriate.

Step 2. This step attempts to find the values of $K[0]$, $K[1]$. The procedure is to write on the tape 60 vectors with the initial four words (X, Y, Z, W), where X, Y are the values recovered in the previous step, $Z = (N - 3) \oplus K[2]$, and with 60 distinct random values of W, and for each such vector, press the output button. This particular initial sequence assures that $S_2[1] = 1$ and $S_2[2] = S_1[0] = K[0]$, and hence the output will be biased towards $K[0]$. Once that has been recovered, $K[1]$ can be computed.

Step 3. This step iteratively recovers individual words of the key. It operates by running a subprocedure that assumes that we have already recovered $(K[0], \ldots, K[A-1])$, and want to learn the value of $K[A]$. The procedure is to write 60 vectors that have the property that, given the known values of $(K[0], \ldots, K[A-1])$, that $S_{A-1}[1] = X < A$ and $X + S_{A-1}[X] = A$. With 60 such vectors, we can use the procedure shown in 7.1 to rederive $K[A]$.

The total time required for the attack is thus (because $2^n \geq \ell$):

$$Step1 + Step2 + (\ell - 3) * Step3 = O(2^{n+8}) + 2^6 + (\ell - 3)2^6 = O(2^{n+8})$$

For a RC4 key with $n = 8$ the time complexity is $O(2^{16})$ and is essentially independent of the key length.

8.3   Comparing the Attacks

Both attacks are able to completely reconstruct the randomly chosen RC4 key[13] with a number of chosen keys and amount of work that is significantly below that of brute force (except for extremely short RC4 keys). The first attack scales upwards as the key grows longer, while the time complexity of the second attack is independent of key length, with a cross-over point at $\ell = 8$.

However, due to the second word weakness, future implementations of RC4 are likely to discard some prefix of the output stream, and in this case the second attack becomes difficult to apply – output word x depends on $2x+1$ permutation elements immediately after KSA, and all the $2x + 1$ elements must occur before r for the resolved condition to hold. On the other hand, the first attack extends well, in that the probability of the output words being patterned drops modestly as the number of discarded words increases.

9   Discussion

Section 3 describes an interesting weakness of RC4 which results from the simplicity of its key scheduling algorithm. We recommend to neutralize this weakness by discarding the first N words of each generated stream. After N rounds, every element of S is swapped at least once and the permutation S and the index j are expected to be "independent" of the initialization process.

Section 6 describes a weakness of RC4 in a common mode of operation in which attacker visible IV's are concatenated with a fixed secret key. It is easy to extend the attack to other simple types of combination operators (e.g., when we XOR the IV and the fixed key) with essentially the same complexity. We recommend to neutralize this weakness by avoiding this mode of operation, or by using a secure hash to form the key presented to the KSA from the IV and secret key.

---

[13] the first attack works only for some key lengths.

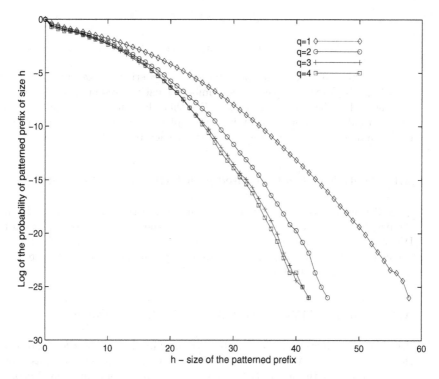

Fig. 8. This graph demonstrates the probabilities of special keys (2^q-exact with $K[0] = 1$, $MSB(K[1] = 1)$) of $RC4_{n=8, \ell=16}$ to produce streams with long patterned prefixes

A Applying the Attack to WEP

The Wired Equivalent Privacy (WEP) protocol is designed to provide privacy to packet based wireless networks based on the 802.11 standard (see [LMSon]). It encrypts by taking a secret key and a per-packet 3 byte IV, and using the IV followed by the secret key as the RC4 key. Then, it transmits the IV, and the RC4 encrypted payload. By using the results from Section 7.1, we can show how, by examining enough ciphertext packets, to reconstruct the secret key for WEP.

We assume that the attacker is able to retrieve the first byte of the RC4 output from each packet[14]. By the analysis done in section 7.1, to recover key byte B, the attacker needs to know the previous key bytes, and then search for IVs that sets up the permutation such that

---

[14] Because of the payload format used with 802.11, the first byte of each plaintext payload is a known constant, and hence the attacker is able to derive the first byte of RC4 output.

$$V = S_{B+3}[1] < B + 3 \qquad (1)$$

$$V + S_{B+3}[V] = B + 3$$

With about 60 such IVs, the attacker can rederive the key byte with reasonable probability of success. The number of packets required to obtain that number of IVs depends on the exact IVs that the sender uses. Although the 802.11 standard does not specify how an implementation should generate these IVs, common practice is to use a counter to generate them.

A.1   Analysis of IVs Generated by a Little Endian Counter

If the IVs are generated by a multibyte counter in little endian order (and hence the first byte of the IV increments the fastest), then the attacker can search for IVs of the form $(B, 255, V)$ for $3 \le B < 8$. If he can collect these for 60 different values of V, then he can derive the secret key with little work. This requires approximately 4,000,000 packets.

A.2   Analysis of IVs Generated by a Big Endian Counter

If the IVs are generated by a multibyte counter in big endian order (and hence the last byte of the IV increments the fastest), then the attacker can, as above, search for IVs of the form $(B, 255, V)$. This requires approximately 1,000,000 packets to collect the requisite IVs, assuming that the counter starts from zero.

However, if the counter doesn't start from zero, the attacker has an alternative strategy available to him. He can assume the first several bytes of secret key, and then search for IVs that set up the permutation as in Equation 1. If the attacker assumes the first two bytes of secret key, then for each initial IV byte, there are approximately 4 settings of the remaining two bytes that set up the permutation as required to rederive a particular key byte. Hence, with approximately 1,000,000 packets, and an additional 2^{16} work factor, he can still rederive the key.

It is common practice in the industry to extend the length of the WEP secret key (which is specified as 40 bit). Because the above attacks recover each key byte individually, the time complexity of the attack grows linearly rather than exponentially with the key length, and the data complexity of the attack remains essentially constant. Consequently, even an extremely long key is not immune to this attack.

Shortly after the publication of a preliminary version of this paper, Stubblefield, Ioannidis and Rubin ([SIR01]) implemented the attack and successfully derived a 128 bit WEP key, by observing the network during a single evening. Several optimization techniques can probably reduce the required amount of data, to the number of packets sent on a fully loaded network, in less than 15 minutes.

B   Ciphertext-Only Distinguishers Based on the Invariance Weakness

The distinguishers we presented in Section 5.1, as well as most of the distinguishers mentioned in the literature (for RC4 and other stream ciphers) assume knowledge of the plaintext in order to isolate the XORed key stream.

However, in practice the only information the attacker has is typically some statistical knowledge about the plaintext, e.g., that it contains English text. Combining the non-random behaviors of the plaintext and the key-stream is not always possible, and there are cases where XORing biased streams result with a totally random stream (e.g. when one stream is biased in its even positions and the other stream is biased in its odd positions). We prove here that if the plaintexts are English texts, it is easy to construct a ciphertext-only distinguisher from our biases. The intuition of this construction is that the biases described in Section 5.1 are in the distribution of the LSBs, and consequently they can be combined with the non-random distribution of the LSBs of English texts.

There are many major biases in the distribution of the LSBs of English texts, and they can be combined with biases of the key-stream words in various ways. In Theorem 3, we show how to combine the distribution of the first LSB of the RC4 output stream, with the first order statistics of English texts[15] :

Theorem 3. *Let C be the ciphertext generated by RC4 from a random key and the ASCII representation of plaintexts, distributed according to the first order statistics of English texts. Let p be the probability of a random key to be special 2-exact. Then C can be distinguished from a random stream by analyzing the first few words of about $\frac{200}{p^2}$ different RC4 streams.*

For example, for $RC4_{n=8}$ with 8 byte keys, $p = 2^{-16}$, which implies a reliable ciphertext-only distinguisher that works with less than 2^{40} data. The proof of Theorem 3 is based on the observation that the LSB of a random English text character is zero with probability of about 55%. The formal proof is omitted due to space limitations.

It is important to note that Theorem 3 does not use all the statistical information which is available in either the key-stream or the plaintext distributions, and consequently does not represent the best possible attack.

C   The Sampling Resistance of RC4

Most of the Time/Memory/Data tradeoff attacks on stream ciphers are based on the following paradigm. The attacker keeps a database of [state,output] pairs (sorted by output) and lookups every subsequence of the output stream in this database. When a (sufficiently long) database sequence is located in the output,

---

[15] Since the purpose of the theorem is only to demonstrate this approach, we ignore the fact that the distribution of the first characters in an English sentence differs from the distribution of mid-text characters.

the attacker can conclude that the actual state is the one stored along with this sequence and predict the rest of the stream.

A drawback of this approach is that the large database must be stored in a hard disk(s) whose random access time is about a million times slower than a computational step. To improve that attack we can keep on disk only states that are guaranteed to produce outputs with some rare but easy recognizable property (e.g., starting with some prefix α). In this case only output sequences that have this property have to be searched in the database, and thus the expected time and the expected number of disk probes is significantly reduced.

In general, producing a pair [state,output] with such a rare property costs much more than producing a random pair. $O(\frac{1}{p})$ *random* states are required to find a single pair, where p is the probability of a random output to have this property. However, if we can efficiently enumerate states that produce such outputs, the number of sampled states decreases dramatically, and this method can be applied without significant additional cost during the preprocessing stage. The sampling resistance of a stream cipher provides a lower bound on the efficiency of such enumeration.

Such an attack can be applied to RC4 in two ways, based on the KSA and PRGA parts. An attack on the generation part constructs a database of pairs [RC4 state, output substring] and analyzes all the substrings along a single output stream. The database construction is very simple since it is easy to enumerate states which produce outputs that have some constant prefix. However, this enumeration seems to be useless due to the huge effective key of this part (1684 bits) which makes such a tradeoff attack completely impractical. A more promising approach is based on the KSA part which uses a key of 40-256 bits and might be vulnerable to tradeoff attacks. In this case, the pairs in the database are [secret key, prefix of the output stream], and the attack requires prefixes from a large number of streams (instead of a single long stream).

The correlation described in Section 4 provides an efficient sampling of keys that are more likely to produce output prefixes of the patterned type specified above (predictable mod b).

For example, consider the problem of sampling M keys which are transformed by the KSA into streams whose first five words are fixed (mod 16). This property of random streams has probability of 2^{-20}, and the expected number of disk probes during the actual attack is reduced by this factor. For stream ciphers with high sampling resistance, such a filter would increase the preprocessing time by a factor of one million, as one would have to sample a million random keys in order to find a single "good" key. For RC4 (due to the invariance weakness), the preprocessing time increases by a factor of less than four, as more than one quarter of the exact special keys produce such streams, which have this fixed pattern. Consequently, the preprocessing stage is accelerated by a factor of 2^{18}.

To summarize this section, we proved that RC4 has relatively low *Sampling Resistance*, which greatly improves the efficiency of tradeoff attacks based on its KSA.

D   Deriving the Secret Key
from an Early Permutation State

Given the values $S_A[0], \ldots, S_A[A-1]$, one method to find all the values of $K[0], \ldots, K[A-1]$ that result in such a permutation is:

$i = j = 0$
$S = [0, 1, \ldots, N-1]$
For $i = 0 \ldots A - 1$
    $X = S^{-1}[S_A[i]]$
    If $i < X < A$
        Branch over all values of $0 \le X < A$ s.r. $X \ge i$ or
            $S[X] \ne S_A[X]$, running the remaining part of this
            algorithm for all such values.
    $K[i] = X - j - S[i]$
    $j = X$
    Swap($S[i]$, $S[j]$)
Verify that $[S[0], \ldots, S[A-1]] = [S_A[0], \ldots, S_A[A-1]]$

The number of times this algorithm will perform an iteration is bounded by $A^{\lambda+1}$, where λ if the number of values $0 \le x < A$ where $S_A[x] < A$. Because λ is typically quite small, this algorithm is typically efficient.

An algorithm with a better run time lower bound could be given by using the values of $S_A[A], \ldots, S_A[N-1]$.

References

[BSW00]   A. Biryukov, A. Shamir, and D. Wagner. Real time cryptanalysis of A5/1 on a PC. In *FSE: Fast Software Encryption*, 2000.

[FM00]   Fluhrer and McGrew. Statistical analysis of the alleged RC4 keystream generator. In *FSE: Fast Software Encryption*, 2000.

[Gol97]   Golić. Linear statistical weakness of alleged RC4 keystream generator. In *EUROCRYPT: Advances in Cryptology: Proceedings of EUROCRYPT*, 1997.

[GW00]   A. L. Grosul and D. S. Wallach. a related-key cryptanalysis of RC4. June 2000.

[KMP+98]   Knudsen, Meier, Preneel, Rijmen, and Verdoolaege. Analysis methods for (alleged) RC4. In *ASIACRYPT: Advances in Cryptology – ASIACRYPT: International Conference on the Theory and Application of Cryptology*. LNCS, Springer-Verlag, 1998.

[LMSon]   Wireless lan medium access control (MAC) and physical layer (PHY) specifications. (IEEE Standard 802.11), 1999 Edition. L. M. S. C. of the IEEE Computer Society.

[MS01]   I. Mantin and A. Shamir. A practical attack on broadcast RC4. In *FSE: Fast Software Encryption*, 2001.

[MT98]   Mister and Tavares. Cryptanalysis of RC4-like ciphers. In *SAC: Annual International Workshop on Selected Areas in Cryptography*. LNCS, 1998.
[Rei01]   Arnold Reinhold. The ciphersaber home page. 2001.
[Roo95]   A. Roos. A class of weak keys in the RC4 stream cipher. September 1995.
[SIR01]   Adam Stubblefield, John Ioannidis, and Aviel D. Rubin. Using the fluhrer, mantin and shamir attack to break WEP. (TD-4ZCPZZ), 2001. AT&T Labs, Technical Report.
[Wag95]   D. Wagner. Re: Weak keys in RC4. September 1995.

A Practical Cryptanalysis of SSC2

Philip Hawkes[1], Frank Quick[2], and Gregory G. Rose[1]

[1] Qualcomm Australia, Level 3, 230 Victoria Rd, Gladesville, NSW 2111, Australia
[2] Qualcomm Inc., 5775 Morehouse Drive, San Diego, CA 92121-1714, USA
{phawkes,fquick,ggr}@qualcomm.com

Abstract. SSC2 is a stream cipher that operates by XORing the output of two "half-ciphers". The first half-cipher is constructed from a linear feedback shift register (LFSR) with a non-linear filter. The second half-cipher is constructed from a lagged Fibonacci generator (LFG) and a multiplexor that chooses values from the Fibonacci register. The second half-cipher has a small cycle length $\pi \approx 2^{52}$. The initial state of the LFSR is derived by performing a fast correlation attack on the sequence resulting when XORing the key-stream at an interval of π words (thus cancelling the effect of the LFG). This attack requires around 2^{25} words of this sequence and a few hours of computation. The initial state of the LFG is then derived from around 15300 outputs using around one second of computation.

Keywords: SSC2, fast correlation attack.

1 Introduction

SSC2 is a stream cipher proposed by Zhang, Carroll and Chan [2]. The cipher is designed for software implementation and is very fast. This paper describes a practical cryptanalysis of SSC2 that requires around 2^{25} words of known key-stream (from a run of 2^{52} words) and a few hours work on a 250 MHz processor with 100 MB of memory.

SSC2 is based on a *linear feedback shift register* (LFSR) and a *lagged Fibonacci generator* (LFG). An LFSR consists of a register that stores a set of bits called the *state*, and a function that is linear modulo 2. This function updates the state bit-by-bit. An LFG consists of a register which stores a set of integers modulo N (once again called the state) and a function that is linear modulo N. This function updates the state integer-by-integer. In SSC2, the modulus is $N = 2^{32}$, and the integers are stored as 32-bit blocks called *words*.

SSC2 achieves its speed by using 32-bit operations. The stream is derived from a 127-bit LFSR, a 17-word LFG and a multiplexor that chooses values from the register of the LFG. The 127-bit register for the LFSR is stored in four 32-bit words (the extra bit is forced to 1 in the filter function). After the states of the LFSR and LFG are initialised, the following steps are repeated to produce each word of output:

S. Vaudenay and A. Youssef (Eds.): SAC 2001, LNCS 2259, pp. 25–37, 2001.

1. 32-bits of the LFSR state are updated simultaneously. A *non-linear filter* (NLF) computes a 32-bit output N_i from the four words in the state of the LFSR.
2. The LFG state is updated. The upper 16-bits and lower 16-bits of the LFG output are swapped to form L_i.
3. The multiplexor uses the four most significant bits (MSBs) of the updated word to choose one of 16 values in the LFG state to be the output M_i.
4. The output of the cipher is $Z_i = (L_i + M_i \bmod 2^{32}) \oplus N_i$, where $\oplus$ denotes XOR.

The value N_i is called the output of the LFSR half-cipher, while $V_i = (L_i + M_i \bmod 2^{32})$ is called the output of the LFG half-cipher.

Previous Results. In the rump session of Crypto 2000, Rose and Hawkes [6] reported on correlations between the least significant bits (LSBs) of certain words output from SSC2. They also noted that the LFG has a small period $\pi = 17 \cdot 2^{31} \cdot (2^{17} - 1) \approx 2^{52}$. Computing $Z_i' \stackrel{\text{def}}{=} Z_i \oplus Z_{i+\pi} = N_i \oplus N_{i+\pi}$, allows the LFSR to be attacked in isolation. The correlation in the LSBs of Z_i' allows an attacker to distinguish the output of SSC2 from a random bit stream. Another analysis by Hawkes and Rose [7] found an attack on the LFSR half-cipher in isolation that requires 382 words and around 2^{42} time. Bleichenbacher and Meier [1] found an attack on the entire cipher that finds the initial state of the LFSR using around 2^{52} words of known key-stream with around 2^{75} time. This attack exploits the small period π. Following this, the initial state of the LFG is found using around 2^{32} known outputs of the LFG half-cipher with around 2^{75} time.

Concurrent Results. Independently, Fluhrer, Crowley and Harvey had also identified a number of correlations in the LFSR half-cipher [4], and give other attacks. They noticed that there are actually two different correlations, apparently equally valid, with the LSB of the N_i.

New Results. The first part of the attack in this paper exploits the small period of the LFG by performing a fast correlation attack on the stream Z_i', based on the correlation noted in [6]. This part of the attack requires around 2^{25} words of known key-stream (from a run of 2^{52} words) with a few hours of processing time on a 250 MHz Sun UltraSPARC (see Section 3). The attack applies simple techniques that increase the accuracy and speed of any fast correlation attack. After the output of the LFSR half-cipher is removed, the attack exploits properties of the LFG noted in [1] to identify when the multiplexor has selected specific words in the LFG register. This information is used to reconstruct the initial state of the LFG (Section 4). This part of the attack requires around 15300 known outputs of the LFG half-cipher (presumed already known from the previous phase) and around a second of processing on a 250 MHz Sun UltraSPARC.

2   A Description of SSC2

LFSR half-cipher. The LFSR state is stored as four 32-bit words denoted $(X_{i+3}, X_{i+2}, X_{i+1}, X_i)$. The state is updated to $(X_{i+4}, X_{i+3}, X_{i+2}, X_{i+1})$ by computing

$$X_{i+4} = X_{i+2} \oplus (X_{i+1} << 31) \oplus (X_i >> 1),$$

where '$<<$' denotes a zero-fill left shift and '$>>$' denotes a zero-fill right shift. The least significant bit of X_i is ignored. If this sequence were converted to a bit-stream b_t, then the bit-sequence would satisfy the linear recursion:

$$b_{t+127} = b_{t+63} + b_t \bmod 2.$$

The corresponding characteristic polynomial is $x^{127} + x^{63} + 1$. This polynomial is irreducible modulo 2, which means that the bit sequence has a period of $(2^{127} - 1)$. The LFSR is implemented using a 4-word array $S[1], \ldots, S[4]$ containing $X_{i+3}, \ldots, X_i$. At each clock, the LFSR computes $A = S[2] \oplus (S[3] << 31) \oplus (S[4] >> 1)$. The values are shifted up ($S[4] \leftarrow S[3]$, $S[3] \leftarrow S[2]$, $S[2] \leftarrow S[1]$) and the value of $S[1]$ is set to A. After the LFSR is updated, the NLF output N_i is computed. The NLF uses a variety of operations: XOR; modular addition; $SWAP(A)$: swaps the upper 16-bits and lower 16-bits of A; and $\widehat{X}_i$: which denotes the word X_i with the LSB forced to 1.

NLF Algorithm
1    $A \leftarrow X_{i+3} + \widehat{X}_i \bmod 2^{32}$, with $c1 \leftarrow$ carry;
2    $A \leftarrow SWAP(A)$;
3   if $(c1 = 0)$ then $A \leftarrow X_{i+2} + A \bmod 2^{32}$ with $c2 \leftarrow$ carry;
4   else $A \leftarrow (X_{i+2} \oplus \widehat{X}_i) + A \bmod 2^{32}$ with $c2 \leftarrow$ carry;
5    $N_i \leftarrow (X_{i+1} \oplus X_{i+2}) + A + c2 \bmod 2^{32}$;

The LFG half-cipher. The LFG state consists of 17 words $(Y_{i+16}, \ldots, Y_i)$. The state is updated to $(Y_{i+17}, \ldots, Y_{i+1})$ using the recurrence:

$$Y_{i+17} = Y_{i+12} + Y_i \bmod 2^{32}. \tag{1}$$

The LFG is implemented using a 17-word array $G[1], \ldots, G[17]$. The key scheduling initialises $G[1], \ldots, G[17]$ to the values $Y_{16}, \ldots, Y_0$, and initialises two pointers r and s to 17 and 5 respectively. The output L_i is defined as $L_i = SWAP(Y_i)$. The LFG state is updated by computing

$$G[r] + G[s] = Y_i + Y_{i+12} = Y_{i+17} \bmod 2^{32},$$

and replacing the value of $G[r]$ (which was Y_i) with the value of Y_{i+17}. The values of r and s are then decreased by 1 (when r or s reaches 0, the value is reset to 17). The output M_i is defined as

$$M_i = G[1 + (s + (Y_{i+17} >> 28) \bmod 16)].$$

As a result of the reduction modulo 16, the formula for M_i in terms of the sequence $\{Y_i\}$ changes according to the value of $i \bmod 17$. Now that L_i, M_i and N_i have been computed, SSC2 outputs $Z_i = ((L_i + M_i \bmod 2^{32}) \oplus N_i)$, increments i and repeats the process. This paper does not address the issue of obtaining the key from the initial states of the LFSR and LFG, so we do not describe the key scheduling algorithm.

3   Attacking the LFSR Half-Cipher

The attack on the LFSR half-cipher is an advanced fast correlation attack, exploiting an observed correlation between the least significant bit of the filtered output words and five of the LFSR state bits. The attack is aided greatly by the fact that the feedback polynomial of the LFSR is only a trinomial: $x^{127} + x^{63} + 1$. Meier and Staffelbach observed in [10] in 1989 "any correlation to an LFSR with less than 10 taps should be avoided".

3.1   Background: Fast Correlation Attacks

The seminal work on Fast Correlation Attacks is [10], and another paper which explains them and explores some heuristic optimisations is [5].

Many stream ciphers have an underlying Linear Feedback Shift Register, and produce output by applying some nonlinear function to the state of the register; many schemes which appear different in structure are equivalent to this formulation. SSC2's LFSR half-cipher is such a construction.

If the nonlinear function is perfect, there should be no (useful) correlation between the output of the generator and any linear function of the state bits. Conversely, if there is a correlation between output bits and any linear combination of the state bits, this may be used by a fast correlation attack to recover the initial state. Consider the output bits of the generator, $\{B_i\}$, to be outputs from an LFSR, $\{A_i\}$, modified by erroneous bits $\{E_i\}$ with some probability $P < 0.5$. The probability of error P is the opposite of the known correlation. Put simply, the technique of a Fast Correlation Attack utilises the recurrence relations obeyed by the X_i to identify particular bits in the output stream which have a high probability of being erroneous, and correct (flip) them. To do this, the attack computes $(B_j + \sum_{i \in T} B_i \mod 2)$, for each recurrence relation $A_j + \sum_{i \in T} A_i \equiv 0 \pmod 2$, (these are also called parity check equations). The error probability for bit j: $P(B_j \neq A_j)$, is computed based on the number of recurrence relations $(B_j + \sum_{i \in T} B_i \equiv 0 \mod 2)$ satisfied and the number of recurrence relations unsatisfied. If there are enough bits in the output stream for the given P, this process will eventually converge until a consistent LFSR output stream remains. Linear algebra is then used to recover the corresponding initial state of the LFSR.

3.2   Fast Correlation Attack on SSC2

Recall that $\pi = 17 \cdot 2^{31} \cdot (2^{17} - 1)$ is the period of the Lagged Fibonacci Generator half-cipher. If two segments of output stream π apart are exclusive-ored together, the contributions from the LFG half-cipher cancel out, leaving the exclusive-or of two filtered LFSR streams to be analysed.

Let $Z'_i = Z_i \oplus Z_{i+\pi} = N_i \oplus N_{i+\pi}$. N_i exhibits a correlation to a linear function of the bits of the four-word state S_i. Define $l(S) = S[1]_{15} \oplus S[1]_{16} \oplus S[2]_{31} \oplus S[3]_0 \oplus S[4]_{16}$, where the subscript indicates a particular bit of the word (with bit 0 being the least significant bit). Then $P(\text{LSB}(Z_i) = l(S_i)) = 5/8$. (Note that this correlation is incorrectly presented in [1]). Intuitively, three of these terms are the bits that are XORed to form the least significant bits of N_i; the other two terms contribute to the carry bits that influence how this result might be inverted or affected by carry propagation. Obviously $N_{i+\pi}$ is similarly correlated to the state $S_{i+\pi}$, but because the state update function is entirely linear, the bits of $S_{i+\pi}$ are in turn linear functions of the bits of S_i. So $\text{LSB}(Z'_i)$ exhibits a correlation to $L(S_i) = l(S_i) \oplus l(S_{i+\pi})$.

Fluhrer [4] shows that there is actually a second linear function $l'(S) = S[4]_{15} \oplus S[1]_{16} \oplus S[2]_{31} \oplus S[3]_0 \oplus S[4]_{16}$ with the same correlation. We find it interesting that in all the test data sets we have used, admittedly a limited number, our program always "homes in" on $l(S)$ and not $l'(S)$. The existence of this second correlation makes it harder for the program to converge to the correct correlation and explains why more input data is required than would be inferred from previous results such as [5]. We are continuing to explore this area.

The words of the LFSR state are updated according to a bitwise feedback polynomial, but since the wordsize (32 bits) is a power of two, entire words of state also obey the recurrence relation, being related by the 32nd power of the feedback polynomial.

If the two streams Z_i and $Z_{i+\pi}$ were independent, then the correlation probability would be $P(\text{LSB}(Z'_i) = L(S_i)) = 17/32$. However these streams are clearly not independent and, experimentally, we have determined that there is a "second order" effect and in practice the error probability is approximately 0.446, rather than the expected 0.46875. This fortuitous occurrence makes the fast correlation attack more efficient, and counters to some extent the confusion caused by the existence of two correlation functions.

The attack on the LFSR half-cipher proceeds by first gathering approximately 32,000,000 words Z'_i, of which only the least significant bits are utilised in the attack. This requires two segments of a single output stream, separated by π. We then perform fast correlation calculations, to attempt to "correct" the output stream, on different amounts of input varying between 29,000,000 bits and 32,000,000 bits. Empirically, about 2/3rds of these trials will terminate and produce the correct output $L(S_i)$; some of the trials might give an incorrect answer, while others will "bog down", performing a large number of iterations without correcting a significant number of the remaining errors. The sections below describe the fast correlation attack itself in some detail. If the attack is thought to have corrected the output, linear algebra is used to relate this back to the initial state S_0. The sequence $Z'_i = Z_i \oplus Z_{i+\pi}$ can be reconstructed from the initial state to verify that S_0 is correct. If S_0 is incorrect or the attack "bogs down", then a different number of input bits will be tried. Thanks to the numerous optimisations discussed below, a single fast-correlation computation when successful takes about an hour on a 250MHz Sun UltraSPARC (not a particularly fast machine by today's standards) and uses about 70MB of memory. When a computation "bogs down" it is arbitrarily terminated after 1000 rounds, and this takes a few hours. For a particular output set, the full initial state is often recovered in as little as one hour, and it is very unlikely that the correct state will not be found within a day.

3.3   Increasing the Accuracy of Fast Correlation Attacks

The discussion below applies mostly to LFSRs with low weight feedback, in particular where a trinomial feedback is in use.

A number of papers have been written since [9] applying heuristic techniques to speeding up or increasing the accuracy of the basic technique of fast correlation attacks. These include [3,5,8,11]. We first spent a lot of time examining some of these techniques, and variation in their basic parameters, to gain an intuitive understanding of what is useful and what is not.

The original technique of [9] distinguished between "rounds" and "iterations", where a round started with each of the bits having the same a priori error probability. A new probability was calculated for each bit based on the probabilities of the other bits involved in parity check equations. Subsequent iterations performed the same calculations based on the updated probabilities, until enough bits had error probabilities exceeding some threshhold, or a predetermined number of iterations had been exceeded. We found the arguments in favour of performing iterations unsatisfying, since it seemed that the new probabilities were just self-reinforcing. Eventually, we made structural changes to our program which made it impossible to do iterations, and found an overall increase in accuracy.

The basic correlation algorithm has the error probability P as an input parameter; P is kept constant throughout the computation, and the bit probabilities are reset to P at the beginning of each round. In reality, the error probabilities decrease with each round (at least initially), so this approach results in inaccurate estimates for the bit probabilities. We found that as the real error probability approaches 0.5, then a constant value of P is unlikely to result in a successful attack. The computation is more likely to be successful if P is estimated at each round. For a given P, it is straightforward to calculate the proportion of parity check equations expected to be satisfied by the data. This process is easily reversible, too; having observed the proportion α of parity check equations satisfied, it is easy to calculate the error probability P:[1]

$$\delta \;=\; 1 - 2\alpha, \qquad P \;=\; \tfrac{1}{2}(1 - \delta^{1/3}).$$

Since each round begins by counting parity check equations, it is a simple matter to calculate P for that round. This technique essentially forbids the use of iterations, and obviates techniques like "fast reset", but nevertheless speeds up the attack and increases the likelihood of success.

We felt that having the greatest possible number of parity check equations for each bit was important to the operation of the algorithm, so we performed a one-time brute force calculation to look for low-weight multiples of the feedback

---

[1] This formula is based on the check equations being trinomials.

polynomial other than the obvious ones (the powers of the basic polynomial). We found a number of them. As well as $x^{127} + x^{64} + 1$, the attack uses

$$x^{16129} + x^{4033} + 1, \qquad x^{12160} + x^{4159} + 1, \qquad x^{12224} + x^{8255} + 1,$$
$$x^{16383} + x^{12288} + 1, \qquad x^{24384} + x^{12351} + 1.$$

and all possible powers of these polynomials. For each bit, the parity checks with that bit at the left, in the middle, and at the right, were all used. For 30,000,000 input bits, an average of 200 parity check equations applied to each bit.

Lastly, we made the observation that relatively early in the computation, a significant number of bits satisfied all of the available parity check equations. We called these *fully satisfied* bits. Experimentally we determined that when more than a few hundred such bits were available, and if the computation was eventually successful, they were almost all correct, so that any subset of 127 of them had a high probability of forming a linearly independent set of equations in the original state bits, which could then be solved in a straightforward manner. Computationally, taking this early opportunity to calculate the answer is a significant performance improvement. In a typical run with 30,000,000 bits of input, 5,040 fully satisfied bits were available after 16 rounds, all of which turned out to be correct, while the full computation required 64 rounds. This is not as great an optimisation as it sounds, because the rounds get faster as the number of bits corrected decreases (see below).

3.4   Increasing the Speed of Fast Correlation Attacks

At the same time as we were analysing the theoretical basis for improvements in the algorithm, we also looked at purely computational optimisations to the algorithm. When the probability of error of individual bits is variable, probability computations are complex and require significant effort for each bit, as well as the requirement to store floating-point numbers for each bit. When the error probability P is assumed the same for all bits at the beginning of a round, the computation is significantly eased. More importantly, the likelihood that a particular bit is in error can be expressed as a threshhold of the number of unsatisfied parity check equations, given the total number of parity check equations for that bit, and the probability P.

The number of parity check equations available for a particular bit is least near the edges of the data set, and increases toward the middle. During the first pass over the data, the number of equations available for each bit is simply counted (this is computationally irrelevant compared to actually checking the equations) and the indexes where this total is different to that for the previous bit is stored. Thus, it requires very little memory to derive the total number of parity checks for a particular bit in subsequent passes. In each round, the first pass over the data calculates (and stores) the number of unsatisfied checks for each bit. From the total proportion of parity checks unsatisfied, P is calculated for this round, and from that, threshhold values above which a bit will be considered to be in error are calculated for each number of parity check equations. When $P <$

0.4 it is approximately correct that more than half of the parity checks unsatisfied implies that the probability of the bit being erroneous is greater than 0.5, and the bit should be corrected. However, when $P > 0.4$, more equations need to be unsatisfied before flipping a bit is theoretically justified. The algorithm's eventual success is known to be very dependent on these early decisions.

A pass is then made through the data, flipping the bits that require it. For each bit that is flipped, the count of unsatisfied parity checks is corrected, not only for that bit, but for each bit involved in a parity check equation with it. The correction factor is accumulated in a separate array so that the correction is applied to all bits atomically. Bits which have no unsatisfied parity checks are noted. In the early rounds, this incremental approach doesn't save very much, but as fewer bits are corrected per round the saving in computation becomes very significant.

Typically another 50% of the overall computation is then saved when the count of fully satisfied bits significantly exceeds the length of the register, and the answer is derived from linear algebra. The net effect of the changes described in this and the previous section is a factor of some hundreds in the time required for data sets of about 100,000 bits over a straightforward implementation. We did not have time to find the speedup for larger data sets, as it would have required too long to run the original algorithm.

4   Attacking the LFG Half-Cipher

This attack derives the initial state $IV = (Y_{16}, \ldots, Y_0)$ of the LFG from outputs of the LFG-half cipher: $V_i = L_i + M_i \bmod 2^{32} = Z_i \oplus N_i$. Much of the analysis is based on dividing the 32-bit words into two 16-bit blocks: $A = A'' \| A'$. Note that

$$Y'_{i+17} - Y'_{i+12} - Y'_i \equiv 0 \bmod 2^{16},$$
$$Y''_{i+17} - Y''_{i+12} - Y''_i \equiv f_i \bmod 2^{16},$$

where $f_i \in \{0, 1\}$, denotes the carry bit to the upper half in the sum $(Y_i + Y_{i+12})$.

The value $\mu_i = (Y_{i+17} >> 28)$ chooses M_i from the set $\{Y_{i+1}, \ldots, Y_{i+17}\}$: $M_i = G[1 + (s + \mu_i \bmod 16)]$. The value α_i such that $M_i = Y_{i+\alpha_i}$, is the *multiplexor difference*. We always write μ_i in hexadecimal form, and α_i in decimal form. The particular word chosen depends on μ_i and s, where s is directly related to value of $\mathbf{i} \equiv i \bmod 17$. For example, if $\mu_i = 0$, then $\alpha_i = 12$ unless $\mathbf{i} \in \{4, 5\}$, in which case $\alpha_i = 11$.

4.1   Motivation

The attack exploits a property of outputs (V_i, V_{i+12}) with $\alpha_i = \alpha_{i+12} = 12$. These are called *good pairs*; all other pairs (V_i, V_{i+12}) are *bad pairs*. The initial state can be derived from good pairs using the following observations.

1. If (V_i, V_{i+12}) is good, then $M_i = Y_{i+12}$, and $M_{i+12} = Y_{i+24}$, so

$$V'_{i+12} - V''_i \equiv (Y'_{i+24} + Y''_{i+12}) - (Y''_{i+12} + Y'_i + g_i)$$
$$\equiv Y'_{i+24} - Y'_i - g_i \bmod 2^{16},$$

where $g_i \in \{0,1\}$ is the carry bit from the lower half to the upper half in the sum $V_i = L_i + M_i$. Note that if an attacker is given a good pair, then $(Y'_{i+24} - Y'_i)$ can be derived from $(V'_{i+12} - V''_i)$ by guessing g_i.

2. Every 16-bit half-word Y'_i is a linear function (mod 2^{16}) of the half-word initial state $IV' = (Y'_{16}, \ldots, Y'_0)$. Thus $(Y'_{i+24} - Y'_i)$ is also a linear function (mod 2^{16}) of IV'. We say that the values $(Y'_{i+24} - Y'_i)$ are *linearly independent* (LI) if the linear equations for $(Y'_{i+24} - Y'_i)$ are linearly independent. If the attacker knows a set of 17 LI values $(Y'_{i+24} - Y'_i)$ then the values of IV' can be determined by solving the system of linear equations.

3. Now, having obtained IV', all values Y'_i in the sequence $\{Y'_i\}$ can be computed. For each of the 17 good pairs, the value of Y'_{i+12} allows

$$Y''_i \equiv V'_i - Y'_{i+12} \bmod 2^{16}$$

to be computed. Computing Y'_i completes the word $Y_i = Y''_i \| Y'_i$. The 17 equations for Y_i (in terms of the complete initial state IV) will also be LI, so this system can be solved to find the initial state, and the attack is complete.

There remain two problems: guessing the 17 carry bits g_i and identifying good pairs.

4.2 Guessing the Carry Bits

The attack will have to try various combinations of values for g_i before the correct carry bits are found. The attack avoids trying all 2^{17} combinations by computing an accurate *prediction* p_i for the value of g_i. Note that if $V'_i < 2^{15}$ then the carry from the sum $(Y''_i + Y'_{i+12} \bmod 2^{16})$ is more likely to be one than zero. That is, we can predict that $g_i = 1$. Conversely, if $V'_i \geq 2^{15}$ then the carry g_i is more likely to be zero than one. Based on this, the attack either sets $p_i = 1$ when $V'_i < 2^{15}$ or sets $p_i = 0$ when $V'_i \geq 2^{15}$. Hence, rather that guessing the carry bits g_i, the attack guesses the 17 *errors* $\epsilon_i = p_i \oplus g_i$. The attack first guesses that there are no errors (all $\epsilon_i = 0$), then one error (one value of $\epsilon_i = 1$), two errors, and so forth. The accuracy of the prediction, $P(p_i = g_i)$, depends on V'_i. Experimental results are shown in Table 1.

Table 1. Experimental approximation to the accuracy of the prediction, $P(p_i = g_i)$, as a function of the four MSBs of V'_i.

The 4 MSBs of V'_i	0,1	2,3	4,5	6,7	8,9	A,B	C,D	E,F
$P(p_i = g_i)$	0.96	0.83	0.7	0.56	0.56	0.69	0.8	0.93

If all the pairs are good then there will be only a small number of errors. When choosing the 17 LI values $(Y'_{i+24} - Y'_i)$, the attack gives preference to values with accurate predictions as there are fewer errors, and the attack will be faster. As shown below, the attack has a small probability of choosing one or more bad pairs. If the correct initial state is not found while the number of errors is small, then this suggests that one of the pairs is bad, so our attack chooses another set of 17 LI values $(Y'_{i+24} - Y'_i)$.

4.3   Identifying Good Pairs

There are 16 possible values for α_i, so we expect good pairs to occur every $16^2 = 256$ words (on average). The problem is identifying good pairs. The trick is to identify *triples* $(V_i, V_{i+12}, V_{i+17})$ with $\alpha_i = \alpha_{i+12} = \alpha_{i+17} = 12$. Bleichenbacher and Meier [1] noted that if $\alpha_i = \alpha_{i+12} = \alpha_{i+17}$, then

$$\Delta_i \overset{\text{def}}{=} V_{i+17} - V_{i+12} - V_i \bmod 2^{32} \in \{0, 1, -2^{16}, 1 - 2^{16}\} = \mathcal{A}.$$

A triple of outputs $(V_i, V_{i+12}, V_{i+17})$ that results in $\Delta_i \in \mathcal{A}$ is said to be *valid*, because $\alpha_i = \alpha_{i+12}(= \alpha_{i+17})$ with probability close to one, (which fulfills part of the requirement for a good pair).

Note that $\mu_{i+17} \equiv \mu_{i+12} + \mu_i + c \pmod{16}$, where $c \in \{0, 1\}$, due to the recurrence (1). Hence, the possible combinations for $(\mu_i, \mu_{i+12}, \mu_{i+17})$ that result in $\alpha_i = \alpha_{i+12} = \alpha_{i+17}$ are those given in Table 2 (these are also noted in [1]).

Table 2. The possible combinations for $(\mu_i, \mu_{i+12}, \mu_{i+17})$ that result in $\alpha_i = \alpha_{i+12} = \alpha_{i+17}$

$(\mu_i, \mu_{i+12}, \mu_{i+17})$	(0,0,0)	(0,F,0)	(1,0,1)	(F,0,F)	(F,F,F)
Values of i	$i \notin \{4, 5, 9, 10\}$	$i = 9$	$i \in \{9, 10\}$	$i = 4$	$i \notin \{4, 9\}$
α_i	12	12	11	12	13

A valid triple that corresponds to a good pair is also said to be *good*; otherwise the triple is said to be *bad*. If $\mathbf{i} = 4$, then all valid triples are good, and they are used in the attack. If $\mathbf{i} \in \{5, 10\}$, then all valid triples are bad so these triples are ignored. We currently do not have an efficient method of distinguishing between the cases when $(\mu_i, \mu_{i+12}, \mu_{i+17}) = (0, F, 0)$ and $(\mu_i, \mu_{i+12}, \mu_{i+17}) = (1, 0, 1)$, so the attack also ignores triples with $\mathbf{i} = 9$.

If $\mathbf{i} \notin \{4, 5, 9, 10\}$, then a valid triple is equally likely to be either good or bad: good when $(\mu_i, \mu_{i+12}, \mu_{i+17}) = (0, 0, 0)$, and bad when $(\mu_i, \mu_{i+12}, \mu_{i+17}) = (F, F, F)$. Most of the bad triples are filtered out by examining the values of V'_i and

$$\delta_i \overset{\text{def}}{=} V''_{i+17} - V''_{i+12} - V''_i \bmod 2^{16},$$

$$\nu_i \overset{\text{def}}{=} ((V''_{i+12} - V'_i \bmod 2^{16}) >> 12)$$

$$= \text{the 4 MSBs of } (V''_{i+12} - V'_i \bmod 2^{16}).$$

Table 3. The probabilities of certain properties being satisfied in the two cases where $(\mu_i, \mu_{i+12}, \mu_{i+17}) \in \{(0,0,0), (F,FF)\}$

$(\mu_i, \mu_{i+12}, \mu_{i+17})$	$P(\nu_i \in \{0,1,F\})$	$P(\delta_i \in \{0,-1\})$	$P(V_i' \geq 2^{15} : \delta_i = 0)$
(0,0,0)	1	0.99	0.85
(F,F,F)	$\frac{3}{16}$	0.51	0.15

The attack discards valid triples with $\mathbf{i} \notin \{4,5,9,10\}$, if

- $\nu_i \notin \{0,1,F\}$, or
- $\delta_i \notin \{0,1\}$, or
- $V_i' < 2^{15}$ and $\delta_i = 0$.

Following this, $0.99 \times 0.85 = 0.84$ of the good triples remain, while only $\frac{3}{16} \times 0.51 \times 0.15 = 0.024$ of the bad triples remain. Thus, $0.024/0.84 = 0.028$ (one in 36) of the remaining valid triples are bad. The bound on V_i' (when $\delta_i = 0$) can be increased to further reduce the fraction of bad triples to good triples. However, this will also reduce the number of good triples that remain so the attack would require more key-stream.

LFG Half-Cipher Attack Algorithm

1. Find a set of triples with $\mathbf{i} \notin \{5,9,10\}$ and $\Delta_i \in \mathcal{A}$ (valid triples). For $\mathbf{i} \neq 4$, discard triples if
   - $\nu_i \notin \{0,1,F\}$,
   - $\delta_i \notin \{0,1\}$, or if
   - $\delta_i = 0$ and $V_i' < 2^{15}$.
   For each remaining triple, set $p_i = 1$ if $V_i' < 2^{15}$; else set $p_i = 0$.
2. From these triples, find 17 LI values $(Y_{i+24}' - Y_i')$, for which $P(p_i = g_i)$ is high.
3. Guess the errors ϵ_i. If the number of errors gets large, then return to Step 2.
4. Compute IV' from $(Y_{i+24}' - Y_i') \equiv V_{i+12}' - V_i'' - (p_i \oplus \epsilon_i) \bmod 2^{16}$.
5. Compute $Y_i'' \equiv V_i' - Y_{i+12}' \bmod 2^{16}$, to obtain Y_i.
6. Compute the entire state IV from Y_i. Return to Step 3 if IV produces the incorrect output.

4.4   Complexity

The number of outputs required for the attack is affected by three factors.

1. **The probability that a triple is valid.** Recall that $\mu_{i+17} \equiv \mu_{i+12} + \mu_i + c \pmod{16}$. To obtain $(\mu_i, \mu_{i+12}, \mu_{i+17}) = (0,0,0)$, it is sufficient to have $\mu_i = 0$, $\mu_{i+12} = 0$ and $c = 0$, so the combination $(0,0,0)$ occurs with probability 2^{-9}. Similarly, $(\mu_i, \mu_{i+12}, \mu_{i+17}) = (F,0,F)$ and $(\mu_i, \mu_{i+12}, \mu_{i+17}) = (F,F,F)$ occur with probability 2^{-9} each.

2. **The probability that a valid triple is good.** Of the good triples with $i \notin \{4, 5, 9, 10\}$, only 0.84 proceed to Step 2, while all of the good triples with $i = 4$ proceed to Step 2. So the probability of a good triple getting to Step 2 is $2^{-9} \times (\frac{13}{17} \times 0.84 + \frac{1}{17} \times 1)$.
3. **Finding 17 LI values of $(Y'_{i+24} - Y'_i)$ from the good triples.** Assuming that 17 good triples get to Step 2 there is no guarantee that the values $(Y'_{i+24} - Y'_i)$ are LI. However, we found that a set of 21 values of $(Y'_{i+24} - Y'_i)$ is typically sufficient to find 17 that are LI.

Therefore, the average number of outputs required for the attack on the LFG half-cipher is around

$$21 \cdot \left[\left(\frac{13}{17} \times 0.84 + \frac{1}{17} \times 1 \right) \times 2^{-9} \right]^{-1} = 15300.$$

There is a large variation in the time/process complexity, as the attacker will have to return to Step 2 if a bad triple has been selected. Our implementation of the attack on a 250MHz Sun UltraSPARC typically takes between 0.1 and 10 seconds.

5   Conclusion

We have demonstrated that attacks on SSC2 are computationally feasible, given a sufficient amount of key-stream. The attack requires portions from a (currently) prohibitive amount of continuous key-stream (around 2^{52} continuous outputs). However, we suggest that the existence of this attack indicates that SSC2 is not sufficiently secure for modern encryption requirements.

References

1. D. Bleichenbacher and W. Meier. Analysis of SSC2. *Fast Software Encryption Workshop, FSE 2001, to be published in the Lecture Notes in Computer Science, Program chair: M. Matsui, Springer-Verlag*, 2001.
2. C. Carroll, A. Chan, and M. Zhang. The software-oriented stream cipher SSC-II. In *Proceedings of Fast Software Encryption Workshop 2000*, pages 39–56, 2000.
3. V. Chepyzhov and B. Smeets. On a fast correlation attack on certain stream ciphers. *Advances in Cryptology, EUROCRYPT'91, Lecture Notes in Computer Science, vol. 547, D. W. Davies ed., Springer-Verlag*, pages 176–185, 1991.
4. S. Fluhrer, P. Crowley, I. Harvey.
   http://www.cluefactory.org.uk/paul/crypto/ssc2/mail1.txt
5. J. Dj. Golić, M. Salmasizadeh, A. Clark, A. Khodkar, and E. Dawson. Discrete optimisations and fast correlation attacks. *Cryptography: Policy and Algorithms, Lecture Notes in Computer Science, vol. 1029, E. Dawson, J. Golić eds., Springer*, pages 186–200, 1996.
6. P. Hawkes and G. Rose. Correlation cryptanalysis of SSC2, 2000. Presented at the Rump Session of CRYPTO 2000.

7. P. Hawkes and G. Rose. Exploiting multiples of the connection polynomial in word-oriented stream ciphers. *Advances in Cryptology, ASIACRYPT2000, Lecture Notes in Computer Science, vol. 1976, T. Okamoto ed., Springer-Verlag*, pages 302–316, 2000.

8. T. Johansson and F Jönsson. Improved fast correlation attacks on stream ciphers via convolutional codes. *Advances in Cryptology, EUROCRYPT'99, Lecture Notes in Computer Science, vol. 1592, J. Stern ed., Springer-Verlag*, pages 347–362, 1999.

9. W. Meier and O. Staffelbach. Fast correlation attacks on certain stream ciphers. *Advances in Cryptology, EUROCRYPT'88, Lecture Notes in Computer Science, vol. 330, C. G. Günther ed., Springer-Verlag*, pages 301–314, 1988.

10. W. Meier and O. Staffelbach. Fast correlation attacks on certain stream ciphers. *Journal of Cryptology*, 1(3):159–176, 1989.

11. M. Mihaljević and J Golić. A comparison of cryptanalytic principles based on iterative error-correction. *Advances in Cryptology, EUROCRYPT'91, Lecture Notes in Computer Science, vol. 547, D. W. Davies ed., Springer-Verlag*, pages 527–531, 1991.

Analysis of the E_0 Encryption System

Scott Fluhrer[1] and Stefan Lucks[2]

[1] Cisco Systems, Inc., 170 West Tasman Drive, San Jose, CA 95134, USA
sfluhrer@cisco.com
[2] University of Mannheim, 68131 Mannheim, Germany
lucks@th.informatik.uni-mannheim.de

Abstract. The encryption system E_0, which is the encryption system used in the Bluetooth specification, is examined. In the current paper, a method of deriving the cipher key from a set of known keystream bits is given. The running time for this method depends on the amount of known keystream available, varying from $O(2^{84})$ if 132 bits are available to $O(2^{73})$, given 2^{43} bits of known keystream.

Although the attacks are of no advantage if E_0 is used with the recommended security parameters (64 bit encryption key), they provide an upper bound on the amount of security that would be made available by enlarging the encryption key, as discussed in the Bluetooth specification.

1 Introduction

We give algorithms for deriving the initial state of the keystream generator used within E_0 given some bits of keystream with less effort than exhaustive search. From this, we derive a method for reconstructing the session encryption key used by E_0 based on some amount of keystream output. E_0 uses a two level rekeying mechanism, using the key to initialialize the level 1 keystream generator to produce the initial state for the level 2 keystream generator, which produces the actual keystream used to encrypt the data.

We use a known keystream to reconstruct the initial state for the level 2 keystream generator, which we then use to reconstruct the initial state for the level 1 keystream generator, from which we can directly deduce the encryption key. Reconstructing the state of the level 2 keystream generator takes an expected $O(2^{76})$ to $O(2^{84})$ work effort (based on the amount of known keystream available). Another attack with even more keystream available takes $O(2^{72})$ work.

By reconstructing the state from either 1 or 2 packets that are encrypted during the same session, we can reconstruct the state of the level 1 keystream generator in an expected $O(2^{81})$ or $O(2^{51})$ time, which gives a total of $O(2^{73})$ to $O(2^{84})$ work effort.

This paper is structured as follows. In Section 2, the E_0 keystream generator, and how it is used within the Bluetooth system is described. In Section 3, previous analysis and results are summarized. Section 4 presents our base attack against the keystream generator, Section 5 describes how to use it against the

S. Vaudenay and A. Youssef (Eds.): SAC 2001, LNCS 2259, pp. 38–48, 2001.
© Springer-Verlag Berlin Heidelberg 2001

level 2 generator. Section 6 deals with another approach to attack the keystream generator and the second level, if a huge amount of known keystream is available. Section 7 describes the basic attack on the first level of E_0 , given one state of the level 2 generator, while Section 2 deals with an attack given two such states. Section 9 comments on attacking the full E_0 system. Section 10 concludes and discusses the ramifications on the Bluetooth system.

2   Description of E_0

E_0 is an encryption protocol that was designed to provide privacy within the Bluetooth wireless LAN specification. When two Bluetooth devices need to communicate securely, they first undergo a key exchange protocol that completes with each unit agreeing on a shared secret, which is used to generate the encryption key (K_C). To encrypt a packet, this private key (K_C) is combined with a publicly known salt value (EN_RAND) to form an intermediate key (K_C')[1]. Then, K_C' is used in a linear manner, along with the publicly known values, the Bluetooth address, and a clock which is distinct for each packet, to form the initial state for a two level keystream generator.

The keystream generator consists of 4 LFSRs with a total length of 128 bits, and a 4 bit finite state machine, refered to as the blender FSM. For each bit of output, each LFSR is clocked once, and their output bits are exclusive-or'ed together with one bit of output from the finite state machine. Then, the 4 LFSR outputs are summed together. The two most significant bits of this 3-bit sum are used to update the state of the finite state machine. We will refer to the 25 bit LFSR as LFSR1, the 31 bit LFSR as LFSR2, the 33 bit LFSR as LFSR3 and the 39 bit LFSR as LFSR4. We will also refer to the finite state machine as the blender FSM. The generator is shown in Figure 1. Note that the least significant bit (LSB) of the sum of the four LFSRs is their bit-wise XOR.

There are logically two such keystream generators. The key of the first level keystream generator is shifted into the LFSRs, while clearing the blender FSM. Then, 200 bits are generated and discarded. Then, the output of this keystream

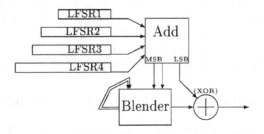

Fig. 1. The E_0 keystream generator.

---

[1] The attacks in the current paper actually provide the value of K_C'.

generator is collected, and is used to initialize the LSFRs of what we call the second level keystream generator, which is structurally identical to the first level keystream generator. This initialization is done by collecting 128 output bits, parallel loading them into the LSFRs, and making the initial second level FSM state be the final first level FSM state.

This output of this second generator is then used as an additive stream cipher to encrypt the packet.

3 Description of Previous Work

In a sci.crypt.research posting [6], Markku-Juhani O. Saarinen showed an attack that rederived the session key. This attack consisted of guessing the states of the 3 smaller LFSRs and the blender FSM, and using those states and the observed keystream to compute whether there is a consistent output from LFSR4 that is consistent with that assumption.

In the original posting, he estimated the attack to have overall complexity of $O(2^{100})$. However, he assumed that only 125 bits of keystream were available, and so he assumed a significant amount of time would be spent checking false hits. Since significantly more keystream is available within a packet, the true complexity is closer to $O(2^{93})$ expected.

Our attacks can be viewed as refinements of Saarinen's attack by taking the same basic approach of guessing the initial states of part of the cipher, and checking for consistency. However, our attacks take advantage of additional relationships within E_0 and use them to gain some performance.

Ekdahl and Johansson have shown in [2] how to extract the initial state from the keystream generator used in E_0 given $O(2^{61})$ time and $O(2^{50})$ known keystream. Their attack works by exploiting some weak linear correlations between the outputs of the LFSRs and the keystream output to verify if a guess on one of the LFSRs is accurate. Previous to that, Hermelin and Nyberg published in [4] an attack which recovered the initial state with $O(2^{64})$ work and $O(2^{64})$ known keystream. However, these are theoretical attacks as they require a far larger amount of consecutive keystream output than is available.

A time-spaces tradeoff attack has been described by Jakobsson and Wetzel [5]. Given N key streams and running time T, it is possible to recover one of the N keys if $N * T > 2^{132}$. A similar attack on the A5 keystream generator has been previously described by Golic [3].

Our attacks resemble a general type of attack, the *linear consistency attack*, which has been described as early as 1989 by Zeng, Yang, and Rao [7].

4 Base Attack on the E_0 Keystream Generator

The base attack rederives the initial settings of the LFSRs, given a limited (132 or so bits) keystream output. We will later show how this attack can be separately optimized for both levels of the keystream generators. For this attack, you assume the initial settings of the blender FSM and the contents of $LFSR1$

and $LFSR2$, and maintain for each state the current settings of the blender FSM, and a set $\mathcal{L}$ of linear equations on the $LFSR3$ and $LFSR4$ output bits. We will refer to those output bits as $LFSR3_n$ and $LFSR4_n$.

First, you initialize the set $\mathcal{L}$ to empty. Then, you perform the below depth-first search:

1. Call the state we are examining n. Compute the exclusive-or of the output n of $LFSR1$ and $LFSR2$, the next output of the blender FSM (based on the current state), and the known keystream bit Z_n. If our assumptions are correct to this point, this must be equal to the exclusive-or of the outputs of $LFSR3$ and $LFSR4$.

2. If the exclusive-or is zero, then we branch and consider the cases that both $LFSR3$ and $LFSR4$ output a zero here, and that they both output a one. When we assume a zero, we include in $\mathcal{L}$ the two linear equations $LFSR3_n = 0$ and $LFSR4_n = 0$, and when we assume a one, we include in $\mathcal{L}$ the two linear equations $LFSR3_n = 1$ and $LFSR4_n = 1$.

3. If the exclusive-or is one, then we include in $\mathcal{L}$ the single linear equation $LFSR3_n \neq LFSR4_n$

4. If $n \geq 33$, then we include in $\mathcal{L}$ the linear equation implied by the $LFSR3$ tap equations. If $n \geq 39$, then we include in $\mathcal{L}$ the linear equation implied by the $LFSR4$ tap equations. In both cases, we check to see if the new equations are inconsistent with the equations already in $\mathcal{L}$. If they are, then some assumption we made is incorrect and we backtrack to consider the next case.

5. Compute the next state of the blender FSM. This is always possible, as the next state depends on the current state (which we know) and the number of LFSRs that output a one, which we know.

6. If n is more than 132, then we have found with high probability the initial state of the encryption engine. If not, then we continue this search for state $n + 1$

There are two ideas behind this algorithm. The first is that the next state function for the blender FSM depends only on the number of LSFRs that output a one. So, when we assume that the outputs of LFSR3 and LFSR4 differ, we need not decide which one outputs a zero and which one outputs a one – instead, we can just note the fact that they differ and continue the search.

The other idea is that systems of linear equations in $GF(2)$ can be quite efficiently examined for contradictions.

How efficient is this attack? We provide some heuristic arguments. First, consider the case that all the assumed bits of LFSRs 1 and 2 and the blender state are correct.

With every step we learn if the sum S of the two output bits is either (a) $S \in \{0, 2\}$ or (b) $S = 1$. Both cases (a) and (b) are equally likely.

Note $\text{Prob}[S = 1] = 0.5$, and $\text{Prob}[S = 0] = \text{Prob}[S = 2] = 0.25$. If $S = 1$, we learn one linear equation on the state bits of LFSRs 3 and 4 (namely the XOR of the two current output bits). If $S \in \{0, 2\}$, we branch and consider both

$S = 0$ and $S = 2$. Both $S = 0$ and $S = 2$ provide us with two linear equations on the state bits of LFSRs 3 and 4.

On the average, we expect to learn 1.5 linear equations and branch 0.5 times for each step. Once we have learned in total 33+39=72 equations, we are in a leaf of the branch tree and know or "have guessed" all bits in the system. The number of such leaves describes the amount of work. (Note that this analysis is based on the heuristic assumption that no equations are redundant or contradictory, or rather, that the effects of redundant and contradictory equations on the amount of work cancel out.)

So, our branch tree has an "average" size determined by $2^{72/3} = 2^{24}$ leaves. We initially assumed 60 bits and can expect to have made a correct assumption after trying 2^{59} times, which gives us a running time of $O(2^{59+24}) = O(2^{83})$ on the average.

Experiments demonstrate that our heuristic arguments on the efficiency of the attack are reasonable, though perhaps a bit optimistic. For a random incorrect guess of initial state, the procedure examines an average of approximately 60 million (2^{26}) states before terminating. Thus we can reconstruct the encryption engine state in

$$O(2^{85}) \text{ expected time.}$$

However, for both the first level and the second level keystream generator, we can take advantage of special conditions that allow us to further optimize the attack.

5   Attack on the Second Level E_0 Keystream Generator

To optimize the attack against the second level keystream generator (which produces the observed keystream directly), we note that the base attack is more efficient if the outputs of LFSR3 and LFSR4 exclusive-or'ed together happens to have a high hamming weight. To take advantage of this, we extend the attack by assuming that, at a specific point in the keystream, the next $n + 1$ bits of LFSR3 exclusive-or'ed with LFSR4 are n ones followed by a zero, where n will be less than the length of the LFSRs. Since LFSR outputs are effectively random and independent with such a length (since both LFSRs can generate any $n + 1$ bit pattern at any time with approximately equal probability if $n < 32$), the probability a $n + k$ length output contains such a sequence is approximately $k \cdot 2^{-n}$ (for $k \ll 2^n$).

If the assumption that the LFSRs produce such an output at the specific point in the keystream is false, we will fail to discover the internal state. However, the amount of work required to make that determination turns out to be rather less than $O(2^{85-n})$, and so if we have 2^n or more starting places to test out, we will find a place where the above procedure discovers the initial state with high probability.

The expected amount of time the base attack will take when we precondition the assumed outputs of LFSR3 and LFSR4 can be experimentally obtained. The results are given in Table 1, together with the expected time for the full search.

Table 1. The expected complexity and plaintext required for various values of n. Base Search Time is the expected number of nodes traversed in a single run of the base attack. Expected Plaintext Required is the expected amount of plaintext we need to prosecute the attack. Expected Search Time is the expected total search time taken.

n	Base Search Time	Expected Plaintext Required	Expected Search Time
5	$2^{24.8}$	165 bytes	$2^{83.8}$
10	$2^{23.5}$	1157 bytes	$2^{82.5}$
15	$2^{22.1}$	33k	$2^{81.1}$
20	$2^{20.5}$	1M	$2^{79.5}$
25	$2^{18.8}$	32M	$2^{77.8}$
30	$2^{17.1}$	1G	$2^{76.1}$

Looking through this table, we can see that modest amounts of keystream reduce the expected work somewhat, however, vast quantities of keystream reduce the expected work only slightly further.

Formally, the algorithm is:

1. Select a position in the known keystream that is the start of more than 132 consecutive known bits.
2. Cycle through all possible combinations of 4 bits of blender FSM state, 25 bits of LFSR1 state and the last $30 - n$ bits of LFSR2 state
3. Compute the initial $n + 1$ bits of LFSR2 state that is consistent with the exclusive-or of LFSR3 and LFSR4 consisting of n ones and then zero.
4. Run the base attack on that setting. Stop if it finds a consistant initial setting.

The above algorithm runs the base attack 2^{59-n} times and has a 2^{-n} probability of success for a single location.

Note that, even though a single packet has a payload with a maximum of 2745 bits, we can have considerably more than 2745 bits of known keystream, if we know the plaintext of multiple packets. All the next phase of the attack needs to know is the initial state of the second level keystream generator for a packet – it does not matter which. If we have multiple packets, we can try all of them, and we will be successful if we manage to find the initial state for any of them.

6   Another Attack on the Second Level Generator

Given a huge amount of known keystream, there is another technique to attack the second level keystream generator more efficiently. The basic attack requires to assume the blender state and the states of both LFSR1 and LFSR2 (i.e. $4 + 25 + 31$ bits $= 60$ bits). Now, we start with assuming only the blender and LFSR1 states (29 bits), at the beginning of the attack. During the course of the attack, we continue to make assumptions on how the blender state is updated.

Denote the sum of the outputs of LFSR2, LFSR3, and LFSR4 by S. Obviously, $S \in \{0, 1, 2, 3\}$. Since we always know (based on previous assumptions) the current blender and LFSR1 state, we only need to know S in order to compute the next blender state. The current output bit tells if S is odd or not. Thus, we know if either (a) S in $\{0, 2\}$ or (b) S in $\{1, 3\}$.

Both cases (a) and (b) are equally likely. And in both cases we learn one linear equation, namely we learn the XOR of the output bits of the LFSRs 2–4.

Now consider the conditional probabilities Prob[$S=2|(a)$] and Prob[$S=1|(b)$]. Assuming the three output bits are independent uniformly distributed random bits (which they are, approximately), we get

$$\text{Prob}[S = 2|(a)] = \text{Prob}[S = 1|(b)] = 0.75.$$

Instead of branching, as we did in the base attack, we simply assume the likely case $S \in \{1, 2\}$, ignoring $S = 0$ and $S = 3$.

We need $31 + 33 + 39 = 103$ linear equations to entirely restore the states of the LFSRs 2–4. The assumptions we get here are linearily independent. If both our initial assumptions on the 29 state bits of blender and LFSR1 and our 103 assumptions on the sum S are correct, we have found restored the correct state. We can check so by computing δ output bits (with $\delta > 29$) and comparing the output stream we get by our assumed E_0 state with the true output stream.

Within these 103 clocks the random variable S takes 103 values $S_1, S_2, \ldots \in \{0, 1, 2, 4\}$ with Prob[$S_i \in \{1, 2\}$ = 0.75]. The attack works if $S_1 \in \{1, 2\}$ and $S_2 \in \{1, 2\}$ and $\ldots$ and $S_{103} \in \{1, 2\}$. Making the heuristic (but apparently plausible) argument that the S_i behave like 103 independent random variables, the probability $p = \text{Prob}[S_1 \in \{1, 2\} \text{ and } \ldots \text{ and } S_{103} \in \{1, 2\}]$ is

$$p = 0.75^{103} \approx 1.35 * 10^{-13} \approx 2^{-42.7}.$$

If the initially assumed 29 bits are correct, the attack requires less than 2^{43} bits of known keystream and less than 2^{43} steps (each step means to solve a system of 103 linear equations). Thus the entire attack needs

$$\text{less than } 2^{43} \text{ bits of known keystream}$$

and

$$\text{less than } 2^{72} \text{ steps.}$$

7   Attack on the First Level E_0 Keystream Generator

To attack the first level keystream generator (which produces the initial LFSR and blender FSM states), we first note that the key setup sets the FSM state of the second level keystream generator to be the final contents of the FSM state after the first level generator has produced the last bit for the LFSR state. We also note that the next-state function of the cipher is invertible – the LFSRs can be run backwards as easily as forwards, and the FSM next state function is invertible given a current LFSR state. We can also test the base attack, and find that it works essentially as well on the backwards cipher as it does the forward cipher.

This suggests this attack: when given one state of the level 2 generator, cycle through all possible combinations of 25 bits of LFSR1 state and 31 bits of LFSR2 state, and use the base attack on the reversed cipher, using as the initial FSM contents the initial contents of the phase 2 FSM. Because we are cycling through an expected $O(2^{55})$ LFSR states, and each check is expected to take $O(2^{26})$ time, we should expect to find the first level initial position in $O(2^{81})$ time.

8   Attack on the First Level E_0 Keystream Generator Given Two Second Level Keystreams

Now, let us consider a possible attack if the attacker has the first level output for two distinct packets that were sent with the same key. In this case, we first note that both keystreams have a clock associated with it, and that the clock is the only thing that differs. We further note that the method of combination is linear, hence if we know the xor differential in the clock (which we do, because we know the actual clock values), we know the xor differential of the first level LFSRs.

We can use this to optimize the attack further, as follows, where we will indicate the two known sides with as x_A and x_B, and where $\mathcal{L}$ is a set of linear equations on the outputs of $LFSR2_A$, $LFSR3_A$, $LFSR4_A$.

Assume the contents of $LFSR1_A$ (which also gives you $LFSR1_B$, because of the known differential between the two).

Initialize the set $\mathcal{L}$ to empty.

Perform the following depth-first search

1. Call the state we are examining n. Compute the output n_A, n_B of $LFSR1_A$, $LFSR1_B$, the previous output of the blender FSMs based on the current state), and the known keystream bit Z_A^n, Z_B^n. If our assumptions are correct to this point, this must be equal to the exclusive-or of the outputs of $LFSR2_A$, $LFSR3_A$, $LFSR4_A$ and of $LFSR2_B$, $LFSR3_B$, $LFSR4_B$.

2. Check the known differential in $LFSR2_A$, $LFSR3_A$, $LFSR4_A$, $LFSR2_B$, $LFSR3_B$, $LFSR4_B$ to see if there is a setting of those bits that satisifies both the known xors and the known differentials. If there is not, then backtrack to consider the next case.

3. If we reach here, there are four possible settings of the outputs of $LFSR2_A$, $LFSR3_A$, $LFSR4_A$ which are consistent with known xors and differentials. At least two of those settings will also update both blender FSMs identically, and will differ in precisely two bits. Here, we branch and consider three cases: one case that corresponds to the two settings which updates both blender FSMs identically, and the other two cases corresponding to the other two settings. For the first case, we include in $\mathcal{L}$

the linear equation implied by the two bits that differ, and the linear equation implied by the third bit setting. For the other two cases, we include in $\mathcal{L}$ three linear equations giving the three bit settings.

4. If $n \geq 31$, then we include in $\mathcal{L}$ the linear equation implied by the $LFSR2_A$ tap equations.

5. If $n \geq 33$, then we include in $\mathcal{L}$ the linear equation implied by the $LFSR3_A$ tap equations. If $n \geq 39$, then we include in $\mathcal{L}$ the linear equation implied by the $LFSR4_A$ tap equations. In all three cases, we check to see if the new equations are inconsistent with the equations already in $\mathcal{L}$. If they are, then some assumption we made is incorrect and we backtrack to consider the next case.

6. Compute the previous state of the blender FSMs. This is always possible, as the next state depends on the current state (which we know) and the number of LFSRs that output a one, which we know.

7. If n is more than 128, then we have found with high probability the initial states of the encryption engines. If not, then we continue this search for state $n + 1$

Experiments show that the above procedure examines an expected $O(2^{51})$ nodes during the search.

9 Attack Against Full E_0

Below is how we can combine these attacks into an attack on the full E_0 encryption system.

Assume we have an amount of known keystream generated with an unknown session key, which may be from a single packet or it may be from multiple packets. We select n based on the amount of known keystream. We can then use the attack shown in Section 5 to find the initial LFSR and blender FSM settings for a packet generated by that session key. If the cost of finding the initial LFSR and blender FSM settings for a second packet is less than $O(2^{81})$, then we find a second one. Then, we either use the attack shown in Section 7 to find all possible initial LFSR settings that generated that initial setting (if we have one initial LFSR setting), or we use the attack shown in Section 8 if we have two initial LFSR settings. Once we find the initial LFSR settings that generates the observed output, we can step the LFSRs back 200 cycles, and use linear transformations to eliminate the Bluetooth address and the block to reconstruct the session key K'_C, and verify that potential key by using to to decrypt other packets.

If we denote the amount of effort to find a LFSR and blender setting given n bytes of known keystream as $F(n)$ (see table 1), then the total effort for this attack is

$$O(min(F(n) + 2^{81}, 2F(n/2) + 2^{51})) \text{ work.}$$

This is $O(2^{84})$ if you have barely enough keystream to uniquely identify the session key (eg., 140 bits), and drops to $O(2^{77})$ if you have a gigabit of known keystream.

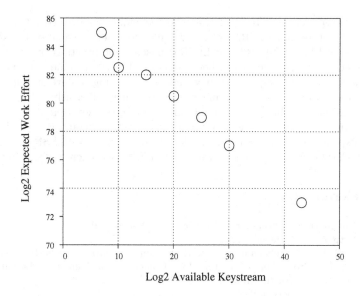

Fig. 2. Expected work effort required to recover session key, versus known keystream.

We can further reduce the effort down to

$$O(2^{73}) \text{ work},$$

if about 14 000 gigabit bits of keystream are available. We simply use the attack from Section 6 twice, to recover two states of the level 2 generator, and then continue with the attack from Section 8.

These results are summarized in Figure 2.

10   Conclusions and Open Problems

We described methods for rederiving the session key for E_0 given a limited amount of known keystream. This session key will allow the attacker to decrypt all messages in that session. We showed that the real security level of E_0 is no more than 73–84 bits (depending the amount of keystream available to the attacker), and that larger key lengths suggested by the Bluetooth specification[2] would not provide additional security.

---

[2]   "For the encryption algorithm, the key size may vary between 1 and 16 octets (8-128 bits). The size of the encryption key shall be configurable for two reasons. [First is export provisions]. The second reason is to facilitate a future upgrade path for the security without a costly redesign of the algorithms and the encryption hardware; increasing the effective key size is the simplest way to combat increased computing power at the opponent side. Currently (1999) it seems that an encryption key size of 64 bits gives satisfying protection for most applications." [1, Section 14, page 148]

We empicically observed that the technique from Section 6 (assume the blender state and LFSR1 only, and build up a set of equations based on the states of LFSR2, LFSR3 and LFSR4) posed some practical problems, because the equations created are rather complex. Also, the technique requires a huge amount of known keystream. It would be interesting to develop improved techniques to handle the set of linear equations more efficiently. Also, it would be interesting to reduce the required amount of known keystream.

Another approach for more practical attacks on E_0 and Bluetooth would be to exploit the weak mixing of the clock into the first level LFSRs, which will, at attacker known times, leave three of the LFSRs with zero differential.

References

1. Bluetooth SIG, "Bluetooth Specification", Version 1.0 B,
   `http://www.bluetooth.com/`
2. P. Ekdahl, T. Johansson, "Some Results on Correlations in the Bluetooth Stream Cipher", Proceedings of the 10th Joint Conference on Communications and Coding, Obertauern, Austria, March 11-18, 2000.
3. J. Golic, Eurocrypt 1997.
4. M. Hermelin, K. Nyberg, "Correlation Properties of the Bluetooth Combiner", proceedings of ICISC '99, LNCS 1787, Springer, 1999.
5. M. Jakobsson, S. Wetzel, "Security Weaknesses in Bluetooth", RSA Conference 2001.
6. M. Saarinen, "Re: Bluetooth und E0", Posting to `sci.crypt.research`, 02/09/00.
7. K. Zeng, C.-H. Yang, T. Rao "On the Linear Consistency Test (LCT) in Cryptanalysis with Applications", Crypto '89, Springer LNCS 435, pp. 164–174.

Boolean Functions with Large Distance to All Bijective Monomials: N Odd Case

Amr Youssef[1] and Guang Gong[2]

[1] Department of Combinatorics & Optimization
[2] Department of Electrical and Computer Engineering,
Center for Applied Cryptographic Research, University of Waterloo,
Waterloo, Ontario N2L 3G1, Canada
{a2youssef,ggong}@cacr.math.uwaterloo.ca

Abstract. Cryptographic Boolean functions should have large distance to functions with simple algebraic description to avoid cryptanalytic attacks based on successive approximation of the round function such as the interpolation attack. Hyper-bent functions achieve the maximal minimum distance to all the coordinate functions of all bijective monomials. However, this class of functions exists only for functions with even number of inputs. In this paper we provide some constructions for Boolean functions with odd number of inputs that achieve large distance to all the coordinate functions of all bijective monomials.

Key words. Boolean functions, hyper-bent functions, extended Hadamard transform, Legendre sequences, nonlinearity.

1   Introduction

Several cryptanalytic attacks on block ciphers are based on approximating the round function (or S-box) with a simpler one. For example, linear cryptanalysis [13] is based on approximating the round function with an affine function. Another example is the interpolation attack [10] on block ciphers using simple algebraic functions as S-boxes and the extended attack in [11] on block ciphers with probabilistic nonlinear relation of low degree.

   Thus, cryptographic functions used in the construction of the round function should have a large distance to functions with simple algebraic description. Along this line of research , Gong and Golomb [9] introduced a new S-box design criterion. By showing that many block ciphers can be viewed as a non linear feedback shift register with input, Gong and Golomb proposed that S-boxes should not be approximated by a bijective monomial. The reason is that, for $gcd(c, 2^N - 1) = 1$, the trace functions $Tr(\zeta x^c)$ and $Tr(\lambda x), x \in GF(2^N)$, are both m-sequences with the same linear span.

For Boolean functions with even number of input variables, bent functions achieve the maximal minimum distance to the set of affine functions. In other words, they achieve the maximal minimum distance to all the coordinate functions of affine monomials (i.e., functions in the form $Tr(\lambda x) + e$). However, this

S. Vaudenay and A. Youssef (Eds.): SAC 2001, LNCS 2259, pp. 49–59, 2001.

doesn't guarantee that such bent functions cannot be approximated by the coordinate functions of bijective monomials (i.e., functions in the form $Tr(\lambda x^c) + e$, $gcd(c, 2^N - 1) = 1$). At Eurocrypt' 2001, Youssef and Gong [19] introduced a new class of bent functions which they called hyper-bent functions. Functions within this class achieve the maximal minimum distance to all the coordinate functions of all bijective monomials.

In this paper we provide some constructions for Boolean functions with odd number of inputs that achieve large distance to all the coordinate functions of all bijective monomials. Unlike the N even case, bounding the nonlinearity (NL) for functions with odd number of inputs, N, is still an open problem. For $N = 1, 3, 5$ and 7, it is known that max $NL = 2^{N-1} - 2^{(N-1)/2}$. However, Patterson and Wiedemann [15], [16] showed that for $N = 15$, max $NL \geq 16276 = 16384 - \frac{27}{32} 2^{\frac{15-1}{2}}$. It should be noted that our task, i.e., finding functions with large distance to all the coordinate functions of all bijective monomials, is far more difficult than finding functions with large nonlinearity. For example, while the (experimental) average nonlinearity for functions with $N = 11$ and 13 is about 941 and 3917 respectively, the (experimental) average minimum distance to the coordinate functions of all bijective monomials is about 916 and 3857 respectively.

We conclude this section with the notation and concepts which will be used throughout the paper.

- $\mathbb{F} = GF(2)$.
- $\mathbb{E} = GF(2^N)$.
- $Tr_M^N(x)$, $M|N$, represents the trace function from $\mathbb{F}_{2^N}$ to $\mathbb{F}_{2^M}$, i.e., $Tr_M^N(x) = x + x^q + \cdots + x^{q^{l-1}}$ where $q = 2^M$ and $l = N/M$. If $M = 1$ and the context is clear, we write it as $Tr(x)$.
- $\mathbf{a} = \{a_i\}$, a binary sequence with period $s|2^N - 1$. Sometimes, we also use a vector of dimension s to represent a sequence with period s. I.e., we also write $\mathbf{a} = (a_0, a_1, \cdots, a_{s-1})$.
- $Per(\mathbf{b})$, the period of a sequence $\mathbf{b}$.
- $\mathbf{a}^{(t)}$ denotes the sequence obtained by decimating the sequence $\mathbf{a}$ by t, i.e., $\mathbf{a}^{(t)} = \{a_{tj}\}_{j \geq 0} = a_0, a_t, a_{2t}, \cdots$.
- $w(s)$: the number of 1's in one period of the sequence s or the number of 1's in the set of images of the function $s(x) : GF(2^N) \rightarrow GF(2)$. This is the so-called the Hamming weight of s whether s is a periodic binary sequence or a function from $GF(2^N)$ to $GF(2)$.
- $\mathcal{S}$ denotes the set of all binary sequences with period $r|2^N - 1$.
- $\mathcal{F}$ denotes the set of all (polynomial) functions from $GF(2^N)$ to $GF(2)$.

2   Preliminaries

The trace representation of any binary sequence with period dividing $2^N - 1$ is a polynomial function from $GF(2^N)$ to $GF(2)$. Any such polynomial function corresponds to a Boolean function in N variables. This leads to a connection among sequences, polynomial functions and Boolean functions. Using this connection, pseudo-random sequences are rich resources for constructing functions with good cryptographic properties.

Any non-zero function $f(x) \in \mathcal{F}$ can be represented as

$$f(x) = \sum_{i=1}^{s} Tr_1^{m_{t_i}}(\beta_i x^{t_i}), \beta_i \in GF(2^{m_{t_i}})^*, \tag{1}$$

where $1 \le s \le |\Omega(2^N - 1)|$, $\Omega(2^N - 1)$ is the set of coset leaders modulo $2^N - 1$, t_i is a coset leader of a cyclotomic coset modulo $2^N - 1$, and $m_{t_i}|N$ is the size of the cyclotomic coset containing t_i. For any sequence $\mathbf{a} = \{a_i\} \in \mathcal{S}$, there exists $f(x) \in \mathcal{F}$ such that

$$a_i = f(\alpha^i), i = 0, 1, \cdots,$$

where α is a primitive element of $\mathbb{E}$. $f(x)$ is called *the trace representation* of $\mathbf{a}$. ($\mathbf{a}$ is also referred to as an s-term sequence.) If $f(x)$ is any function from $\mathbb{E}$ to $\mathbb{F}$, by evaluating $f(\alpha^i)$, we get a sequence over $\mathbb{F}$ with period dividing $2^N - 1$. Thus

$$\delta : \mathbf{a} \leftrightarrow f(x) \tag{2}$$

is a one-to-one correspondence between $\mathcal{F}$ and $\mathcal{S}$ through the trace representation in (1). We say that $f(x)$ is the *trace representation* of $\mathbf{a}$ and $\mathbf{a}$ is the *evaluation* of $f(x)$ at α. In this paper, we also use the notation $\mathbf{a} \leftrightarrow f(x)$ to represent the fact that $f(x)$ is the trace representation of $\mathbf{a}$. The set consisting of the exponents that appear in the trace terms of $f(x)$ is said to be the *null spectrum set* of $f(x)$ or $\mathbf{a}$.

If $s = 1$, i.e.,

$$a_i = Tr_1^N(\beta \alpha^i), i = 0, 1, \cdots, \beta \in \mathbb{E}^*,$$

then $\mathbf{a}$ is an m-sequence over $\mathbb{F}$ of period $2^N - 1$ of degree N. (For a detailed treatment of the trace representation of sequences, see [14]).

3 Extended Transform Domain Analysis for Boolean Functions

The Hadamard transform of $f : \mathbb{E} \to \mathbb{F}$ is defined by [1]

$$\hat{f}(\lambda) = \sum_{x \in \mathbb{E}} (-1)^{f(x) + Tr(\lambda x)}, \lambda \in \mathbb{E}. \tag{3}$$

The Hadamard transform spectrum of f exhibits the nonlinearity of f. More precisely, the nonlinearity of f is given by

$$NL(f) = 2^{N-1} - \frac{1}{2} \max_{\lambda \in \mathbb{E}} |\hat{f}(\lambda)|,$$

which indicates that the absolute value of $\hat{f}(\lambda)$ reflects the difference between agreements and disagreements of $f(x)$ and the linear function $Tr(\lambda x)$. Only bent functions [17] have a constant spectrum of their Hadamard transform. Gong and Golomb [9] showed that many block ciphers can be viewed as a non linear feedback shift register with input. In the analysis of shift register sequences [4], all m-sequences are equivalent under the decimation operation on elements in a sequence. The same idea can be used to approximate Boolean functions, i.e., we can use monomial functions instead of linear functions to approximate Boolean functions.

Gong and Golomb [9] introduced the concept of extended Hadamard transform (*EHT*) for a function from $\mathbb{E}$ to $\mathbb{F}$. The extended Hadamard transform is defined as follows.

Definition 1. *Let $f(x)$ be a function from $\mathbb{E}$ to $\mathbb{F}$. Let*

$$\hat{f}(\lambda, c) = \sum_{x \in \mathbb{E}} (-1)^{f(x) + Tr(\lambda x^c)} \tag{4}$$

where $\lambda \in \mathbb{E}$ and c is a coset leader modulo $2^N - 1$ co-prime to $2^N - 1$. Then we call $\hat{f}(\lambda, c)$ an extended Hadamard transform of the function f.

Notice that the Hadamard transform of f, defined by (3), is $\hat{f}(\lambda, 1)$. The numerical results in [9] show that, for all the coordinate functions $f_i, i = 1, \cdots, 32$ of the DES s-boxes, the distribution of $\hat{f}_i(\lambda, c)$ in λ is invariant for all c.

Thus a new generalized nonlinearity measure can be defined as

$$NLG(f) = 2^{N-1} - \frac{1}{2} \max_{\substack{\lambda \in \mathbb{E}, \\ c \, : \, gcd(c, 2^N - 1) = 1}} |\hat{f}(\lambda, c)|.$$

This leads to a new criterion for the design of Boolean functions used in conventional cryptosystems. The *EHT* of Boolean functions should not have any large component.

In what follows we will provide constructions for Boolean functions with large distance to all the coordinate functions of bijective monomials. The construction method depends on whether N is a composite number or not.

4 Case 1: N Is a Composite Number

Let $N = nm$ where $n, m > 1$. Let $\underline{b} = \{b_j\}_{j \geq 0}$ be a binary sequence with $per(\underline{b}) = d = \frac{q^n - 1}{q - 1}$, $q = 2^m$, and $w(\underline{b}) = v$. Let $g(x) \leftrightarrow \underline{b}$. In the following, we derive some bounds on $NLG(g)$ in terms of v.

Write $a_i = Tr_1^{nm}(\alpha^i)$, $i = 0, 1, \cdots$. Thus $\underline{a} = \{a_i\}$ is an m-sequence of period $2^N - 1$. Let

$$\delta(\tau) = |\{0 \leq i < d | b_i = 1, Tr_m^N(\alpha^{i+\tau}) = 0\}|.$$

Lemma 1. *With the above notation, we have*

$$w(Tr(\alpha^\tau x^r) + g(x)) = 2^{nm-1} - v + q\delta(\tau). \tag{5}$$

Proof. Throughout the proof, we will write $\delta(\tau)$ as δ for simplicity. The sequence $\underline{a}$ can be arranged into a $(q-1, d)$-interleaved sequence [8]. Thus $\underline{a}$ can be arranged into the following array

$$A = \begin{bmatrix} a_0 & a_1 & \cdots & a_{d-1} \\ a_d & a_{d+1} & \cdots & a_{2d-1} \\ \vdots & \vdots & \vdots & \vdots \\ a_{d(q-2)} & v_{d(q-2)+1} & \cdots & v_{(q-1)d-1} \end{bmatrix} = [A_0, A_1, \cdots, A_{d-1}],$$

where A_i's are columns of the matrix. Similarly we can arrange the sequence $\underline{b}$ in the following array

$$B = \begin{bmatrix} b_0 & b_1 & \cdots & b_{d-1} \\ b_d & b_{d+1} & \cdots & b_{2d-1} \\ \vdots & \vdots & \vdots & \vdots \\ b_{d(q-2)} & b_{d(q-2)+1} & \cdots & b_{(q-1)d-1} \end{bmatrix}.$$

Note that $w(A) = |\{(i,j)|a_{ij} = 1\}, 0 \le i < q-1, 0 \le j < d\}|$. Thus

$$w(A + B) = \sum_{b_i = 0} w(A_i) + \sum_{b_i = 1} w(A_i + 1)$$

$$= \sum_{b_i = 0} w(A_i) + \sum_{b_i = 1} (q - 1 - w(A_i)).$$

In the array A, there are

$$r = \frac{q^{n-1} - 1}{q - 1} \tag{6}$$

zero columns (See Lemma 1 in [18]). If there are δ zero columns corresponding to the indices of the 1's in $\{b_i\}$, then they contribute $\delta(q-1)$ 1's. Thus we have

$$w(A + B) =$$

$$\sum_{b_i = 0, A_i \ne 0} w(A_i) + \sum_{b_i = 0, A_i = 0} w(A_i) + \sum_{b_i = 1, A_i \ne 0} (d - w(A_i)) + \sum_{b_i = 1, A_i = 0} (q - 1 - w(A_i)).$$

Since A_i's are m-sequences, then for all the non-zero A_i's we have $w(A_i) = 2^{m-1}$. Let

$$N_{ij} = |\{b_k = i, char(A_k) = j, 0 \le k < d\}|,$$

where $i, j \in \{0, 1\}$ and

$$char(A_i) = \begin{cases} 0 \text{ if } A_i = 0, \\ 1 \text{ if } A_i \ne 0. \end{cases}$$

Note that

$$N_{1,0} = \delta,$$
$$N_{1,0} + N_{0,0} = r, \tag{7}$$
$$N_{0,0} = r - \delta.$$

Hence we have

$$N_{1,0} + N_{1,1} = v \Rightarrow N_{1,1} = v - N_{1,0} = v - \delta,$$

$$N_{0,1} + N_{1,1} = d - r \Rightarrow N_{0,1} = d - r - N_{1,1} = d - r - (v - \delta) = d - r - v - \delta.$$

Thus

$$\begin{aligned}
w(A + B) &= 2^{m-1}N_{0,1} + 0N_{0,0} + (2^{m-1}N_{1,1} + (2^m - 1)N_{1,0} \\
&= 2^{m-1}(d - r - v) + \delta 2^{m-1} + v(2^{m-1} - 1) - \delta 2^{m-1} + \delta + 2^m \delta - \delta \\
&= 2^{m-1}(d - r) - v2^{m-1} + v2^{m-1} - v + 2^m \delta \\
&= 2^{m-1}(d - r) - v + 2^m \delta = 2^{m-1}(d - r) - v + 2^m \delta.
\end{aligned}$$

(8)

By noting that $d - r = q^{n-1}$ then we have

$$w(A + B) = 2^{nm-1} - v + 2^m \delta,$$

which proves the lemma.

Theorem 1. *With the notation above, if $v = \frac{d-1}{2}$ then*

$$NLG(g) \geq 2^{nm-1} - \frac{d-1}{2}.$$

Proof.

$$\begin{aligned}
\widehat{g}(0, c) &= \sum_{x \in \mathbb{E}}(-1)^{g(x)} = 1 + \sum_{x \in E^*}(-1)^{g(x)} = 1 + (q - 1)\sum_{k=0}^{d-1}(-1)^{b_k} \\
&= 1 + (q - 1)(d - 2wt(\mathbf{b})) = q.
\end{aligned}$$

For $\lambda \neq 0$,

$$\begin{aligned}
\widehat{g}(\lambda, c) &= \sum_{x \in \mathbb{E}}(-1)^{Tr(\lambda x^c) + g(x)} = 1 + \sum_{i=0}^{2^{nm}-1}(-1)^{a_i + b_i} \\
&= 2^{nm} - 2wt(A + B).
\end{aligned}$$

(9)

Note that $\delta \leq r = \frac{q^{n-1}-1}{q-1}$. Thus

$$w(A + B) \leq 2^{nm-1} - \frac{d-1}{2},$$

and

$$w(A + B) \geq -2^{nm-1} - \frac{d-1}{2} + q\frac{q^{n-1}-1}{q-1} = -2^{nm-1} + \frac{d-1}{2}.$$

By noting that $n > 1$ then $d - 1 > q$ and hence

$$|\widehat{g}(\lambda, c)| \leq (d - 1)$$

which proves the theorem.

Using the construction above for $N = 9$, $m = n = 3$ we get $NLG = 220$. It is clear that, in order to maximize NLG, we should minimize $d = \frac{2^{nm}-1}{2^m-1}$. Thus we should choose m to be the large factor of $N = n \times m$. For example, let $N = 15 = 3 \times 5$. If we choose $m = 5$, then we have $NLG = 15856$. However, if we picked $m = 3$, then we get $NLG = 14044$.

5   Case 2: N Is a Prime Number

If N is a prime number then the above sub-field construction is not applicable. This case is further divided into two cases depending on whether $2^N - 1$ is a prime number or not.

5.1   Case 2.1: $2^N - 1$ Is a Prime Number

In this case, we base our construction on the Legendre sequence. Let γ be a primitive root of a prime p, then the Legendre sequence (also called quadratic residue sequence) of period p, $p \equiv 3 \pmod 4$, is defined by

$$a_i = \begin{cases} 1 \text{ or } 0, \text{ if } i = 0 \\ 1, \quad i \text{ is a residue } (i \equiv \gamma^{2s} \bmod p \\ 0, \quad i \text{ is a non-residue } (i \not\equiv \gamma^{2s} \bmod p) \end{cases}$$

Note that for $N \geq 2$ we always have $2^N - 1 \equiv 3 \bmod 4$. The properties of Legendre sequences have been extensively studied (e.g., [2], [5], [6], [12]). In here we are concerned with the following fact:

Fact 1 *If a Legendre sequence of period $p \equiv 3 \bmod 4$ is decimated with d then the original sequence is obtained if d is a quadratic residue mod p, and the reverse sequence is obtained if d is non-quadratic residue mod p.*

This fact can be easily explained by noting that the Boolean function corresponding to Legendre sequence has the following trace representation [12]

$$f(x) = \sum_{c \in QR} Tr(x^c),$$

where QR denotes the set of quadratic residue mod $2^N - 1$.

Example 1. Let p=7, then $\underline{a} = \{1110100\}$ The sequences $\underline{a}^{(d)}$ obtained by decimating $\underline{a}$ with d are given by

$$\begin{aligned} \underline{a}^{(1)} &= \{1110100\}, \\ \underline{a}^{(2)} &= \{1110100\}, \\ \underline{a}^{(3)} &= \{1001011\}, \\ \underline{a}^{(4)} &= \{1110100\}, \\ \underline{a}^{(5)} &= \{1001011\}, \\ \underline{a}^{(6)} &= \{1001011\}. \end{aligned} \tag{10}$$

Note that $\underline{a}^{(1)} = \underline{a}^{(2)} = \underline{a}^{(3)}$ since $1, 2, 4$ are quadratic residue mod 7. Also $\underline{a}^{(3)} = \underline{a}^{(5)} = \underline{a}^{(6)}$ are the same since since $3, 5, 6$ are quadratic non-residue mod 7.

The following property follows directly from Fact 1.

Property 1. Let $f \leftrightarrow \underline{\mathbf{a}}$ where $\underline{\mathbf{a}}$ is a Legendre sequence. Then we have

$$NLG(f) = \min \{NL(f), NL(g)\}, \tag{11}$$

where $g \leftrightarrow \underline{\mathbf{a}}^{(c)}$ and c is any quadratic non-residue modulo $2^N - 1$.

Example 2. For $N = 5$, $\underline{\mathbf{b}} = \{11101101111000101011110000100100\}$. If we let $f \leftrightarrow \underline{\mathbf{b}}$ with $f(0) = 1$ then we have $\hat{f}(\lambda, c) \in \{-2, -6, -10, 2, 6, 10\}$ for $c \in$ set of quadratic residue mod 31. $\hat{f}(\lambda, c) \in \{-2, -6, 2, 10\}$ for $c \notin$ set of quadratic residue mod 31. Thus we have $NLG(f) = 11$.

Table 1 shows NLG of the functions obtained from this construction. In this case, we set $f(0) = 1$. If we set $f(0) = 0$ then we obtain balanced functions for which NLG is 1 less than the values shown in the table.

Table 2 shows NLG versus NL distribution for $N = 5$. It is clear that our Legendre sequence construction achieves the maximum possible NLG. Table 3 shows the same distribution for balanced functions. For $N = 7$ we searched all functions in the form [7]

$$f(x) = \sum_{c \in \Omega(2^N - 1)} Tr_1^{n_c}(x^c),$$

where $\Omega(2^N - 1)$ is the set of coset leaders mod $2^N - 1$ and n_c is the size of the coset containing c. Table 4 shows NLG versus NL distribution for this case. Table 5 shows the same distribution for the balanced functions of the same form. Again, it's clear that the construction above achieves the best possible NLG. For larger values of N, our construction is no longer optimum. For example, for $N = 13$, $g(x) \leftrightarrow \underline{\mathbf{b}} = \{i \bmod 2, i = 0, 1, \cdots\}$ have $NLG = 3972$.

Table 1.

N	3	5	7	13	17	19
NLG	1	11	55	3964	64816	259882

Table 2. $N = 5$

NLG NL	0	1	2	3	4	5	6	7	8	9	10	11
0	64	0	0	0	0	0	0	0	0	0	0	0
1	0	2048	0	0	0	0	0	0	0	0	0	0
2	0	0	31744	0	0	0	0	0	0	0	0	0
3	0	0	0	317440	0	0	0	0	0	0	0	0
4	0	0	0	0	2301440	0	0	0	0	0	0	0
5	0	0	0	0	0	12888064	0	0	0	0	0	0
6	0	0	0	0	13020	0	57983268	0	0	0	0	0
7	0	0	0	7440	0	3919392	0	211487952	0	0	0	0
8	0	0	2790	0	2396610	0	74021180	0	571246300	0	0	0
9	0	620	0	923180	0	39040780	0	544800200	0	777687700	0	0
10	62	0	149668	0	8474160	0	189406218	0	1022379070	0	191690918	0
11	0	9300	0	606980	0	19419516	0	232492250	0	302968890	0	911896
12	248	0	1302	0	263810	0	3803018	0	20035610	0	3283148	0

Table 3. $N = 5$ balanced case

NLG	0	2	4	6	8	10
NL						
0	62	0	0	0	0	0
2	0	15872	0	0	0	0
4	0	0	892800	0	0	0
6	0	0	6200	19437000	0	0
8	0	1550	1074150	27705010	167500130	0
10	62	77128	3274220	62085560	276057170	34259588
12	248	682	109430	1536050	6312220	735258

Table 4. $N = 7$

NLG	0	2	8	14	16	22	28	30	36	42	44	46	48	50	52	54
NL																
2	0	2	0	0	0	0	0	0	0	0	0	0	0	0	0	0
8	0	0	72	0	0	0	0	0	0	0	0	0	0	0	0	0
14	0	0	0	306	0	0	0	0	0	0	0	0	0	0	0	0
16	0	0	0	0	306	0	0	0	0	0	0	0	0	0	0	0
22	0	0	0	0	0	3264	0	0	0	0	0	0	0	0	0	0
28	0	0	0	0	90	0	6030	0	0	0	0	0	0	0	0	0
30	0	0	0	90	0	0	1269	4761	0	0	0	0	0	0	0	0
36	0	0	72	0	0	3156	4032	2916	23088	0	0	0	0	0	0	0
42	0	6	0	280	460	2448	4715	4408	12927	7012	0	0	0	0	0	0
44	6	0	0	460	280	2448	6696	2427	12927	6248	764	0	0	0	0	0
46	0	0	0	4	1	121	157	174	326	119	0	8	0	0	0	0
48	0	0	1	25	46	578	1232	757	2486	1052	3	0	50	0	0	0
50	0	0	326	918	948	10187	16267	9632	32340	16288	33	90	401	742	0	0
52	0	1	120	549	504	4746	6409	4236	12167	5781	15	25	170	272	5	0
54	2	10	46	228	164	1281	2557	1295	4548	1727	1	21	5	84	1	1
56	10	0	47	47	108	619	1270	570	2241	926	15	0	13	27	0	1

5.2 Case 2.2: N Is a Prime Number and $2^N - 1$ Is a Composite Number

Let $2^N - 1 = dr$, $d > r > 1$. In this case we a use construction similar to case 1, i.e., we let $f \leftrightarrow \underline{\mathbf{b}}$ where $per(\underline{\mathbf{b}}) = d$ and $w(\underline{\mathbf{b}}) = \frac{d-1}{2}$. However, unlike case 1, there is no easy way to determine the weight distribution of A_i's because they are no longer m-sequences. Using this approach for $N = 11$, $d = 89$ we obtained several functions with $NLG(f) = 980 = 2^{N-1} - \frac{d-1}{2}$.

Table 5. $N = 7$ balanced case

NLG	0	2	14	16	28	30	42	44	52	54
NL										
0	1	0	0	0	0	0	0	0	0	0
2	0	1	0	0	0	0	0	0	0	0
14	0	0	81	0	0	0	0	0	0	0
16	0	0	0	81	0	0	0	0	0	0
28	0	0	0	54	1242	0	0	0	0	0
30	0	0	54	0	561	681	0	0	0	0
42	0	6	160	232	2144	1997	2517	0	0	0
44	6	0	232	160	3067	1074	2510	7	0	0
46	0	0	4	1	66	76	28	0	0	0
48	0	0	21	39	904	561	747	3	0	0
50	0	0	82	115	1220	544	908	1	0	0
52	0	1	549	504	6409	4236	5781	15	5	0
54	2	10	228	164	2557	1295	1727	1	1	1
56	10	0	47	108	1270	570	926	15	0	1

6   Conclusions and Open Problems

In this paper we presented some methods to construct functions with odd number of inputs which achieve large minimum distance to the set of all bijective monomials. However, since a a general upper bound on NLG is not known, it is interesting to search for other functions that outperform the constructions presented in this paper. Finding NLG of functions corresponding to the Legendre sequences is another interesting open problem.

References

1. R.E. Blahut, *Theory and practice of error control codes*, Addison-Wesley Publishing Company, 1983.
2. I. B. Damgard, *On the randomness of Legendre and Jacobi sequences*, Advances in Cryptology - CRYPTO '88, pp. 163-172.
3. J. F. Dillon, *Elementary Hadamard difference sets*, Ph.D. Dissertation, University of Maryland, 1974.
4. S.W. Golomb, *Shift register sequences*, Aegean Park Press, Laguna Hills, California, 1982.
5. C. Ding, T. Helleseth and W. Shan, *On the linear complexity of Legendre sequences*, IEEE Trans. Inf. Theory, vol. IT-44, no.3, pp. 1276-1278, May 1998.
6. P. Fan and M. Darnell, *Sequence design for communications applications*, John Wiley and Sons Inc., 1996.
7. C. Fontaine, *The Nonlinearity of a class of Boolean functions with short representation*, Proc. of Pragocrypt ' 96, pp. 129-144, 1996.

8. G. Gong, Theory and applications of q-ary interleaved sequences, *IEEE Trans. on Inform. Theory*, vol. 41, No. 2, 1995, pp. 400-411.

9. G. Gong and S. W. Golomb, *Transform domain analysis of DES*, IEEE transactions on Information Theory. Vol. IT-45, no. 6, pp. 2065-2073, September, 1999.

10. T. Jakobsen and L. Knudsen, *The interpolation attack on block ciphers, LNCS 1267*, Fast Software Encryption, pp. 28-40. 1997.

11. T. Jakobsen, *Cryptanalysis of block ciphers with probabilistic non-linear relations of low degree*, Proceedings of Crypto'99, LNCS 1462, pp. 213-222, 1999.

12. J. No; H. Lee; H. Chung; H. Song, *Trace representation of Legendre sequences of Mersenne prime period*, IEEE Trans. Inf. Theory, vol. IT-42, no.6, pt.2, pp.2254-2255.

13. M. Matsui, *Linear cryptanalysis method for DES cipher* Advances in Cryptology, Proceedings of Eurocrypt'93, LNCS 765, pp. 386-397, Springer-Verlag, 1994.

14. R. J. McEliece, *Finite fields for computer scientists and engineers* , Kluwer Academic Publishers, Dordrecht, 1987.

15. N.J. Patterson, D.H. and Wiedemann, *The covering radius of the* $(2^{15}, 16)$ *Reed-Muller code is at least 16276*, IEEE Trans. Inf. Theory, vol.IT-29, no.3, pp. 354-356, May 1983.

16. N.J. Patterson, D.H. and Wiedemann, *Correction to: The covering radius of the* $(2^{15}, 16)$ *Reed-Muller code is at least 16276*, IEEE Trans. Inf. Theory, vol.IT-36, no.2, pp. 343, March 1990.

17. O.S. Rothaus , On bent functions, *J. Combinatorial Theory*, vol. 20(A), 1976, pp.300-305.

18. R.A. Scholtz and L.R. Welch, *GMW sequences*, IEEE Trans. Inf. Theory, vol.IT-30, no.3, pp. 548-553, May 1984.

19. A. Youssef and G. Gong, Hyper-bent functions, *Advances in Cryptology, Proc. of Eurocrypt'2001*, LNCS 2045, pp. 406-419, Springer-Verlag, 2001.

Linear Codes in Constructing Resilient Functions with High Nonlinearity

Enes Pasalic[1] and Subhamoy Maitra[2]

[1] Department of Information Technology, Lund University,
P.O. Box 118, 221 00 Lund, Sweden
enes@it.lth.se

[2] Computer & Statistical Service Center, Indian Statistical Institute,
203, B.T. Road, Calcutta 700 035, India
subho@isical.ac.in

Abstract. In this paper we provide a new generalized construction method of highly nonlinear t-resilient functions, $F : \mathbb{F}_2^n \mapsto \mathbb{F}_2^m$. The construction is based on the use of linear error correcting codes together with multiple output bent functions. Given a linear $[u, m, t + 1]$ code we show that it is possible to construct n-variable, m-output, t-resilient functions with nonlinearity $2^{n-1} - 2^{\lceil \frac{n+u-m-1}{2} \rceil}$ for $n \geq u + 3m$. The method provides currently best known nonlinearity results.

Keywords: Resilient functions, Nonlinearity, Correlation Immunity, Stream Ciphers, Linear Codes.

1 Introduction

A well known method for constructing a running key generator exploits several linear feedback shift registers (LFSR) combined by a nonlinear Boolean function. This method is used in design of stream cipher system where each key stream bit is added modulo two to each plaintext bit in order to produce the ciphertext bit. The Boolean function used in this scenario must satisfy certain properties to prevent the cipher from common attacks, such as Siegenthaler's correlation attack [18], linear synthesis attack by Berlekamp and Massey [14] and different kinds of approximation attacks [7]. If we use multiple output Boolean function instead of single output one, it is possible to get more than one bits at each clock and this increases the speed of the system. Such a multiple output function should possess high values in terms of order of resiliency, nonlinearity and algebraic degree.

Research on multiple output binary resilient functions has received attention from mid eighties [6,1,8,19,2,9,21,12,11,4,5]. The initial works on multiple output binary resilient functions were directed towards linear resilient functions. The concept of multiple output resilient functions was introduced independently by Chor et al [6] and Bennett et al [1]. A similar concept was introduced at the same time for single output Boolean functions by Siegenthaler [17]. Besides its importance in random sequence generation for stream cipher systems, these

S. Vaudenay and A. Youssef (Eds.): SAC 2001, LNCS 2259, pp. 60–74, 2001.
© Springer-Verlag Berlin Heidelberg 2001

resilient functions have applications in quantum cryptographic key distribution, fault tolerant distributed computing, etc.

The nonlinearity issue for such multiple output resilient functions was first discussed in [20]. After that, serious attempts towards construction of nonlinear resilient functions have been taken in [21,12,11,5]. We here work in that direction and provide better results than the existing work. For given number of input variables n, number of output variables m, and order of resiliency t, we can construct functions $F : \mathbb{F}_2^n \mapsto \mathbb{F}_2^m$ that achieve higher nonlinearity values than existing constructions for almost all choices of n, m and t.

The paper is organized as follows. Section 2 provides basic definitions and notations both for 1-output and m-output functions, $m > 1$. In Section 3, we review some important techniques and results used towards the new construction of t-resilient functions. Section 4 provides the new construction based on the use of linear error-correcting codes together with bent functions. Some numerical values for the constructed functions and comparison with previous constructions are presented in Section 5. Section 6 concludes this paper.

2 Preliminaries

For binary strings S_1, S_2 of the same length λ, we denote by $\#(S_1 = S_2)$ (respectively $\#(S_1 \neq S_2)$), the number of places where S_1 and S_2 are equal (respectively unequal). The *Hamming distance* between S_1, S_2 is denoted by $d(S_1, S_2)$, i.e.,

$$d(S_1, S_2) = \#(S_1 \neq S_2).$$

Also the *Hamming weight* or simply the weight of a binary string S is the number of ones in S. This is denoted by $wt(S)$.

By $\mathbb{F}_2^n$ we denote the vector space corresponding to the finite field $\mathbb{F}_{2^n}$. The addition operator over $\mathbb{F}_2$ is denoted by $\oplus$ (the XOR operation, which is basically addition modulo 2). By V_n we mean the set of all Boolean functions on n-variables, i.e., V_n corresponds to all possible mappings $\mathbb{F}_2^n \mapsto \mathbb{F}_2$. We interpret a Boolean function $f(x_1, \ldots, x_n)$ as the output column of its truth table, that is, a binary string of length 2^n,

$$[f(0, 0, \ldots, 0), f(1, 0, \ldots, 0), f(0, 1, \ldots, 0), \ldots, f(1, 1, \ldots, 1)].$$

An n-variable function f is said to be *balanced* if its output column in the truth table contains equal number of 0's and 1's (i.e., $wt(f) = 2^{n-1}$).

An n-variable Boolean function $f(x_1, \ldots, x_n)$ can be considered to be a multivariate polynomial over $\mathbb{F}_2$. This polynomial can be expressed as a sum of products representation of all distinct k-th order product terms $(0 \leq k \leq n)$ of the variables. More precisely, $f(x_1, \ldots, x_n)$ can be written as

$$f(x_1, \ldots, x_n) = a_0 \oplus (\bigoplus_{i=1}^{i=n} a_i x_i) \oplus (\bigoplus_{1 \leq i \neq j \leq n} a_{ij} x_i x_j) \oplus \ldots \oplus a_{12\ldots n} x_1 x_2 \ldots x_n,$$

where the coefficients $a_0, a_{ij}, \ldots, a_{12\ldots n} \in \{0, 1\}$. This representation of f is called the *algebraic normal form* (ANF) of f. The number of variables in the highest order product term with nonzero coefficient is called the *algebraic degree*, or simply degree of f.

Functions of degree at most one are called affine functions. An affine function with constant term equal to zero is called a linear function. The set of all n-variable affine (respectively linear) functions is denoted by A_n (respectively L_n). The *nonlinearity* of an n variable function f is

$$nl(f) = min_{g \in A_n}(d(f, g)),$$

i.e., the distance from the set of all n-variable affine functions.

Let $x = (x_1, \ldots, x_n)$ and $\omega = (\omega_1, \ldots, \omega_n)$ both belong to $\mathbb{F}_2^n$. The *dot product* of x and ω is defined as

$$x \cdot \omega = x_1\omega_1 \oplus \ldots \oplus x_n\omega_n.$$

For a Boolean function $f \in V_n$ the *Walsh transform* of $f(x)$ is a real valued function over $\mathbb{F}_2^n$ that can be defined as

$$W_f(\omega) = \sum_{x \in \mathbb{F}_2^n} (-1)^{f(x) \oplus x \cdot \omega}.$$

Next we define *correlation immunity* in terms of the characterization provided in [10]. A function $f(x_1, \ldots, x_n)$ is m-th order correlation immune (CI) iff its Walsh transform W_f satisfies

$$W_f(\omega) = 0, \text{ for all } \omega \in \mathbb{F}_2^n \text{ s.t. } 1 \leq wt(\omega) \leq m.$$

If f is balanced then $W_f(\overline{0}) = 0$. Balanced m-th order correlation immune functions are called *m-resilient* functions. Thus, a function $f(x_1, \ldots, x_n)$ is m-resilient iff its Walsh transform W_f satisfies

$$W_f(\omega) = 0, \text{ for all } \omega \in \mathbb{F}_2^n \text{ s.t. } 0 \leq wt(\omega) \leq m.$$

Given all these definitions we now start the definitions with respect to the multiple output Boolean functions $\mathbb{F}_2^n \mapsto \mathbb{F}_2^m$. That is, in this case we provide the truth table of m different columns of length 2^n. Let us consider the function $F(x) : \mathbb{F}_2^n \mapsto \mathbb{F}_2^m$ such that $F(x) = (f_1(x), \ldots, f_m(x))$. Then the nonlinearity of F is defined as,

$$nl(F) = \min_{\tau \in \mathbb{F}_2^{m*}} nl(\bigoplus_{j=1}^m \tau_j f_j(x)).$$

Here, $\mathbb{F}_2^{m*} = \mathbb{F}_2^m \backslash 0$ and $\tau = (\tau_1, \ldots, \tau_m)$. Similarly the algebraic degree of F is defined as,

$$deg(F) = \min_{\tau \in \mathbb{F}_2^{m*}} deg(\bigoplus_{j=1}^m \tau_j f_j(x)).$$

Now we define an n-variable, m-output, t-resilient function, denoted by (n, m, t), as follows. A function F is an (n, m, t) resilient function, iff $\bigoplus_{j=1}^{m} \tau_j f_j(x)$ is an $(n, 1, t)$ function (n variable, t-resilient Boolean function) for any choice of $\tau \in \mathbb{F}_2^{m*}$. Since we are also interested in nonlinearity, we provide the notation (n, m, t, w) for an (n, m, t) resilient function with nonlinearity w. In this paper we concentrate on the nonlinearity value. Thus, for given size of input parameters n, m, t, we construct the functions with currently best known nonlinearity.

3 Useful Techniques

In this section we will describe a few existing techniques that will be used later. First we recapitulate one result related to linear error correcting codes. The following lemma was proved in [11]. We will use it frequently in our construction, and therefore it is stated with the proof.

Proposition 1. *Let $c_0, \ldots, c_{m-1}$ be a basis of a binary $[u, m, t+1]$ linear code C. Let β be a primitive element in $\mathbb{F}_{2^m}$ and $(1, \beta, \ldots, \beta^{m-1})$ be a polynomial basis of $\mathbb{F}_{2^m}$. Define a bijection $\phi : \mathbb{F}_{2^m} \mapsto C$ by*

$$\phi(a_0 + a_1\beta + \cdots a_{m-1}\beta^{m-1}) = a_0 c_0 + a_1 c_1 + \cdots a_{m-1} c_{m-1}.$$

Consider the matrix

$$A^* = \begin{pmatrix} \phi(1) & \phi(\beta) & \cdots & \phi(\beta^{m-1}) \\ \phi(\beta) & \phi(\beta^2) & \cdots & \phi(\beta^m) \\ \vdots & \vdots & \ddots & \vdots \\ \phi(\beta^{2^m-2}) & \phi(1) & \cdots & \phi(\beta^{m-2}) \end{pmatrix}.$$

For any linear combination of columns (not all zero) of the matrix A^*, each nonzero codeword of C will appear exactly once.

Proof. Since ϕ is a bijection, it is enough to show that the matrix

$$\begin{pmatrix} 1 & \beta & \cdots & \beta^{m-1} \\ \beta & \beta^2 & \cdots & \beta^m \\ \vdots & \vdots & \ddots & \vdots \\ \beta^{2^m-2} & 1 & \cdots & \beta^{m-2} \end{pmatrix}$$

has the property that each element in $\mathbb{F}_{2^m}^*$ will appear once in any nonzero linear combination of columns of the above matrix.

Any nonzero linear combination of columns can be written as

$$(c_0 + c_1\beta + \cdots + c_{m-1}\beta^{m-1}) \begin{pmatrix} 1 \\ \beta \\ \vdots \\ \beta^{2^m-2} \end{pmatrix},$$

for some $c_0, c_1, \ldots, c_{m-1} \in \mathbb{F}_2$, and this gives the proof. $\qquad\square$

There are $2^m - 1$ rows in the matrix A^*. Let us only concentrate on the first 2^{m-1} rows of this matrix. That is, we consider each column to be of length 2^{m-1}. It is clear that for any nonzero linear combination of the columns, a nonzero codeword of C will appear exactly once in it. Hence, in the resulting column of length 2^{m-1}, no codeword will appear more than once. In this direction, we update Proposition 1 with the following result.

Proposition 2. *Let $c_0, \ldots, c_{m-1}$ be a basis of a binary $[u, m, t+1]$ linear code C. Let β be a primitive element in $\mathbb{F}_{2^m}$ and $(1, \beta, \ldots, \beta^{m-1})$ be a polynomial basis of $\mathbb{F}_{2^m}$. Define a bijection $\phi : \mathbb{F}_{2^m} \mapsto C$ by*

$$\phi(a_0 + a_1\beta + \cdots a_{m-1}\beta^{m-1}) = a_0 c_0 + a_1 c_1 + \cdots a_{m-1} c_{m-1}.$$

For $0 \leq q \leq m-1$, consider the matrix

$$D = \begin{pmatrix} \phi(1) & \phi(\beta) & \cdots & \phi(\beta^{m-1}) \\ \phi(\beta) & \phi(\beta^2) & \cdots & \phi(\beta^m) \\ \vdots & \vdots & \ddots & \vdots \\ \phi(\beta^{2^q-1}) & \phi(\beta^{2^q}) & \cdots & \phi(\beta^{2^q+m-2}) \end{pmatrix}.$$

For any linear combination of columns (not all zero) of the matrix D, each nonzero codeword of C will either appear exactly once or not appear at all.

Note that the entries of D are elements from $\mathbb{F}_2^u$. For convenience, we use a standard index notation to identify the elements of D. That is, $d_{i,j}$ denotes the element in i-th row and j-th column of D, for $i = 1, \ldots, 2^q$, and $j = 1, \ldots, m$.

Throughout the paper we consider C to be a binary linear $[u, m, t+1]$ code with a set of basis vectors $c_0, c_1, \ldots, c_{m-1}$. To each codeword $c_i \in C$, $i = 0, \ldots, 2^m - 1$, we can associate a linear function $l_{c_i} \in L_u$, where

$$l_{c_i} = c_i \cdot x = \bigoplus_{k=1}^{u} c_{i,k} x_k.$$

This linear function is uniquely determined by c_i. Since the minimum distance of C is $t+1$, any function l_{c_i} for $c_i \in C$ will be nondegenerate on at least $t+1$ variables, and hence t-resilient.

According to Proposition 1, any column of the matrix A^* can be seen as a column vector of $2^m - 1$ distinct t-resilient linear functions on u variables. In [11], it was proved that the existence of a set $\mathcal{C}$ of linear $[u, m, t+1]$ *nonintersecting codes* of cardinality $|\mathcal{C}| = \lceil 2^{n-u}/2^m - 1 \rceil$ was sufficient and necessary requirement in construction of an $(n, m, t, 2^{n-1} - 2^{u-1})$ function. A set of linear $[u, m, t+1]$ codes $\mathcal{C} = \{C_1, C_2, \ldots, C_s\}$ such that $C_i \cap C_j = \{0\}, 1 \leq i < j \leq s$, is called a set of linear $[u, m, t+1]$ nonintersecting codes.

The results in [11] were obtained using a computer search for the set $\mathcal{C}$. Good results could be obtained only for small size of $n \leq 20$, thus not providing a good construction for arbitrary n.

In this initiative our approach is different. We do not try to search for non-intersecting linear codes. We only consider a single linear code with given parameters and use a repetition of the codewords in a specific manner. If we look into the matrix D of Proposition 2, and consider each column as concatenation of 2^q $(0 \leq q \leq m-1)$ linear functions on u variables, then each column can be seen as a Boolean function on $u + q$ variables, i.e., $g_j \in V_{u+q}$, $j = 1, \ldots, m$. In the ANF notation the functions $g_j \in V_{u+q}$ will be given by,

$$g_j(y, x) = \bigoplus_{\tau \in \mathbb{F}_2^q} (y_1 \oplus \tau_1) \cdots (y_q \oplus \tau_q)(d_{[\tau]+1,j} \cdot x),$$

where $[\tau]$ denotes the integer representation of vector τ. Once again note that we have denoted the elements of D matrix as $d_{i,j}$, for $i = 1, \ldots, 2^q$, and $j = 1, \ldots, m$. Since each of the constituent linear functions is nondegenerate on $t+1$ variables, they are all t-resilient. Thus, each of the $(u+q)$-variable Boolean function g_j is t-resilient. Next we have the following result on nonlinearity.

Proposition 3. *Any nonzero linear combination of the functions $g_1, \ldots, g_m$ has the nonlinearity $2^{u+q-1} - 2^{u-1}$.*

Proof. From [16], we have, $nl(g_j) = 2^{u+q-1} - 2^{u-1}$ for $j = 1, \ldots, m$. Moreover, from Proposition 2, it is clear that any nonzero linear combination of these functions $g_1, \ldots, g_m$ will have the same property. $\qquad\square$

Hence we get the following result related to multiple output functions.

Proposition 4. *Given a $[u, m, t+1]$ linear code, it is possible to construct $(u + q, m, t, 2^{u+q-1} - 2^{u-1})$ resilient functions, for $0 \leq q \leq m - 1$.*

A simple consequence of Proposition 4 is that for given m and t our goal is to use a linear code of minimum length, i.e., u should be minimized, since the nonlinearity is maximized in that case. Throughout this paper the functions constructed by means of Proposition 4 will be denoted by g_j. We immediately get the following corollary concerning the construction of 1-resilient functions.

Corollary 1. *It is possible to construct an $(n = 2m, m, 1, nl(F) = 2^{n-1} - 2^{\frac{n}{2}})$ function $F(x)$.*

Proof. It is possible to construct $[m + 1, m, 2]$ linear code. Putting $u = m + 1$ and $q = m - 1$, we get $(n, m, 1, 2^{n-1} - 2^m)$ resilient functions. $\qquad\square$

Thus, using Corollary 1 with $m = 16$, we can construct 1-resilient function $F(x) : \mathbb{F}_2^{32} \mapsto \mathbb{F}_2^{16}$ with nonlinearity $N_F = 2^{n-1} - 2^{\frac{n}{2}} = 2^{31} - 2^{16}$. This function can be used in a stream cipher system where at each clock it is possible to get 2-byte output.

Next we look into a more involved technique. For this we need a set of m bent functions such that any nonzero linear combination of these bent functions will also be a bent function.

The following proposition is well known and therefore stated without proof (for proof see [16]).

Proposition 5. *Let $h(y) \in V_k$ and $g(x) \in V_{n_1}$. Then the nonlinearity of $f(y, x) = h(y) \oplus g(x)$ is given by, $nl(f) = 2^k nl(g) + 2^{n_1} nl(h) - 2nl(g)nl(h)$.*

Next we present the following Corollaries which will be useful in the sequel.

Corollary 2. *Let $h(y)$ be a bent function on V_k, $k = 2m$. Let $g(x) \in V_{n_1}$ with $nl(g) = 2^{n_1-1} - 2^{u-1}$, for $u \leq n_1$. Then the nonlinearity of $f(y, x) = h(y) \oplus g(x)$ is given by, $nl(f) = 2^{n_1+k-1} - 2^{\frac{k}{2}} 2^{u-1}$.*

Proof. Put $nl(h) = 2^{k-1} - 2^{\frac{k}{2}-1}$ in Proposition 5. ☐

Corollary 3. *Let $h'(y')$ be a bent functions on V_k, $k = 2r$, and let $h(y)$ be a function on V_{k+1} given by $h(y) = x_{k+1} \oplus h'(y')$. Let $g(x) \in V_{n_1}$ with $nl(g) = 2^{n_1-1} - 2^{u-1}$, for $u \leq n_1$. Then the nonlinearity of $f(y, x) = h(y) \oplus g(x)$ is given by, $nl(f) = 2^{n_1+k-1} - 2^{\frac{k+1}{2}} 2^{u-1}$.*

Proof. Put $nl(h) = 2^{k-1} - 2^{\frac{k+1}{2}-1}$ in Proposition 5. ☐

Corollary 4. *Let $h(y)$ be a constant function on V_k, $k > 0$. Let $g(x) \in V_{n_1}$ with $nl(g) = 2^{n_1-1} - 2^{u-1}$, for $u \leq n_1$. Then the nonlinearity of $f(y, x) = h(y) \oplus g(x)$ is given by, $nl(f) = 2^{n_1+k-1} - 2^k 2^{u-1}$.*

Proof. Put $nl(h) = 0$ in Proposition 5. ☐

Thus, using the composition of bent functions with resilient functions, one may construct highly nonlinear resilient Boolean functions on higher number of variables. The question is if we may use the same technique for construction of multiple output functions. In other words, we want to find a set of bent functions of cardinality $2^m - 1$, say $B = \{b_1, \ldots, b_{2^m-1}\}$, with basis $b_1, \ldots, b_m$, such that $\bigoplus_{j=1}^m \tau_j b_j \in B$, for $\tau \in \mathbb{F}_2^{m*}$.

Now we discuss the construction in more detail [15]. Let A be of size $2^m \times m$ given by $A = (\frac{0}{A^*})$, where A^* is a matrix constructed by means of Proposition 1 using $c_0, \ldots, c_{m-1}$, that spans an $[m, m, 1]$ code C with the unity matrix I as the generator matrix. Now consider each column of the matrix A, which can be seen as concatenation of 2^m distinct linear functions on m variables. This is a Maiorana-McFarland type bent function in $2m$-variables. Also using Proposition 1, it is clear that any nonzero linear combination of these bent functions will provide a bent function. The algebraic degree of this class of bent functions is equal to m. Thus, we have the following result.

Proposition 6. *It is possible to get m distinct bent functions on $2m$-variables, say $b_1, \ldots, b_m$, such that any nonzero linear combination of these bent functions will provide a bent function. Also, $\deg(\bigoplus_{i=1}^m \tau_i b_i) = m$, for $\tau \in \mathbb{F}_2^{m*}$.*

Example 1. Let $m = 2$ and $c_0 = (01)$, $c_1 = (10)$. We use an irreducible polynomial $p(z) = z^2 + z + 1$ to create the field $\mathbb{F}_{2^2}$. Then it can be shown that the matrix A is given by,

$$A = \begin{pmatrix} 0 & 0 \\ c_0 & c_1 \\ c_1 & c_0 + c_1 \\ c_0 + c_1 & c_0 \end{pmatrix}.$$

In the truth table notation, let us consider the 4-variable bent function $g_1(x)$ as the concatenation of the 2-variable linear functions $0, x_1, x_2, x_1 \oplus x_2$ and similarly, $g_2(x)$ as concatenation of $0, x_2, x_1 \oplus x_2, x_1$. Then the function $g_1(x) \oplus g_2(x)$ is also bent, which is a concatenation of $0, x_1 \oplus x_2, x_1, x_2$.

Also note the following updation of Proposition 6.

Proposition 7. *It is possible to get m distinct bent functions on $2p$-variables $(p \geq m)$, say $b_1, \ldots, b_m$, such that any nonzero linear combination of these bent functions will provide a bent function. Also, $\deg(\bigoplus_{i=1}^m \tau_i b_i) = p$, for $\tau \in \mathbb{F}_2^{m*}$.*

With these results we present our construction method in the following section.

4   New Construction

In this section we will first provide the general construction idea using a $[u, m, t+1]$ linear code and then we will use specific codes towards construction of resilient functions of specific orders. Let us first discuss the idea informally. We take the matrix D as described in Proposition 2. Now it is clear that each column of D can be seen as a $u+q$ variable function with order of resiliency t and nonlinearity $2^{u+q-1} - 2^{u-1}$. Let us name these functions as $g_1, \ldots g_m$. From Proposition 4, it is known that any nonzero linear combination of these functions will provide $u+q$ variable function g with order of resiliency t and nonlinearity $2^{u+q-1} - 2^{u-1}$.

Now we concentrate on n-variable functions. It is clear that the $(u + q)$-variable function need to be repeated 2^{n-u-q} times to make an n-variable function. We will thus use an $(n - u - q)$-variable function and XOR it with the $(u+q)$-variable function to get an n-variable function. Also to get the maximum possible nonlinearity in this method, the $(n - u - q)$-variable function must be of maximum possible nonlinearity. We will use m different functions $h_1, \ldots, h_m$ and use the compositions $f_1 = h_1 \oplus g_1, \ldots, f_m = h_m \oplus g_m$, to get m different n-variable functions. Thus any nonzero linear combination of $f_1, \ldots, f_m$ can be seen as the XOR of linear combinations of $h_1, \ldots, h_m$ and linear combinations of $g_1, \ldots, g_m$. In order to get a high nonlinearity of the vector output function we will need high nonlinearity of the functions $h_1, \ldots, h_m$ and also high nonlinearity for their linear combinations.

If $(n - u - q)$ is even, we can use bent functions $h_1, \ldots, h_m$. Importantly, we require m different bent functions (as in Proposition 6) such that the nonzero linear combinations will also produce bent functions. For this we need $n - u - q \geq$

$2m$ (see Proposition 7). If $(n - u - q)$ is odd, we can use bent functions b_j of $(n - u - q - 1)$ variables and take $h_j = x_n \oplus b_j$. This requires the condition $n - u - q - 1 \geq 2m$ to get m distinct bent functions as in Proposition 7.

It may very well happen that the value of $n - u - q$ may be less than $2m$ and in such a scenario it may not be possible to get $2m$ bent functions with desired property. In such a situation we may not get very good nonlinearity. We formalize the results in the following theorem.

Theorem 1. *Given a linear* $[u, m, t+1]$ *code, for* $n \geq u$ *it is possible to construct* $(n, m, t, nl(F))$ *function* $F = (f_1, \ldots, f_m)$, *where*

$$nl(F) = \begin{cases} 2^{n-1} - 2^{u-1}, & u \leq n < u + m; & (1) \\ 2^{n-1} - 2^{n-m}, & u + m \leq n < u + 2m; & (2) \\ 2^{n-1} - 2^{u+m-1}, & u + 2m \leq n < u + 3m; & (3) \\ 2^{n-1} - 2^{\frac{n+u-m-1}{2}}, & n \geq u + 3m - 1, \ n - u - m + 1 \ even; & (4) \\ 2^{n-1} - 2^{\frac{n+u-m}{2}}, & n \geq u + 3m, \ n - u - m + 1 \ odd. & (5) \end{cases}$$

Proof. We consider different cases separately. We will use functions $g_1, \ldots, g_m$ on $u + q$ variables which are basically concatenation of q distinct linear functions on u variables. These linear functions are nondegenerate on at least $t + 1$ variables. From Proposition 3, we get that for any $\tau \in \mathbb{F}_2^{m*}$, $nl(\bigoplus_{j=1}^m \tau_j g_j) = 2^{u+q-1} - 2^{u-1}$. Next we consider m different functions $h_1, \ldots, h_m$ on $(n - u - q)$ variables. We will choose those functions in such a manner so that, for any $\tau \in \mathbb{F}_2^{m*}$, $nl(\bigoplus_{j=1}^m \tau_j h_j)$ is high. Mostly we will use bent functions as in Proposition 6 and Proposition 7 in our construction. Now we construct the vector output function $F = (f_1, \ldots, f_m)$ where, $f_j = h_j \oplus g_j$. For any $\tau \in \mathbb{F}_2^{m*}$, $\bigoplus_{j=1}^m \tau_j f_j(x)$ can be written as $\bigoplus_{j=1}^m \tau_j h_j \oplus \bigoplus_{j=1}^m \tau_j g_j$. This can be done since the set of variables are distinct. The input variables of g_j's are $x_1, \ldots, x_{u+q}$ and the input variables of h_j's are $x_{u+q+1}, \ldots, x_n$.

1. Here, $u \leq n < u + m$. By Proposition 4, we construct $(n = u + q, m, t, 2^{n-1} - 2^{u-1})$ function F.

2. Let $u + m \leq n < u + 2m$. Here we take $q = m - 1$ in Proposition 2. The functions g_j's are of $u + m - 1$ variables. Thus we need to repeat each function $\frac{2^n}{2^{u+m-1}}$ times. We will use functions h_j's of $(n-u-m+1)$ variables which are constant functions. We know, $nl(g_j) = 2^{u+m-2} - 2^{u-1}$. Hence, $nl(f_j) = 2^{n-u-m+1}(2^{u+m-2} - 2^{u-1}) = 2^{n-1} - 2^{n-m}$ as in Corollary 4.

3. Let $u + 2m \leq n < u + 3m$. We take q such that $n - u - q = 2m$. In this case g_j's are of $u + q$ variables. We take m bent functions h_j's, each of $2m$-variables as in Proposition 6. We know, $nl(g_j) = 2^{u+q-1} - 2^{u-1}$ and $nl(h_j) = 2^{2m-1} - 2^{m-1}$. Thus, if we consider the function $F = (f_1, \ldots, f_m)$, we get, $nl(F) = 2^{n-1} - 2^{u+m-1}$ as in Corollary 2.

4. For $n \geq u + 3m - 1$, $n - u - m + 1$ even, we use $q = m - 1$ and a set of bent functions on $n - u - m + 1$ variables. Note that in this case $n - u - m + 1 \geq 2m$. Thus we will get a set of m bent functions as in Proposition 7. Here, $nl(g_j) = 2^{u+m-1} - 2^{u-1}$ and $nl(h_j) = 2^{(n-u-m+1)-1} - 2^{\frac{n-u-m+1}{2}-1}$. Thus we get, $nl(F) = 2^{n-1} - 2^{\frac{n+u-m-1}{2}}$ as in Corollary 2.

5. For $n \geq u + 3m$, $n - u - m + 1$ odd, we use $q = m - 1$ and a set of bent functions on $n - u - m$ variables, say $b_1, \ldots, b_m$ as in Proposition 7. Note that in this case, $n - u - m \geq 2m$. We construct $h_j = x_n \oplus b_j$. Thus we get, $nl(g_j) = 2^{u+m-1} - 2^{u-1}$ and $nl(h_j) = 2^{(n-u-m+1)-1} - 2^{\frac{(n-u-m+1)-1}{2}}$. In this case, the nonlinearity is $nl(F) = 2^{n-1} - 2^{\frac{n+u-m}{2}}$ as in Corollary 3. □

Note that Corollary 1 in Section 3 is a special case of the item 1 in the above theorem. Next we consider the algebraic degree of functions constructed by means of Theorem 1.

Theorem 2. *In reference to Theorem 1, the algebraic degree of the function F is given by,*

$$2 \leq deg(F) \leq n - u + 1, \quad u \leq n < u + m; \tag{1}$$

$$2 \leq deg(F) \leq m, \quad u + m \leq n < u + 2m; \tag{2}$$

$$deg(F) = \begin{cases} m, & u + 2m \leq n < u + 3m; & (3) \\ \frac{n-u-m+1}{2}, & n \geq u + 3m - 1, \ n - u - m + 1 \ \ even; & (4) \\ \frac{n-u-m}{2}, & n \geq u + 3m, \ n - u - m + 1 \ \ odd. & (5) \end{cases}$$

Proof. Let us consider any nonzero linear combination f of $(f_1, \ldots, f_m)$. Also we denote any nonzero linear combination of h_j's as h and that of g_j's as g. It is clear that $deg(F) = deg(f) = \max(deg(h), deg(g))$, as h, g are functions on distinct set of input variables.

1. Here f can be seen as the concatenation of 2^q linear functions ($0 \leq q < m$) of u variables each. The exact calculation of algebraic degree will depend in a complicated way on the choice of the codewords from C. However, it is clear that the function is always nonlinear and hence the algebraic degree must be ≥ 2. Also the function f will have degree at most $q + 1$. Here $q = n - u$, which gives the result.
2. In this case $q = m - 1$. Now f can be seen as the 2^{n-u-q} times repetition of function g, where g is the concatenation of 2^q linear functions ($0 \leq q < m$) of u variables each. The exact calculation of algebraic degree will depend in a complicated way on the choice of the codewords from C. However, it is clear that the function is always nonlinear and hence $deg(f) \geq 2$. Furthermore, the function g will have degree at most $q + 1$. Thus the result.
3. In this case $deg(f) = \max(deg(h), deg(g))$. Now, $deg(h) = m$ as we consider $2m$ variable bent functions with property as described in Proposition 6. Also, $deg(g)$ is at most $q + 1$. Now, $u + 2m \leq n < u + 3m$, which gives $q < m$. Hence $deg(f) = m$.
4. In this case $deg(h) = \frac{n-u-m+1}{2}$ (from Proposition 7) and $deg(g) \leq q+1 = m$. Here $n \geq u + 3m - 1$, i.e., $n - u - m + 1 \geq 2m$, which gives $\frac{n-u-m+1}{2} \geq m$. Thus $deg(f) = \frac{n-u-m+1}{2}$.
5. In this case $deg(h) = \frac{n-u-m}{2}$ and $deg(g) \leq q + 1 = m$. Here $n \geq u + 3m$, i.e., $n - u - m \geq 2m$, which gives $\frac{n-u-m}{2} \geq m$. Thus $deg(f) = \frac{n-u-m}{2}$. □

At this point, let us comment on construction of resilient functions of order 1 and 2. First we concentrate on 1-resilient functions. Let C_1 be an $[m+1, m, 2]$ linear code in systematic form, i.e., $C_1 = (I|\mathbf{1})$, where I is an identity matrix of size $m \times m$, and $\mathbf{1}$ is a column vector of all ones. In this case, we have $u = m+1$. Then we can apply Theorem 1 on this $[m+1, m, 2]$ code.

Next we look into the construction of 2-resilient functions. From the theory of error correcting codes we know that for any $l \geq 3$ there exists a linear $[u = 2^l - 1, m = 2^l - l - 1, 3]$ Hamming code. The codewords from such a code provide the construction of $(n, m, 2, nl(F))$ nonlinear resilient functions F. Also, given l, it is possible to obtain a sequence of linear codes of different length and dimension. In other words, given a linear $[2^l - 1, 2^l - l - 1, 3]$ Hamming code the generated sequence of codes is $[2^l - 1 - j, 2^l - l - 1 - j, 3]$, for $j = 0, 1, \ldots, 2^{l-1} - 1$. This code with Theorem 1 can be used to construct 2-resilient functions with high nonlinearity. Note that this construction of 2-resilient functions is not the best using this technique due to the existence of better linear $[n, m, 3]$ codes than those provided by the Hamming design.

The construction of resilient functions using simplex code has been discussed in [5]. A simplex code [13] is a $[2^m - 1, m, 2^{m-1}]$ linear code, whose minimal distance is maximal. By concatenating each codeword v times, one can get a $[v(2^m - 1), m, v2^{m-1}]$ linear code. Given Theorem 1, one can use such codes for construction of functions with order of resiliency $v2^{m-1} - 1$.

Given a linear $[u, m, t+1]$ code, where fixing u, m the maximum possible $t+1$ value can be achieved, will obviously be the most well suited for our construction as this will maximize the order of resiliency. Such table for $u, m \leq 127$ is available in [3].

5   Results and Comparison

In this section we compare the results obtained using the techniques presented in the previous section with the existing results. It was demonstrated that for a low order of resiliency and a moderate number of input variables the construction in [11] was superior to the other constructions, namely the constructions in [12,21]. However, the main disadvantage of the construction in [11] is the necessity of finding a set of nonintersecting linear codes of certain dimension. This may cause a large complexity for the search programs, since there is no theoretical basis for finding such a set. Next we show that our results are superior in comparison to [21,12,5]. Note that the construction of [12] gives higher nonlinearity than [21], whereas the construction of [21] provides larger order of resiliency than [12].

Theorem 3. [21, Corollary 6] *If there exists a linear (n, m, t) resilient function, then there exists a nonlinear $(n, m, t, 2^{n-1} - 2^{n-\frac{m}{2}})$ whose algebraic degree is $m - 1$.*

Note that given any $[u, m, t+1]$ code, it is easy to construct a linear (u, m, t) function. Thus, using the method of [21] it is possible to construct a nonlinear

(u, m, t) function also. Consequently, for $n = u$, the result of [21] provides the presently best known parameters. Note that there are some cases (when the value of n is very close to u, which falls under item 1 of Theorem 1) where the results of [21] are better than ours. This is when $u - 1 > n - \frac{m}{2}$, i.e., $n < u + \frac{m}{2} - 1$. However, if we fix the values of m, t, then for the values of n that falls under items 2, 3, 4 and 5 of Theorem 1 (and also under item 1 when $n \geq u + \frac{m}{2} - 1$), our nonlinearity supersedes that of [21]. Hence, as we choose n comparatively larger than u, $n \geq u + \frac{m}{2} - 1$, the advantage of [21] decreases and our method provides better result. Moreover, the items 3, 4, 5 of Theorem 2 show that the algebraic degree of our construction is better than $(m - 1)$ given in [21]. We present an example here for the comparison.

We know the existence of a $[36, 8, 16]$ linear code. Hence, it is easy to get a linear $(36, 8, 15)$ resilient function. Using the method of [21] it is possible to get a $(36, 8, 15, 2^{36-1} - 2^{36-\frac{8}{2}} = 2^{35} - 2^{32})$ function. Moreover, it has been mentioned in [12, Proposition 19] how to get a $(36, 8, 15, 2^{35} - 2^{31})$ function using the technique of [21]. Our method can not provide a function with these parameters. Let us now construct a function on larger number of input variables, say $n = 43$, for same m and t. For $n = 43$ and $t = 15$ the best known linear code have the parameters $[43, 12, 16]$. Then, with construction in [21], it is possible to construct a $(43, 12, 15, 2^{42} - 2^{37})$ and consequently a $(43, 8, 15, 2^{42} - 2^{37})$ function using less number of output columns. In our construction we start with a $[36, 8, 16]$ code and applying item 1 of Theorem 1 we obtain a $(43, 8, 15, 2^{42} - 2^{35})$ function which provides better nonlinearity.

Theorem 4. [12, Theorem 18] *For any even l such that $l \geq 2m$, if there exists an $(n - l, m, t)$ function $\Phi(x)$, then there exists an $(n, m, t, 2^{n-1} - 2^{n-\frac{l}{2}-1})$ resilient function.*

Note that if there exists a linear $[u = n - l, m, t + 1]$ code, then by the above theorem [12] it is possible to get the nonlinearity $2^{n-1} - 2^{n-\frac{n-u}{2}-1} = 2^{n-1} - 2^{\frac{n+u}{2}-1}$. Items 4 and 5 of our Theorem 1 provide better nonlinearity than [12]. Also a closer look reveals that our construction outperforms the result of [12] for any $n > u$, with same quality result for $n = u + 2m$.

Next we compare our result with a very recent work [5].

Theorem 5. [5, Theorem 5] *Given a linear $[u, m, t + 1]$ code $(0 < m \leq u)$, for any nonnegative integer Δ, there exists a $(u + \Delta + 1, m, t)$ resilient function with algebraic degree Δ, whose nonlinearity is greater than or equal to $2^{u+\Delta} - 2^u \lfloor \sqrt{2^{u+\Delta+1}} \rfloor + 2^{u-1}$.*

Thus it is clear that given a linear $[u, m, t + 1]$ code, the above construction provides $(n, m, t, 2^{n-1} - 2^{\frac{n+2u}{2}} + 2^{u-1})$ resilient function. Note that the construction provides some nonlinearity only when $n - 1 \geq \frac{n+2u}{2}$, i.e., $n \geq 2u + 2$. It is very clear that our construction of $(n, m, t, 2^{n-1} - 2^{\lfloor \frac{n+u-m-1}{2} \rfloor})$ resilient functions for $n \geq u + 3m$ presents much better nonlinearity than that of [5]. However, comparing our result in Theorem 2 with [5, Theorem 5], it is clear that in terms of algebraic degree the result of [5] is superior to our result. It will be of interest to construct functions with nonlinearity as good as our results with better algebraic degree as given in [5].

5.1   Examples

Next we compare the results with specific examples. Let us start with the construction of a $(24, 4, 2, nl(F))$ function $F(x)$. Given $m = 4$, it is possible to construct a nonlinear function $F(x)$ using the technique in [21] with $nl(F) \geq 2^{23} - 2^{22}$. We know the existence of $[7, 4, 3]$ linear Hamming code [13]. This gives $(7, 4, 2)$ resilient function. Using the construction in [12], we obtain a function $F(x)$ with $nl(F) > 2^{23} - 2^{15}$.

In our notation, $u = 7, m = 4, t = 2$. In this case, $n - u - m + 1 = 24 - 7 - 4 + 1 = 14$ and $n = 24 \geq u + 3m - 1 = 18$. Thus from Theorem 1, we get the nonlinearity $2^{23} - 2^{13}$. Thus, our technique provides the currently best known nonlinearity.

Starting with a $[7, 4, 3]$ code, if we use the construction of [5], we will get $(24, 4, 2, 2^{23} - 2^{19} + 2^6)$ resilient function. To obtain the same value of nonlinearity using the construction in [11], one is forced to find $|\mathcal{C}| = \lceil 2^{n-u'} / (2^m - 1) \rceil = \lceil 2^{10} / 15 \rceil$ nonintersecting linear $[14, 4, 3]$ codes, and this is computationally an extremely hard problem to solve.

In [12] the construction of a $(36, 8, 5, nl(F))$ function was discussed. Using a linear $[18, 8, 6]$ code the authors proved the existence of $(36, 8, 5, nl(F))$ function, where $nl(F) \geq 2^{35} - 2^{26}$. We use a linear $[17, 8, 6]$ code [3] to construct a $(36, 8, 5, 2^{35} - 2^{24})$ function (here $n \geq u + 2m$) by means of Theorem 1. Using the same linear code, we can obtain a $(40, 8, 5, 2^{39} - 2^{24})$ function (here $n \geq u + 3m - 1$).

Nonlinearity of $(36, 8, t)$ resilient functions has been used as important examples in [12,11]. We here compare our results with existing ones.

In this table the results of [12] are the existing best known construction results and our results clearly supersede these [12]. The results of [11] are not the construction results. They show that resilient functions with such parameters exist. However, the construction of functions with such parameters are not available in [11]. Note that, for resiliency of orders 3, 2 and 1 our construction provides better results than the existential bound in [11]. In the last row of Table 1, we describe the linear codes [3] which we use for our construction.

Table 1. Nonlinearity of $(36, 8, t)$ resilient functions.

Order of resiliency t	7	5	4	3	2	1
Nonlinearity of [12]	$2^{35} - 2^{27}$	$2^{35} - 2^{26}$	$2^{35} - 2^{25}$	$2^{35} - 2^{24}$	$2^{35} - 2^{23}$	$2^{35} - 2^{22}$
Nonlinearity of [11]	$2^{35} - 2^{22}$	$2^{35} - 2^{23}$	$2^{35} - 2^{22}$	$2^{35} - 2^{22}$	$2^{35} - 2^{21}$	$2^{35} - 2^{21}$
Our nonlinearity	$2^{35} - 2^{27}$	$2^{35} - 2^{24}$	$2^{35} - 2^{23}$	$2^{35} - 2^{20}$	$2^{35} - 2^{19}$	$2^{35} - 2^{18}$
The codes	$[20, 8, 8]$	$[17, 8, 6]$	$[16, 8, 5]$	$[13, 8, 4]$	$[12, 8, 3]$	$[9, 8, 2]$

6   Conclusion

A new generalized construction of highly nonlinear resilient multiple output functions has been provided. The construction is based on the use of linear codes together with a specific set of bent functions. We show that our construction outperforms all previous constructions for almost all choices of input parameters n, m, t. Many examples are provided demonstrating the better nonlinearity attained using this new construction in comparison to the previous ones. It will be of interest to construct functions with better nonlinearity than our method or to show that some of our constructions provide optimized nonlinearity which can not be improved further.

References

1. C. H. Bennet, G. Brassard, and J. M. Robert. Privacy amplification by by public discussion. *SIAM Journal on Computing*, 17:210–229, 1988.
2. J. Bierbrauer, K. Gopalakrishnan, and D. R. Stinson. Bounds on resilient functions and orthogonal arrays. In *Advances in Cryptology - CRYPTO'94*, number 839 in Lecture Notes in Computer Science, pages 247–256. Springer Verlag, 1994.
3. A. Brouwer and T. Verhoeff. An updated table of minimum-distance bounds for binary linear codes. *IEEE Transactions on Information Theory*, 39(2):662–677, 1993.
4. J. H. Cheon and S. Chee. Elliptic Curves and Resilient Functions. In *ICISC 2000*, number 2015 in Lecture Notes in Computer Science, pages 64–72. Springer Verlag, 2000.
5. J. H. Cheon. Nonlinear Vector Resilient Functions. In *Advances in Cryptology - CRYPTO 2001*, Lecture Notes in Computer Science. Springer Verlag, 2001.
6. B. Chor, O. Goldreich, J. Hastad, J. Friedman, S. Rudich, and R. Smolensky. The bit extraction problem or t-resilient functions. In *26th IEEE Symposium on Foundations of Computer Science*, pages 396–407, 1985.
7. C. Ding, G. Xiao, and W. Shan, The stability theory of stream ciphers, *Number 561, Lecture Notes in Computer Science*, Springer-Verlag, 1991.
8. J. Friedman. On the bit extraction problem. In *33rd IEEE Symposium on Foundations of Computer Science*, pages 314–319, 1982.
9. K. Gopalakrishnan. A study of Correlation-immune, resilient and related cryptographic functions. *PhD thesis, University of Nebraska*, 1994.
10. X. Guo-Zhen and J. Massey. A spectral characterization of correlation immune combining functions. *IEEE Transactions on Information Theory*, 34(3):569–571, May 1988.
11. T. Johansson and E. Pasalic, A construction of resilient functions with high nonlinearity, In *IEEE International Symposium on Information Theory*, ISIT, June 2000, full version available at *Cryptology ePrint Archive, eprint.iacr.org, No.2000/053*.
12. K. Kurosawa, T. Satoh, and K. Yamamoto Highly nonlinear t-Resilient functions. *Journal of Universal Computer Science*, vol. 3, no. 6, pp. 721–729, Springer Publishing Company, 1997.
13. F. J. MacWilams and N. J. A. Sloane. *The Theory of Error Correcting Codes.* North Holland, 1977.
14. A. Menezes, P. Van Oorschot, and S. Vanstone, *Handbook of applied cryptography*, CRC Press, 1997.

15. K. Nyberg. Constructions of bent functions and difference sets. In *Advances in Cryptology - EUROCRYPT 1990*, number 473 in Lecture Notes in Computer Science, pages 151–160. Springer Verlag, 1991.
16. P. Sarkar and S. Maitra. Construction of nonlinear Boolean functions with important cryptographic properties. In *Advances in Cryptology - EUROCRYPT 2000*, number 1807 in Lecture Notes in Computer Science, pages 485–506. Springer Verlag, 2000.
17. T. Siegenthaler. Correlation-immunity of nonlinear combining functions for cryptographic applications. *IEEE Transactions on Information Theory*, IT-30(5):776–780, September 1984.
18. T. Siegenthaler, Decrypting a class of stream ciphers using ciphertext only. *IEEE Trans. Comput.*, vol. C-34, pp. 81–85, 1985.
19. D. R. Stinson. Resilient functions and large sets of orthogonal arrays. *Congressus Numerantium*, 92:105–110, 1993.
20. D. R. Stinson and J. L. Massey. An infinite class of counterexamples to a conjecture concerning non-linear resilient functions. *Journal of Cryptology*, 8(3):167–173, 1995.
21. X. M. Zhang and Y. Zheng. Cryptographically resilient functions. *IEEE Transactions on Information Theory*, 43(5):1740–1747, 1997.

New Covering Radius of Reed-Muller Codes for *t*-Resilient Functions

Tetsu Iwata[1], Takayuki Yoshiwara[1], and Kaoru Kurosawa[2]

[1] Department of Communications and Integrated Systems,
Tokyo Institute of Technology,
2-12-1 O-okayama, Meguro-ku, Tokyo 152-8552, Japan
`tez@ss.titech.ac.jp`
[2] Department of Computer and Information Sciences,
Ibaraki University,
4-12-1 Nakanarusawa, Hitachi, Ibaraki, 316-8511, Japan
`kurosawa@cis.ibaraki.ac.jp`

Abstract. In stream ciphers, we should use a *t*-resilient Boolean function $f(X)$ with large nonlinearity to resist fast correlation attacks and linear attacks. Further, in order to be secure against an *extension* of linear attacks, we wish to find a *t*-resilient function $f(X)$ which has a large distance even from low degree Boolean functions. From this point of view, we define a new covering radius $\hat{\rho}(t, r, n)$ as the maximum distance between a *t-resilient* function $f(X)$ and the *r*-th order Reed-Muller code $RM(r, n)$. We next derive its lower and upper bounds. Finally, we present a table of numerical bounds for $\hat{\rho}(t, r, n)$.

Keywords: Nonlinearity, *t*-resilient function, Reed-Muller code, covering radius, stream cipher.

1 Introduction

Nonlinearity and resiliency are two of the most important cryptographic criteria of Boolean functions which are used in stream ciphers and block ciphers. The nonlinearity of a Boolean function $f(X)$, denoted by $nl(f)$, is the distance between $f(X)$ and the set of affine (linear) functions. It must be large to avoid linear attacks.

$f(X)$ is said to be balanced if $\#\{X \mid f(X) = 0\} = \#\{X \mid f(X) = 1\} = 2^{n-1}$, where $X = (x_1, \ldots, x_n)$. Suppose that $f(X)$ is balanced even if any t variables $x_{i_1}, \ldots, x_{i_t}$ are fixed to any t values $b_{i_1}, \ldots, b_{i_t}$. Then $f(X)$ is called a *t*-resilient function. $f(X)$ should be *t*-resilient for large t to resist fast correlation attacks in stream ciphers such as combination generators and nonlinear filter generators.

Therefore, $f(X)$ should satisfy both large nonlinearity $nl(f)$ and large resiliency. Recently, Sarkar and Maitra derived an upper bound on $nl(f)$ of *t*-resilient functions [5].

We further observe that $f(X)$ should not be approximated even by low degree Boolean functions $g(X)$ in order to be secure against an *extension* of linear

S. Vaudenay and A. Youssef (Eds.): SAC 2001, LNCS 2259, pp. 75–86, 2001.

attacks [3]. Note that the set of n variable Boolean functions $g(X)$ such that $\deg(g) \leq r$ is identical to an error correcting code known as the r-th order Reed-Muller code $RM(r, n)$.

Consequently, we wish to find a t-resilient function $f(X)$ which has a large distance even from $RM(r, n)$ for small r. On the other hand, the covering radius of $RM(r, n)$, denoted by $\rho(r, n)$, is defined as the maximum distance between $f(X)$ and $RM(r, n)$, where the maximum is taken over *all* n variable Boolean functions $f(X)$. That is,

$$\rho(r, n) \stackrel{\text{def}}{=} \max_{f(X)} d(f(X), RM(r, n)).$$

In this paper, we introduce a new definition of covering radius of $RM(r, n)$ from this point of view. We define *t-resilient covering radius* of $RM(r, n)$, denoted by $\hat{\rho}(t, r, n)$, as the maximum distance between a t-resilient function $f(X)$ and $RM(r, n)$, where the maximum is taken over all *t-resilient functions* $f(X)$. That is,

$$\hat{\rho}(t, r, n) \stackrel{\text{def}}{=} \max_{t\text{-resilient } f(X)} d(f(X), RM(r, n)).$$

We then derive lower bounds and upper bounds on $\hat{\rho}(t, r, n)$. The result of Sarkar and Maitra [5] is obtained as a special case of one of our upper bounds. Finally, we present a table of numerical bounds for $\hat{\rho}(t, r, n)$ which are derived from our bounds.

2   Preliminaries

Let $X = (x_1, \ldots, x_n)$.

2.1   Nonlinearity of Boolean Functions

Define the distance between two Boolean functions $f(X)$ and $g(X)$ as

$$d(f(X), g(X)) \stackrel{\text{def}}{=} \#\{X \mid f(X) \neq g(X)\} .$$

Define the weight of $f(X)$ as

$$w(f) \stackrel{\text{def}}{=} \#\{X \mid f(X) = 1\} .$$

A Boolean function such that $a_0 \oplus a_1 x_1 \oplus \cdots \oplus a_n x_n$ is called an affine function. Let A_n denote the set of n variable affine functions. That is,

$$A_n \stackrel{\text{def}}{=} \{a_0 \oplus a_1 x_1 \oplus \cdots \oplus a_n x_n\} .$$

The nonlinearity of $f(X)$, denoted by $nl(f)$, is defined as the distance between $f(X)$ and A_n. That is,

$$nl(f) \stackrel{\text{def}}{=} \min_{g(X) \in A_n} d(f(X), g(X)) .$$

Cryptographically secure Boolean functions should have large nonlinearity to resist linear attacks. Then the following upper bound is known.

$$nl(f) \leq 2^{n-1} - 2^{\frac{n}{2}-1} .$$

It is tight if $n = even$. $f(X)$ which satisfies the above equality is called a bent function.

2.2 t-Resilient Function and its Nonlinearity

$f(X)$ is said to be balanced if

$$\#\{X \mid f(X) = 1\} = \#\{X \mid f(X) = 0\} = 2^{n-1} .$$

Suppose that $f(X)$ is balanced even if any t variables $x_{i_1}, \ldots, x_{i_t}$ are fixed to any values $b_{i_1}, \ldots, b_{i_t}$. Then $f(X)$ is called a t-resilient function. Boolean functions used in stream ciphers should be t-resilient for large t to resist fast correlation attacks.

Therefore, $f(X)$ should satisfy both large nonlinearity $nl(f)$ and large resiliency. Sarkar and Maitra derived an upper bound on $nl(f)$ of t-resilient functions [5].

Proposition 2.1. Let $f(X)$ be a t-resilient function and $l(X)$ be an affine function. Then

$$d(f(X), l(X)) \equiv 0 \bmod 2^{t+1}.$$

Proposition 2.2. Suppose that $f(X)$ is a t-resilient function. If $n = even$, then

$$nl(f) \leq \begin{cases} 2^{n-1} - 2^{t+1} & \text{if } t+1 > n/2 - 1 \\ 2^{n-1} - 2^{\frac{n}{2}-1} - 2^{t+1} & \text{if } t+1 \leq n/2 - 1 \end{cases}$$

They derived a similar bound for $n = odd$.

3 Reed-Muller Code and Its Covering Radius

Any Boolean function is written as the algebraic normal form such that

$$g(X) = a_0 \oplus \bigoplus_{1 \leq i \leq n} a_i x_i \oplus \bigoplus_{1 \leq i < j \leq n} a_{i,j} x_i x_j \oplus \cdots \oplus a_{1,2,\ldots,n} x_1 x_2 \cdots x_n$$

The degree of $g(X)$, denoted by $\deg(g)$, is the degree of the highest degree term in the algebraic normal form. The r-th order Reed-Muller code $RM(r, n)$ is identical to the set of n-variable Boolean function $g(X)$ such that $\deg(g) \leq r$.

The covering radius of $RM(r, n)$, denoted by $\rho(r, n)$, is defined as the maximum distance between $f(X)$ and $RM(r, n)$, where the maximum is taken over all n variable Boolean functions $f(X)$. That is,

$$\rho(r, n) \stackrel{\text{def}}{=} \max_{f(X)} d(f(X), RM(r, n)),$$

where

$$d(f(X), RM(r, n)) \overset{\text{def}}{=} \min_{\deg(g) \le r} d(f(X), g(X)).$$

Note that $\rho(1, n)$ is equal to the maximum nonlinearity of n-variable Boolean functions.

In the following table, the best known numerical bounds for $\rho(r, n)$ with $n \le 7$ are presented.

n	1	2	3	4	5	6	7
$r = 1$	0	1	2	$6^{[4]}$	12	28	56
$r = 2$		0	1	2	$6^{[4]}$	$18^{[6]}$	$40^{[1]}$-$44^{[2]}$
$r = 3$			0	1	2	$8^{[4]}$	$20^{[1]}$-$23^{[1]}$
$r = 4$				0	1	2	$8^{[4]}$
$r = 5$					0	1	2
$r = 6$						0	1
$r = 7$							0

It is easy to see the following propositions.

Proposition 3.1. *Any Boolean function $f(x_1, \ldots, x_n)$ such that $\deg(f) \le r$ is written as*

$$f(X) = f_1(x_1, \ldots, x_{n-1}) \oplus x_n \cdot f_2(x_1, \ldots, x_{n-1}) ,$$

where $\deg(f_1) \le r$ and $\deg(f_2) \le r - 1$.

Proposition 3.2. $d(f, g \oplus h) \ge d(f, g) - w(h)$.

Proof.

$$\begin{aligned}
d(f, g \oplus h) &= w(f \oplus g \oplus h) \\
&\ge w(f \oplus g) - w(h) \\
&= d(f, g) - w(h)
\end{aligned}$$

$\square$

4   New Covering Radius for t-Resilient Functions

4.1   New Covering Radius

Boolean functions $f(X)$ used in stream ciphers and block ciphers should not be approximated by affine (linear) functions to resist linear attacks. This leads to the notion of the nonlinearity $nl(f)$ which is defined as the distance between $f(X)$ and the set of affine (linear) functions.

We also observe that $f(X)$ should not be approximated even by low degree Boolean functions to resist an extension of linear attacks [3]. Remember that

$RM(r, n)$ is identical to the set of $g(X)$ such that $\deg(g) \leq r$, and the covering radius of $RM(r, n)$ is the maximum distance between $f(X)$ and $RM(r, n)$. That is,

$$\rho(r, n) = \max_{f(X)} d(f(X), RM(r, n)).$$

Further, $f(X)$ should be t-resilient to be secure against fast correlation attacks in stream ciphers.

In this section, we introduce a new definition of covering radius of $RM(r, n)$ from this point of view. We define t-*resilient covering radius* of $RM(r, n)$, denoted by $\hat{\rho}(t, r, n)$, as the maximum distance between a t-resilient function $f(X)$ and $RM(r, n)$, where the maximum is taken over all t-*resilient functions* $f(X)$. That is,

$$\hat{\rho}(t, r, n) \stackrel{\text{def}}{=} \max_{t\text{-resilient } f(X)} d(f(X), RM(r, n)).$$

Note that $\hat{\rho}(t, r, n) = 0$ if $n - t - 1 \leq r$. This follows immediately from Siegenthalar's inequality on resilient functions [7].

We then derive lower bounds and upper bounds on $\hat{\rho}(t, r, n)$.

4.2   Lower Bounds on $\hat{\rho}(t, r, n)$

In this subsection, we derive lower bounds on $\hat{\rho}(t, r, n)$.

Theorem 4.1.

$$\hat{\rho}(t, r, n) \geq \begin{cases} 2\rho(r, n-1) & \text{if } t = 0 \\ 2\hat{\rho}(t-1, r, n-1) & \text{if } t \geq 1 \end{cases}$$

Proof. **(1)** $t = 0$.   Suppose that $\rho(r, n-1)$ is achieved by $f'(x_1, \ldots, x_{n-1})$. That is,

$$d(f', RM(r, n-1)) = \rho(r, n-1) .$$

Let $f(x_1, \ldots, x_n) = f'(x_1, \ldots, x_{n-1}) \oplus x_n$. Then it is easy to see that $f(x_1, \ldots, x_n)$ is balanced. Therefore, $f(X)$ is a 0-resilient function. Further,

$$\begin{aligned} \hat{\rho}(t, r, n) &\geq d(f, RM(r, n)) \\ &= d(f', RM(r, n-1)) + d(f', RM(r, n-1)) \\ &= 2\rho(r, n-1) \end{aligned}$$

(2) $t \geq 1$.   Suppose that $\hat{\rho}(t-1, r, n-1)$ is achieved by a $(t-1)$-resilient function $f'(x_1, \ldots, x_{n-1})$. That is,

$$d(f', RM(r, n-1)) = \hat{\rho}(t-1, r, n-1) .$$

Let $f(x_1, \ldots, x_n) = f'(x_1, \ldots, x_{n-1}) \oplus x_n$. Then it is easy to see that $f(x_1, \ldots, x_n)$ is a t-resilient function. The rest of the proof is similar to the above.

$\square$

Corollary 4.1. $\hat{\rho}(t, r, n) \geq 2^{t+1}\rho(r, n - t - 1)$.

Theorem 4.2. *Suppose that there exists* $f(x_1, \ldots, x_n)$ *such that*

$$d(f, RM(r, n)) \geq k$$

and

$$f(x_1, \ldots, x_n) = f_1(x_1, \ldots, x_m) \oplus f_2(x_l, \ldots, x_n)$$

for some f_1 *and* f_2, *where* $1 \leq m \leq n - 1$, $2 \leq l \leq n - 1$. *Let*

$$t = \min(n - m - 1, l - 2).$$

Then

$$\hat{\rho}(t, r + 1, n + 1) \geq k.$$

Proof. Let

$$\begin{cases} h_1(x_1, \ldots, x_n) \stackrel{\text{def}}{=} f_1(x_1, \ldots, x_m) \oplus x_{m+1} \oplus \cdots \oplus x_n \\ h_2(x_1, \ldots, x_n) \stackrel{\text{def}}{=} x_1 \oplus \cdots \oplus x_{l-1} \oplus f_2(x_l, \ldots, x_n) \end{cases}$$

It is easy to see that $h_1(X)$ is $(n - m - 1)$-resilient and $h_2(X)$ is $(l - 2)$-resilient. Then define

$$h(X, x_{n+1}) \stackrel{\text{def}}{=} h_1(X) \oplus x_{n+1} \cdot (h_1(X) \oplus h_2(X)) \ ,$$

where $X = (x_1, \ldots, x_n)$.

We first show that h is t-resilient. For $x_{n+1} = 0$,

$$h(X, 0) = h_1(X)$$

which is $(n - m - 1)$-resilient. For $x_{n+1} = 1$,

$$h(X, 1) = h_2(X)$$

which is $(l - 2)$-resilient. Therefore, $h(X, x_{n+1})$ is t-resilient, where $t = \min(n - m - 1, l - 2)$.

We next prove that $d(h, RM(r + 1, n + 1)) \geq k$. Choose $g(X, x_{n+1})$ such that $\deg(g) \leq r + 1$ and

$$d(h, g) = d(h, RM(r + 1, n + 1)) \ .$$

From Proposition 3.1, g is written as

$$g(X, x_{n+1}) = g_1(X) \oplus x_{n+1} \cdot g_2(X)$$

for some $g_1 \in RM(r + 1, n)$ and $g_2 \in RM(r, n)$. Then from Proposition 3.2,

$$\begin{aligned} d(h, g) &= d(h, g)|_{x_{n+1}=0} + d(h, g)|_{x_{n+1}=1} \\ &= d(h_1, g_1) + d(h_2, g_1 \oplus g_2) \\ &= d(h_1, g_1) + d(h_1 \oplus h_2, h_1 \oplus g_1 \oplus g_2) \\ &\geq d(h_1, g_1) + d(h_1 \oplus h_2, g_2) - w(h_1 \oplus g_1) \\ &= d(h_1 \oplus h_2, g_2) \end{aligned}$$

Let $l(X) \stackrel{\text{def}}{=} x_1 \oplus \cdots \oplus x_{l-1} \oplus x_{m+1} \oplus \cdots \oplus x_n$. Then

$$d(h, g) \geq d(h_1 \oplus h_2, g_2)$$
$$= d(f_1 \oplus f_2 \oplus l, g_2)$$
$$= d(f_1 \oplus f_2, g_2 \oplus l)$$
$$\geq d(f, RM(r, n))$$

because $g_2 \in RM(r, n)$ and $g_2 \oplus l \in RM(r, n)$. Hence

$$d(h, RM(r + 1, n + 1)) = d(h, g)$$
$$\geq d(f, RM(r, n))$$
$$\geq k$$

$\square$

Corollary 4.2. $\hat{\rho}(0, 3, 7) \geq 18$.

Proof. Let

$$f(x_1, \ldots, x_6) = (x_1 x_2 x_3 \oplus x_1 x_4 x_5) \oplus (x_2 x_3 x_6 \oplus x_2 x_4 x_6 \oplus x_3 x_5 x_6) .$$

Then it is known that [6]

$$d(f, RM(2, 6)) = 18 .$$

Let $r = 2$, $n = 6$, $m = 5$ and $l = 2$ in Theorem 4.2. Then we obtain this corollary. $\square$

Corollary 4.3. *Suppose that* $n = 4k + s$, *where* $0 \leq s \leq 3$ *and* $k \geq 1$. *Let* $t = 2k - 1$. *Then*

$$\hat{\rho}(t, 2, n + 1) \geq \begin{cases} 2^{n-1} - 2^{\frac{n}{2}-1} & \text{if } n = \text{even} \\ 2^{n-1} - 2^{\frac{n-1}{2}} & \text{if } n = \text{odd} \end{cases}$$

Proof. For $n = \text{even}$, let

$$f(x_1, \ldots, x_n) = x_1 x_2 \oplus x_3 x_4 \oplus \cdots \oplus x_{n-1} x_n .$$

Then it is known that

$$d(f, RM(1, n)) = 2^{n-1} - 2^{\frac{n}{2}-1}$$

(f is a bent function). In Theorem 4.2, let

$$\begin{cases} f_1(x_1, \ldots, x_{2k}) = x_1 x_2 \oplus \cdots \oplus x_{2k-1} x_{2k}, \\ f_2(x_{2k+1}, \ldots, x_n) = x_{2k+1} x_{2k+2} \oplus \cdots \oplus x_{n-1} x_n \end{cases}$$

Then $m = 2k$ and $l = 2k + 1$. Hence

$$t = \min(n - 2k - 1, 2k + 1 - 2)$$
$$= \min(4k + s - 2k - 1, 2k - 1)$$
$$= 2k - 1$$

because $s \geq 0$.

For $n = odd$, let

$$f(x_1, \ldots, x_n) = x_1 x_2 \oplus x_3 x_4 \oplus \cdots \oplus x_{n-2} x_{n-1} .$$

Then for any $g(x_1, \ldots, x_n)$ such that $\deg(g) \leq 1$,

$$
\begin{aligned}
d(f, g) &= d(f, g)|_{x_n=0} + d(f, g)|_{x_n=1} \\
&\geq d(f, RM(1, n-1)) + d(f, RM(1, n-1)) \\
&= 2 \left(2^{n-2} - 2^{\frac{n-1}{2}-1} \right) \\
&= 2^{n-1} - 2^{\frac{n-1}{2}}
\end{aligned}
$$

Hence

$$d(f, RM(1, n)) \geq 2^{n-1} - 2^{\frac{n-1}{2}} .$$

Finally similarly to $n = even$, we have $t = 2k - 1$.

Therefore, this corollary holds from Theorem 4.2. $\square$

4.3   Upper Bounds on $\hat{\rho}(t, r, n)$

In this subsection, we derive upper bounds on $\hat{\rho}(t, r, n)$.

Theorem 4.3. *For $t \geq 1$,*

$$\hat{\rho}(t, r, n) \leq \hat{\rho}(t-1, r, n-1) + \rho(r-1, n-1) .$$

Proof. Any $f(x_1, \ldots, x_n)$ and $g(x_1, \ldots, x_n)$ are written as

$$
\begin{cases}
f(x_1, \ldots, x_n) = f_1(x_1, \ldots, x_{n-1}) \oplus x_n \cdot f_2(x_1, \ldots, x_{n-1}), \\
g(x_1, \ldots, x_n) = g_1(x_1, \ldots, x_{n-1}) \oplus x_n \cdot g_2(x_1, \ldots, x_{n-1}).
\end{cases}
$$

Then

$$
\begin{aligned}
d(f, g) &= d(f, g)|_{x_n=0} + d(f, g)|_{x_n=1} \\
&= d(f_1, g_1) + d(f_1 \oplus f_2, g_1 \oplus g_2) \\
&= d(f_1, g_1) + d(f_1 \oplus f_2 \oplus g_1, g_2)
\end{aligned}
$$

Now let f be any t-resilient function such that

$$d(f, RM(r, n)) = \hat{\rho}(t, r, n) . \tag{1}$$

Choose g_1 such that $\deg(g_1) \leq r$ and

$$d(f_1, g_1) = d(f_1, RM(r, n-1))$$

arbitrarily. Choose g_2 such that $\deg(g_2) \leq r - 1$ and

$$d(f_1 \oplus f_2 \oplus g_1, g_2) = d(f_1 \oplus f_2 \oplus g_1, RM(r-1, n-1))$$

arbitrarily. Then

(1). $\deg(g) \leq r$. Therefore,

$$d(f, g) \geq d(f, RM(r, n)) = \hat{\rho}(t, r, n) .$$

(2). f_1 is $(t-1)$-resilient. Therefore,

$$d(f_1, g_1) = d(f_1, RM(r, n-1)) \leq \hat{\rho}(t-1, r, n-1) .$$

(3). It is easy to see

$$d(f_1 \oplus f_2 \oplus g_1, g_2) \leq \rho(r-1, n-1) .$$

Therefore,

$$\begin{aligned} \hat{\rho}(t, r, n) &\leq d(f, g) \\ &= d(f_1, g_1) + d(f_1 \oplus f_2 \oplus g_1, g_2) \\ &\leq \hat{\rho}(t-1, r, n-1) + \rho(r-1, n-1) . \end{aligned}$$

$\square$

Lemma 4.1. *Suppose that $f(X)$ is balanced and $\deg(g(X)) \leq n-1$, where $X = (x_1, \ldots, x_n)$. Then*

$$d(f, g) \equiv 0 \bmod 2 .$$

Proof. Note that

$$d(f, g) = w(f) + w(g) - 2w(f \times g) .$$

Since $\deg(g) \leq n-1$, it holds that $w(g) \equiv 0 \bmod 2$. Therefore, it holds that $d(f, g) \equiv 0 \bmod 2$. $\square$

We finally generalize Proposition 2.1 [5] and Proposition 2.2 [5].

Theorem 4.4. *Let $1 \leq r \leq n-2$ and $0 \leq t \leq n-r-2$. If $f(x_1, \ldots, x_n)$ is a t-resilient function, then*

$$d(f, RM(r, n)) \equiv 0 \bmod 2^{\lfloor \frac{t}{r} \rfloor + 1} .$$

Proof. We show that

$$d(f(X), g(X)) \equiv 0 \bmod 2^{\lfloor \frac{t}{r} \rfloor + 1} \tag{2}$$

for any $g(X)$ such that $\deg(g) \leq r$, where $X = (x_1, \ldots, x_n)$. Let $\alpha(g, r)$ be the number of degree r terms $x_{i_1} \cdots x_{i_r}$ involved in g.

Base step on r. If $r = 1$, then the theorem follows from Proposition 2.1.

Inductive step on r. Assume that (2) is true for $r = r_0$. We will show that it is true for $r = r_0 + 1$.

Base step on $\alpha(g, r_0 + 1)$. If $\alpha(g, r_0 + 1) = 0$, then $g(x_1, \ldots, x_n) \in RM(r_0, n)$. By an induction hypothesis on r, we have

$$\begin{aligned} d(f, g) &\equiv 0 \bmod 2^{\lfloor \frac{t}{r_0} \rfloor + 1} \\ &\equiv 0 \bmod 2^{\lfloor \frac{t}{r_0 + 1} \rfloor + 1} . \end{aligned}$$

Inductive step on $\alpha(g, r_0+1)$. Assume that (2) is true for $\alpha(g, r_0+1) \leq \alpha_0$. We show that (2) is true for $\alpha(g, r_0 + 1) = \alpha_0 + 1$. Without loss of generality, we assume that

$$g(x_1, \ldots, x_n) = x_1 \cdots x_{r_0+1} \oplus g^*(x_1, \ldots, x_n)$$

for some g^* such that $\alpha(g^*, r_0 + 1) = \alpha_0$.

Define

$$\begin{cases} f_{b_1 \ldots b_{r_0+1}} \stackrel{\text{def}}{=} f(b_1, \ldots, b_{r_0+1}, x_{r_0+2}, \ldots, x_n) \\ g^*_{b_1 \ldots b_{r_0+1}} \stackrel{\text{def}}{=} g^*(b_1, \ldots, b_{r_0+1}, x_{r_0+2}, \ldots, x_n) \\ d_{b_1 \ldots b_{r_0+1}} \stackrel{\text{def}}{=} d(f_{b_1 \ldots b_{r_0+1}}, g^*_{b_1 \ldots b_{r_0+1}}) \end{cases}$$

Then we have

$$\begin{cases} d(f, g^*) = d_{0 \ldots 0} + \cdots + d_{1 \ldots 10} + d_{1 \ldots 1} = 2^{\lfloor \frac{t}{r_0+1} \rfloor + 1} k \\ d(f, g) = d_{0 \ldots 0} + \cdots + d_{1 \ldots 10} + 2^{n-(r_0+1)} - d_{1 \ldots 1} \end{cases}$$

for some integer k by an induction hypothesis on $\alpha(g, r_0 + 1)$. Therefore we have

$$d(f, g) = 2^{\lfloor \frac{t}{r_0+1} \rfloor + 1} k + 2^{n-(r_0+1)} - 2d_{1 \ldots 1} .$$

From our condition on the parameters, it holds that

$$t \leq n - (r_0 + 1) - 2 .$$

Therefore, we have

$$n - (r_0 + 1) \geq t + 2 \geq \lfloor \frac{t}{r_0 + 1} \rfloor + 1$$

Hence

$$2^{n-(r_0+1)} \equiv 0 \bmod 2^{\lfloor \frac{t}{r_0+1} \rfloor + 1} .$$

Further, from the induction hypothesis on $\alpha(g, r_0 + 1)$, we have

$$d_{1 \ldots 1} \equiv 0 \bmod 2^{\lfloor \frac{t-(r_0+1)}{r_0+1} \rfloor + 1}$$

$$\equiv 0 \bmod 2^{\lfloor \frac{t}{r_0+1} \rfloor} .$$

since $f_{1 \ldots 1}$ is a $(t-(r_0+1))$-resilient function and $\alpha(g^*_{1 \ldots 1}, r_0+1) \leq \alpha_0$. Therefore,

$$2d_{1 \ldots 1} \equiv 0 \bmod 2^{\lfloor \frac{t}{r_0+1} \rfloor + 1} .$$

Finally, putting all things together, we have

$$d(f, g) \equiv 0 \bmod 2^{\lfloor \frac{t}{r} \rfloor + 1}$$

for any g such that $\deg(g) \leq r$. Therefore, this Theorem holds. $\square$

Corollary 4.4. *If $r \leq n - t - 2$, then*

$$\hat{\rho}(t, r, n) \leq \rho(r, n) - \left(\rho(r, n) \bmod 2^{\lfloor \frac{t}{r} \rfloor + 1}\right) .$$

Proof. It is clear that $\hat{\rho}(t, r, n) \leq \rho(r, n)$. Then apply Theorem 4.4   □

Corollary 4.5. *Let $Y \overset{\text{def}}{=} \hat{\rho}(t - 1, r, n - 1) + \rho(r - 1, n - 1)$. Then*

$$\hat{\rho}(t, r, n) \leq Y - \left(Y \bmod 2^{\lfloor \frac{t}{r} \rfloor + 1}\right) .$$

Proof. From Theorem 4.3 and Theorem 4.4.   □

5 Numerical Result

We present a table of numerical values of $\hat{\rho}(t, r, n)$ which are obtained from our bounds and the previous bounds. The entry α-β means that $\alpha \leq \hat{\rho}(t, r, n) \leq \beta$.

	n	1	2	3	4	5	6	7
	$r = 1$		0	2^a	$4^{a,b}$	12^a	24^a-26^b	56^a
	$r = 2$			0	2^a	6^c	12^a-18	36^a-44
$t = 0$	$r = 3$				0	2^a	4^a-8	18^d-22^e
	$r = 4$					0	2^a	4^a-8
	$r = 5$						0	2^a
	$r = 6$							0
	n	1	2	3	4	5	6	7
	$r = 1$			0	$4^{a,g}$	8^a-12	$24^{a,b}$	56^a
	$r = 2$				0	6^f	12^a-18	28^f-44
$t = 1$	$r = 3$					0	4^a-8	8^a-22^e
	$r = 4$						0	4^a-8
	$r = 5$							0
	n	1	2	3	4	5	6	7
	$r = 1$				0	$8^{a,g}$	16^a-24^g	48^a-56
$t = 2$	$r = 2$					0	12^a-16^e	24^a-44
	$r = 3$						0	8^a-22^e
	$r = 4$							0

(a) is obtained from Theorem 4.1, (b) is obtained from Proposition 2.2, (c) is obtained from Theorem 4.2, (d) is obtained from Corollary 4.2, (e) is obtained from Corollary 4.4, (f) is obtained from Corollary 4.3, and (g) is obtained from Proposition 2.1. Unmarked values are obtained from $\rho(r, n)$.

References

1. X.D.Hou. Some results on the covering radii of Reed-Muller codes. *IEEE Transactions on Information Theory*, IT-39:366-378, 1993.
2. X.D.Hou. Further results on the covering radii of the Reed-Muller codes. *Designs, Codes and Cryptography*, vol.3, pages 167–177, 1993.
3. X.Lai. Higher order derivatives and differential cryptanalysis. In Proceedings of Symposium on Communication, Coding and Cryptography, in honor of James L.Massey on the occasion of his 60'th birthday, February 10-13, 1994, Monte-Verita, Ascona Switzerland, 1994.
4. A.M.MacLoughlin. The covering radius of the $(m-3)$-rd order Reed-Muller codes and lower bounds on the $(m-4)$-th order Reed-Muller codes. *SIAM Journal of Applied Mathematics*, vol. 37, no. 2, October 1979.
5. P.Sarkar and S.Maitra. Nonlinearity bounds and constructions of resilient Boolean functions. *Advances in Cryptology — CRYPTO 2000, LNCS 1880*, pages 515–532, 2000.
6. J.R.Schatz. The second order Reed-Muller code of length 64 has covering radius 18. *IEEE Transactions on Information Theory*, IT-27(5):529-530 September 1981.
7. T.Siegentharler. Correlation-immunity of nonlinear combining functions for cryptographic applications. *IEEE Transactions on Information Theory*, IT-30(5):776-780 September 1984.

Generalized Zig-zag Functions
and Oblivious Transfer Reductions

Paolo D'Arco[1] and Douglas Stinson[2]

[1] Dipartimento di Informatica ed Applicazioni, Università di Salerno,
84081 Baronissi (SA), Italy
paodar@dia.unisa.it
[2] Department of Combinatorics and Optimization,
University of Waterloo, Waterloo Ontario, N2L 3G1, Canada
dstinson@uwaterloo.ca

Abstract. In this paper we show some *efficient* and *unconditionally secure* oblivious transfer reductions. Our main tool is a class of functions that *generalizes* the Zig-zag functions, introduced by Brassard, Crepéau, and Sántha in [6]. We show necessary and sufficient conditions for the existence of such generalized functions, and some characterizations in terms of well known combinatorial structures. Moreover, we point out an interesting relation between these functions and ramp secret sharing schemes where each share is a single bit.

Keywords: Oblivious Transfer, Zig-zag Functions, Ramp Schemes.

1 Introduction

The oblivious transfer is a well known cryptographic primitive. Introduced by Rabin in [24], and subsequently defined in different forms in [16,5], it has found many applications in cryptographic studies and protocol design. One of the most common forms in which the oblivious transfer is used is the following[1] [5]: Let S, the Sender, and let R, the Receiver, be two players. Assume that S holds n secrets of ℓ bits and R is interested in one of them, say the i-th one. An oblivious transfer protocol enables R to receive the i-th secret out of the n S holds in such a way that

- S does not know which of the n secrets R has received
- R does not receive any information on the other secrets S holds

We will refer to such a protocol as to an $\binom{n}{1}$-OT^ℓ. All the oblivious transfer definitions [24,16,5] were shown to be equivalent [12,4,13,6]. Moreover, Kilian, in [21], showed that the oblivious transfer is *complete*; in other words, it can be used to construct *any* other cryptographic protocol. Due to the importance of the oblivious transfer many papers [6,12,11,13,14,22,23], assuming that an

---

[1] Recently, it has been pointed out that Wiesner independently developed a similar concept in 1970, unpublished until [27].

S. Vaudenay and A. Youssef (Eds.): SAC 2001, LNCS 2259, pp. 87–102, 2001.

$\binom{n}{1}$-OT^{ℓ} is available, have been focusing on designing protocols that realize an $\binom{N}{1}$-OT^{L}, where $N \geq n$ and $L \geq \ell$, using in an efficient way the given $\binom{n}{1}$-OT^{ℓ}. Such kind of protocols are usually referred to as oblivious transfer reductions.

In [14], *unconditionally secure* oblivious transfer reductions have been studied. Lower bounds on the number of times an $\binom{n}{1}$-OT^{ℓ} oblivious transfer protocol must be called to realize an $\binom{N}{1}$-OT^{L} one, as well as on the number of random bits needed to implement such a reduction, have been proven. The bounds were shown to be tight when the parameter $L = \ell$. Unfortunately, when $L > \ell$, the trivial extension of the described protocol leaks some information. Actually, a cheating receiver is able to obtain pieces of different secrets.

In this paper we focus our attention on unconditionally secure reductions of $\binom{N}{1}$-OT^{L} to $\binom{n}{1}$-OT^{ℓ} . We show how to modify the protocol proposed in [14] in order to avoid information leakage. To this aim, we investigate the properties of a class of functions that *generalizes* the Zig-zag function class introduced by Brassard, Crepéau, and Sántha in [6] in order to reduce in an unconditionally secure way $\binom{2}{1}$-OT^{ℓ} to $\binom{2}{1}$-OT^{1}. Using these generalized Zig-zag functions we set up an unconditionally secure oblivious transfer reduction of $\binom{N}{1}$-OT^{L} to $\binom{n}{1}$-OT^{ℓ}, which is optimal up to a small multiplicative constant with respect to the number of invocations of the smaller oblivious transfer needed to implement such a reduction [14].

Zig-zag functions have been deeply studied in the last years. The authors of [6] showed that linear Zig-zag functions are equivalent to a special class of codes, the self-intersecting codes [9]. Moreover, they described several efficient methods to construct these codes. On the other hand, Stinson, in [25], found bounds and combinatorial characterizations both for linear and for non-linear Zig-zag functions. Applying techniques developed in [25,26], we show necessary and sufficient conditions for the existence of generalized Zig-zag functions, and some characterizations in terms of orthogonal arrays and large set of orthogonal arrays as well.

Then, we show that the reduction presented in [14] can be viewed as a two-stage process, and using a ramp secret sharing scheme [1] in the first stage, we set up a reduction of $\binom{N}{1}$-OT^{L} to $\binom{n}{1}$-OT^{ℓ}, which is optimal with respect to the number of invocations of the available $\binom{n}{1}$-OT^{ℓ}, up to a factor 2.

Finally, we point out an interesting relation between generalized Zig-zags and ramp secret sharing schemes where the size of each share is exactly one bit.

2   Oblivious Transfer

The following definitions were given by Brassard, Crepéau, and Sántha in [6] and were used, in a slightly simplified form[2] in [14]. We refer the reader to [6] for more details.

---

[2] The goal of that paper was to find out lower bounds and the awareness condition does not influence them in any way

Assume that $\mathcal{S}$ and $\mathcal{R}$ hold two programs, S and R respectively, which specify the computations to be performed by the players to achieve $\binom{N}{1}$-OT^L. These programs encapsulate, as black box, *ideal* $\binom{n}{1}$-OT^ℓ. Hence, during the execution, $\mathcal{S}$ and $\mathcal{R}$ are able to carry out many times unconditionally secure $\binom{n}{1}$-OT^ℓ. In order to model dishonest behaviours, where one of the player tries to obtain unauthorized information from the other, we assume that a cheating $\mathcal{S}$ (resp. $\mathcal{R}$) holds a modified version of the program, denoted by $\overline{S}$ (resp. $\overline{R}$).

Let $[\mathbf{P}_0, \mathbf{P}_1](a)(b)$ be the random variable representing the *output* obtained by $\mathcal{S}$ and $\mathcal{R}$ when they execute together their own programs, P_0 held by $\mathcal{S}$ and P_1 held by $\mathcal{R}$, with private inputs a and b, respectively. Moreover, let $[\mathbf{P}_0, \mathbf{P}_1]^*(a)(b)$ be the random variable that describes the *total information* acquired during the execution of the protocol on input a and b, and let $[\mathbf{P}_0, \mathbf{P}_1]^*_{\mathcal{S}}(a)(b)$ (resp. $[\mathbf{P}_0, \mathbf{P}_1]^*_{\mathcal{R}}(a)(b)$) be the random variable obtained by restricting $[\mathbf{P}_0, \mathbf{P}_1]^*(a)(b)$ to $\mathcal{S}$ (resp. to $\mathcal{R}$). These restrictions are the *view* each player has while running the protocol.

Finally, let W be the set of all length N sequences of L-bit secrets, and, for any $w \in W$, let w_i be the i-th secret of the sequence. Denoting by $\mathbf{W}$ the random variable that represents the choice of an element in W, and by $\mathbf{T}$ the random variable representing the choice of an index i in $T = \{1, \ldots, N\}$, we can define the conditions that an $\binom{N}{1}$-OT^L oblivious transfer protocol must satisfy as follows:

Definition 1. *The pair of programs $[S, R]$ is correct for $\binom{N}{1}$-OT^L if for each $w \in W$ and for each $i \in T$*

$$\mathcal{P}([\mathbf{S}, \mathbf{R}](w)(i)) \neq (\epsilon, w_i)) = 0, \tag{1}$$

and, for any program $\overline{S}$, there exists a probabilistic program Sim such that, for each $w \in W$ and $i \in T$

$$([\overline{\mathbf{S}}, \mathbf{R}](w)(i)|\mathcal{R} \text{ accepts }) = ([\mathbf{S}, \mathbf{R}](Sim(w))(i)|\mathcal{R} \text{ accepts }). \tag{2}$$

Notice that condition (1) means that two honest players always complete successfully the execution of the protocol. More precisely, $\mathcal{R}$ receives w_T, the secret in which he is interested, while $\mathcal{S}$ receives nothing. The output pair (ϵ, w_i), where ϵ denotes the empty string, describes this situation. On the other hand, condition (2), referred to as the *awareness condition*, means that, when $\mathcal{R}$ does not abort, a dishonest $\mathcal{S}$ cannot induce on $\mathcal{R}$'s output a distribution that he could not induce by changing the input $(Sim(w))$ and being honest. As explained in [6], this condition is necessary for future uses of the output of the protocol.

Assuming that both $\mathcal{S}$ and $\mathcal{R}$ are aware of the joint probability distribution $\mathcal{P}_{W,T}$ on W and T, the probability with which $\mathcal{S}$ chooses the secrets in W and $\mathcal{R}$ chooses an index $i \in T$, and using the *mutual information*[3] between two random variables, the privacy property of $\binom{N}{1}$-OT^L can be defined as follows:

---

[3] The reader is referred to Appendix A for the definition and some basic properties of the concept of mutual information.

Definition 2. *The pair of programs* $[S, R]$ *is* private *for* $\binom{N}{1}$-OT^L *if for each* $w \in W$ *and* $i \in T$, *for any program* $\overline{S}$

$$I(\mathbf{T}; [\overline{\mathbf{S}}, \mathbf{R}]_S^*(w)(i) | \mathbf{W}) = 0, \tag{3}$$

while, for any program $\overline{R}$, *there exists a random variable* $\overline{\mathbf{T}} = f(\mathbf{T})$ *such that*

$$I(\mathbf{W}; [\mathbf{S}, \overline{\mathbf{R}}]_{\mathcal{R}}^*(w)(i) | \mathbf{T}, \mathbf{W}_{\overline{T}}) = 0. \tag{4}$$

These two conditions ensure that a dishonest S does not gain information about $\mathcal{R}$'s index; and a dishonest $\mathcal{R}$ infers at most one secret among the ones held by S.

3 Unconditionally Secure Reductions

In the literature can be found many unconditionally secure reductions of more "complex" OT to "simpler" ones [11,12,4,14]. The efficiency of such reductions has been careful analyzed in [14]. Therein, the authors considered two types of reductions: reductions for *strong* $\binom{N}{1}$-OT^L, where condition (4) of Definition 2 holds, and reductions for *weak* $\binom{N}{1}$-OT^L, where condition (4) is substituted by the following condition:

for any program $\overline{R}$ *and* $i \in T$, *it holds that*

$$I(\mathbf{W}; [\mathbf{S}, \overline{\mathbf{R}}]_{\mathcal{R}}^*(w)(i)) \leq L. \tag{5}$$

Roughly speaking, in a weak reduction, a dishonest $\mathcal{R}$ can gain partial information about several secrets, but at most L bits overall. Besides, they termed *natural* reductions the reductions where the receiver $\mathcal{R}$ sends no messages to the sender S. This automatically implies that condition (3) of Definition 2 is satisfied. Using the above terminology, they showed the following lower bounds on the number α of invocations the $\binom{N}{1}$-OT^L protocol must do of the ideal $\binom{n}{1}$-OT^ℓ sub-protocol, and on the number of random bits required to implement the $\binom{N}{1}$-OT^L.

Theorem 1. *[14] Any information-theoretical secure reduction of weak* $\binom{N}{1}$-OT^L *to* $\binom{n}{1}$-OT^ℓ *must have* $\alpha \geq \frac{L}{\ell} \cdot \frac{N-1}{n-1}$

Theorem 2. *[14] In any information-theoretic natural reduction of weak* $\binom{N}{1}$-OT^L *to* $\binom{n}{1}$-OT^ℓ *the sender must flip at least* $\frac{L(N-n)}{n-1}$ *random bits.*

When $L = \ell$, the bounds are tight both for the strong and the weak case, since they showed a protocol realizing $\binom{N}{1}$-OT^ℓ where $N > n$ which makes exactly $\frac{N-1}{n-1}$ invocations of the $\binom{n}{1}$-OT^ℓ and flips exactly $\frac{L(N-n)}{n-1}$ random bits [14]. However, for the case $L > \ell$, they gave a protocol (see Table 1), which is optimal with respect to condition (5), but which does not meet condition (4). The idea is

Table 1. Basic protocol for a weak reduction

Protocol weakly reducing $\binom{N}{1}$-OT^L (with $L > \ell$) to $\binom{n}{1}$-OT$^\ell$.
Assume that $\ell | L$.

- Let $w = w_1, \ldots, w_N$ be the length N sequence of secrets $\mathcal{S}$ holds. For each $i = 1, \ldots, N$, w_i is a string of L bits.
- Split the strings into $\frac{L}{\ell}$ pieces. More precisely, let $w_i = w_i^1, \ldots, w_i^{\frac{L}{\ell}}$, where, $w_i^j \in \{0,1\}^\ell$, for each $j = 1, \ldots, \frac{L}{\ell}$.
- For $j = 1, \ldots, \frac{L}{\ell}$, execute an $\binom{N}{1}$-OT$^\ell$ oblivious transfer of *the j-th piece of* $w = w_1, \ldots, w_N$. In other words, compute

$$\binom{N}{1}\text{-}OT^\ell \text{ on } (w_1^j, \ldots, w_N^j)$$

where the $\binom{N}{1}$-OT$^\ell$ is the reduction of $\binom{N}{1}$-OT$^\ell$ to $\binom{n}{1}$-OT$^\ell$ described in [14].

---

simply to split each of the N secret strings in L/ℓ pieces of ℓ bits, and to run the available $\binom{N}{1}$-OT$^\ell$, optimal with respect to the use of the $\binom{n}{1}$-OT$^\ell$ black box, exactly $\frac{L}{\ell}$ times.

An honest $\mathcal{R}$ always obtains the secret in which he is interested in, recovering the "right" pieces at each execution. On the other hand, a *cheating* $\mathcal{R}$ is able to recover $\frac{L}{\ell}$ pieces of possibly *different* secrets among $w = w_1, \ldots, w_N$. We would like to modify this basic construction in order to achieve condition (4) without losing too much in efficiency.

Brassard, Crepéau, and Sántha solved a similar problem in [6]. They studied how to reduce $\binom{2}{1}$-OT$^\ell$ to $\binom{2}{1}$-OT^1 in an information theoretic secure way. Starting from the observation that trivial serial executions of ℓ $\binom{2}{1}$-OT^1 oblivious transfer, one for each bit of the two secret strings w_0 and w_1, didn't work, they pursued the idea of finding a function f where, given x_0 and x_1 such that $f(x_0) = w_0$ and $f(x_1) = w_1$, from two disjoint subsets of bits of x_0 and x_1 it is possible to gain information on *at most one* of w_0 and w_1. Using such a (public) function, the reduction would have been simple to implement (see Table 2).

Table 2. Protocol for two secrets of ℓ bits

Protocol strongly reducing $\binom{2}{1}$-OT$^\ell$ to $\binom{2}{1}$-OT^1

- $\mathcal{S}$ picks random $x_0, x_1 \in \{0,1\}^n$ such that $f(x_0) = w_0$ and $f(x_1) = w_1$
- For $i = 1, \ldots, n$, $\mathcal{S}$ performs a $\binom{2}{1}$-OT^1 on the pair (x_0^i, x_1^i)
- $\mathcal{R}$ recovers w_0 or w_1 by computing $f(x_0)$ or $f(x_1)$.

The property of f ensures that an honest receiver is always able to recover one of the secrets, while a dishonest receiver can obtain information on at most one of the secrets. They called such functions *Zig-zag functions*.

Notice that we have to solve a very close problem: in our scenario, a cheating receiver is able to obtain *partial* information about *many* secrets. Our aim is to find out a class of functions where disjoint subsets of strings $x_1, x_2, \ldots$ give information about at most one of the secrets $w_1, w_2, \ldots$

4   Generalized Zig-zag Functions

Let $X = GF(q)$, and let $X^n = \{(x_1, \ldots, x_n) : x_i \in X, \text{ for } 1 \le i \le n\}$. Moreover, for each $I = \{i_1, \ldots, i_{|I|}\} \subseteq \{1, \ldots, n\}$, denote by $x^I = (x_{i_1}, \ldots, x_{i_{|I|}})$ the subsequence of $x \in X^n$ indexed by I. Finally, let X^I be the set of all possible subsequences x^I for a given I.

A function is *unbiased* with respect to a subset I if the knowledge of the value of x^I does not give *any* information about $f(x)$. More formally, we have the following definition

Definition 3. *Suppose that* $f : X^n \to X^m$, *where* $n \ge m$. *Let* $I \subseteq \{1, \ldots, n\}$. *We say that f is* unbiased *with respect to I if, for all possible choices of* $x^I \in X^I$, *and for every* $(y_1, \ldots, y_m) \in X^m$, *there are exactly* $q^{n-m-|I|}$ *choices for* $x^{\{1, \ldots, n\} \setminus I}$ *such that* $f(x_1, \ldots, x_n) = (y_1, \ldots, y_m)$.

This concept has been introduced in [6]. Actually, the form in which it is stated here is the same as [25]. Since we are going to follow the same approach applied in [25] to study the properties of linear and nonlinear Zig-zag functions, we prefer this definition. The definition of Zig-zag functions relies on the unbiased property.

Definition 4. *A function* $f : X^n \to X^m$ *is said to be a* Zig-zag *function if, for every* $I \subseteq \{1, \ldots, n\}$, *$f$ is unbiased with respect to at least one of I and* $\{1, \ldots, n\} \setminus I$.

We would like some "generalized" Zig-zag property, holding for different disjoint subsets of indices. Roughly speaking, a generalized Zig-zag function should be unbiased with respect to at least $s - 1$ of the subsets $I_1, \ldots, I_s$ into which $\{1, \ldots, n\}$ is partitioned (for all possible partitions). More formally, we can state the following

Definition 5. *Let s be an integer such that* $2 \le s \le n$. *A function* $f : X^n \to X^m$ *is said to be an* s-Zig-zag *function if, for every set of s subsets* $I_1, \ldots, I_s \subseteq \{1, \ldots, n\}$, *such that* $\cup_i I_i = \{1, \ldots, n\}$, *and* $I_j \cap I_j = \emptyset$ *if* $i \ne j$, *f is unbiased with respect to at least $s - 1$ of* $I_1, \ldots, I_s$.

In an s-Zig-zag function, if $\mathcal{R}$ collects information about s x_i's, for some s, then he can get information on at most one w_i. If the above property is satisfied *for every* $2 \le s \le n$, then we say that f is *fully Zig-zag* (see Appendix B for an example of such a function). Fully Zig-zag functions enable us to apply the same approach developed in [6] in order to substitute the real secrets w_i with some pre-images x_i of w_i. The generalized property of the function *ensures* the privacy of the transfer.

Note: The functions $f : X^n \rightarrow X^m$ we are looking for must be *efficient to compute*. Moreover, there must exist an *efficient procedure* to compute a *random pre-image* $x \in f^{-1}(y)$, for each $y \in X^m$.

4.1   Zig-zag and Fully Zig-zag Functions

We briefly review some definitions and known results about Zig-zag. A Zig-zag (resp. s-Zig-zag, fully Zig-zag) function is said to be *linear* if there exists an $m \times n$ matrix M with entries from $GF(q)$ such that $f(x) = xM^T$ for all $x \in GF(q)^n$.

The following results have been shown in [25] and are recalled here since they will be used in the following subsection. The next lemma shows an upper bound on the size of the set of index I with respect to a function can be unbiased.

Lemma 1. *[25] If $f : X^n \rightarrow X^m$ is unbiased with respect to I, then $|I| \leq n - m$.*

As a consequence, it is possible to show a lower bound on the size n of the domain of the function, given the size m of the codomain.

Lemma 2. *[25] If $f : X^n \rightarrow X^m$ is a Zig-zag function, then $n \geq 2m - 1$.*

The following theorem establishes that a Zig-zag function is unbiased with respect to all the subsets of size $m - 1$.

Theorem 3. *[25] If $f : X^n \rightarrow X^m$ is a Zig-zag function, then f is unbiased with respect to I for all I such that $|I| = m - 1$.*

Moreover, notice that it is not difficult to prove the following result

Lemma 3. *If $f : X^n \rightarrow X^m$ is unbiased with respect to I, then f is unbiased with respect to all $I' \subseteq I$.*

Using the above results, we can prove our main result of this section: an equivalence between certain fully Zig-zag functions and Zig-zag functions.

Theorem 4. *Let $n \geq 2m - 1$. Then $f : X^n \rightarrow X^m$ is a fully Zig-zag function if and only if f is a Zig-zag function.*

Proof. We give the proof for $n = 2m - 1$. The if part is straightforward. Indeed, if f is fully Zig-zag, then *for each* partition $I_1, \ldots, I_s$ of $\{1, \ldots, n\}$ f is unbiased with respect to at least $s - 1$ subsets out of the s in the partition. Hence, it is unbiased with respect to at least 1 subset out of the 2 for any possible bipartition of $\{1, \ldots, n\}$. Therefore, f is Zig-zag.

Assume now that f is Zig-zag. Hence, by definition, for each $I \subseteq \{1, \ldots, n\}$, f is unbiased with respect to at least one of I and $\{1, \ldots, n\} \setminus I$.

Let $I_1, \ldots, I_s$ be a partition of $\{1, \ldots, n\}$. We can consider two cases.

a) There exists a subset I_i of the partition such that $|I_i| > n - m$. Consider this subset. Since f is Zig-zag, by Lemma 1, f is unbiased with respect to $\{1, \ldots, n\} \setminus I_i$. But $\{1, \ldots, n\} \setminus I_i = \cup_{j \neq i} I_j$. Hence, applying Lemma 3, we can conclude that f is unbiased with respect to all I_j, for $j \neq i$.

b) For each $i = 1, \ldots, s$, $|I_i| \leq n - m$. Notice that, since $n = 2m - 1$,

$$|I_i| \leq n - m \Leftrightarrow |I_i| \leq 2m - 1 - m \Leftrightarrow |I_i| \leq m - 1.$$

Since f is a Zig-zag function, applying Theorem 3, we can say that f is unbiased with respect to all $I_i : |I_i| = m - 1$. Therefore, by Lemma 3, we can conclude that f is unbiased with respect to all of $I_1, \ldots, I_s$.

Therefore, f is fully Zig-zag. □

The proof for $n > 2m - 1$ is similar. Therefore, we can conclude saying that Zig-zag and fully Zig-zag definitions define the *same class of functions*. Therefore, the known constructions for Zig-zag functions enable us to improve the protocol described in Table 1 by substituting the secrets with the pre-images of a Zig-zag functions, as done in the protocol described in Table 2 for two secrets. A complete description of our protocol can be found in Table 3. Moreover, since both in [6] and in [25], has been shown that for each m there exist functions $f : X^n \to X^m$, where $n = \Theta(m)$ and the asymptotic notation hides a small constant, the modified protocol is still efficient and optimal with respect to the bound obtained in [14] up to a small multiplicative constant [4].

Table 3. General protocol, depending on f.

Protocol strongly reducing $\binom{N}{1}$-OT^L to $\binom{n}{1}$-OT$^\ell$

Let $f : X^P \to X^L$ be a fully Zig-zag function such that $\ell | P$.

- S picks random $x_0, x_1, \ldots, x_{N-1} \in \{0, 1\}^P$ such that, for $i = 0, \ldots, N-1$, $f(x_i) = w_i$.
- S performs the protocol described in Table 1, using $x_0, x_1, \ldots, x_{N-1}$ instead of the real secrets $w_0, \ldots, w_{N-1}$.
- R recovers x_i, and computes $w_i = f(x_i)$.

4.2   On the Existence of s-Zig-zags

A question coming up to mind now is the following: Zig-zag functions are equivalent to fully Zig-zag functions. But these functions, according to Lemma 2, exist

---

[4] After the submission of this extended abstract to the conference, we found out that Dodis and Micali, working on the journal version of the paper presented at Eurocrypt '99, have independently obtained the same reduction, which will appear in the full version of their paper.

only if $n \geq 2m - 1$. Do s-Zig-zag functions exist when $n < 2m - 1$? The example reported in Appendix C shows that the answer is again affirmative. It is interesting to investigate some necessary and sufficient conditions for the existence of such generalized functions. The following lemma extends Lemma 2:

Lemma 4. *If an s-Zig-zag function $f : X^n \to X^m$ exists, then*

$$n \geq \begin{cases} 2m - s + 2, & \text{if } n \text{ and } s \text{ are both odd or both even} \\ 2m - s + 1, & \text{otherwise.} \end{cases}$$

Proof. Notice that, by definition, f must be unbiased with respect to at least $s - 1$ subsets of each possible s-partition. It is not difficult to check that the worst case we have to consider is when a partition has $s - 2$ subsets of size 1 and two subsets of essentially the same size. Therefore, f must be unbiased with respect to at least one of the two "big" subsets. Hence, applying Lemma 1, it follows that

$$\lfloor \frac{n - (s - 2)}{2} \rfloor \leq n - m. \tag{6}$$

The result follows by simple algebra.   □

An interesting relation between s-Zig-zag and t-Zig-zag, where $t \geq s$, is stated by the following lemma, whose proof can be obtained essentially noticing that a t-partition is a refinement of an s-partition.

Lemma 5. *If $f : X^n \to X^m$ is s-Zig-zag, then f is t-Zig-zag for every $s < t \leq n$.*

4.3   A Combinatorial Characterization

Let t be an integer such that $1 \leq t \leq k$ and $v \geq 2$. An *orthogonal array* $OA_\lambda(t, k, v)$ is a $\lambda v^t \times k$ array A of v symbols, such that within any t columns of A, every possible t-tuple of symbols occurs in exactly λ rows of A. An orthogonal array is *simple* if it does not contain two identical rows. A *large set* of orthogonal arrays $OA_\lambda(t, k, v)$, denoted $LOA_\lambda(t, k, v)$, is a set of v^{k-t}/λ simple $OA_\lambda(t, k, v)$, such that every possible k-tuple occurs as a row in exactly one of the orthogonal arrays in the set (see [20] for the theory and applications of these structures).

Theorem 5. *If $f : X^n \to X^m$ is an s-Zig-zag function where n and s have different parity, and $m > \lfloor \frac{n}{2} \rfloor + \lfloor \frac{s-2}{2} \rfloor$ then f is unbiased with respect to all the subsets of size $\lfloor \frac{n - (s-2)}{2} \rfloor$.*

Proof. Notice that, our assumptions imply $\lceil \frac{n - (s-2)}{2} \rceil > \lfloor \frac{n - (s-2)}{2} \rfloor$. By definition, f is unbiased with respect to at least $s - 1$ subsets of each s-partition of $\{1, \ldots, n\}$. Suppose there exists a subset I_i such that $|I_i| = \lfloor \frac{n - (s-2)}{2} \rfloor$ with respect to f is biased. Then, it would be possible to define an s-partition having $s - 2$ subsets of size 1, the subset I_i, and a subset R having size

$$|R| = n - (s - 2) - \lfloor \frac{n - (s - 2)}{2} \rfloor = \lceil \frac{n - (s - 2)}{2} \rceil.$$

Since f is biased with respect to I_i, then f must be unbiased with respect to R. This is possible only if

$$|R| = \lceil \frac{n - (s - 2)}{2} \rceil \leq n - m \iff m \leq n - \lceil \frac{n - (s - 2)}{2} \rceil.$$

Since $\lceil \frac{n-(s-2)}{2} \rceil = \lceil \frac{n}{2} \rceil - \lfloor \frac{s-2}{2} \rfloor$ the above inequality is satisfied only if $m \leq \lfloor \frac{n}{2} \rfloor + \lfloor \frac{s-2}{2} \rfloor$. But $m > \lfloor \frac{n}{2} \rfloor + \lfloor \frac{s-2}{2} \rfloor$ and, hence, we have a contradiction. □

The following theorem establishes a necessary and sufficient condition for the existence of certain s-Zig-zag functions.

Theorem 6. *An s-Zig-zag function $f : X^n \rightarrow X^m$, where n and s have different parity, and $m > \lfloor \frac{n}{2} \rfloor + \lfloor \frac{s-2}{2} \rfloor$ exists if and only if a large set of orthogonal arrays $LOA_\lambda(\lfloor \frac{n-(s-2)}{2} \rfloor, n, q)$ with $\lambda = q^{n-m-\lfloor \frac{n-(s-2)}{2} \rfloor}$ exists.*

Proof. The necessity of the condition derives from Theorem 5, analyzing the arrays containing the pre-images of f, as done in [25]. The sufficiency can be proved as follows: label each of the q^m arrays of the large set with a different element of $y \in X^m$. Denote such array with A_y. Then, define a function $f : X^n \rightarrow X^m$ as

$$f(x_1, \ldots, x_n) = y \iff (x_1, \ldots, x_n) \in A_y.$$

The properties of the arrays and the condition $m > \lfloor \frac{n}{2} \rfloor + \lfloor \frac{s-2}{2} \rfloor$ assure that f is s-Zig-zag. □

On the other hand, using the same proof technique, it is possible to show a *sufficient* condition for the existence of an s-Zig-zag for *any* n and $2 \leq s \leq n$. More precisely, we can state the following

Theorem 7. *If a large set of orthogonal arrays $LOA_\lambda(\lfloor \frac{n-(s-2)}{2} \rfloor, n, q)$ with $\lambda = q^{n-m-\lfloor \frac{n-(s-2)}{2} \rfloor}$ exists, then an s-Zig-zag function exists.*

5   Towards a General Reduction

The protocol described before can be conceptually divided in two phases: a first phase in which x_i is split into several pieces and $\mathcal{R}$ needs all the pieces to retrieve x_i; and a second phase where, once having obtained x_i, $\mathcal{R}$ recovers the secret by computing $y_i = f(x_i)$ for some function f. Since each piece gives partial knowledge of x_i, f needs to hide the value of y_i according to the definition of a correct and private reduction (i.e., the Zig-zag property). In this section, we show that using in the first phase an appropriate *ramp secret sharing scheme* [1] (see Appendix D for a brief review of the definition and some basic properties) to share x_i then, in the second phase the function f needs *weaker requirements* than the Zig-zag property. In this case, the pieces that $\mathcal{R}$ recovers from each transfer are not *substrings* of the value x_i he needs to compute the real secret $y_i = f(x_i)$, but *shares* that he has to combine according to the given ramp scheme in order to recover x_i.

Actually, notice that the splitting of the strings can be seen as a sharing according to a $(0, \frac{p}{\ell}, \frac{p}{\ell})$-RS, where p is $|x_i|$ and ℓ is the size of each share/piece. The questions therefore are: is it possible to design an overall better protocol, using in the first phase some *non trivial* ramp scheme to share x_i. Does there exist a *trade-off* between what we pay in the first phase and what we pay in the second phase? Using a generic (t_1, t_2, n)-RS, what properties does f need to satisfy in order to hide y_i from partial knowledge of x_i as required by our problem? It is not difficult to check that the condition f needs is the following.

Definition 6. *A function* $f : X \to Y$ *realizes an unconditionally secure oblivious transfer reduction if and only if, for each set of shares* $\{x_1, \ldots, x_n\}$ *for a secret* $x \in X$ *generated by a given* (t_1, t_2, n)-RS, *for every sequence of subsets* $I_1, \ldots, I_s \subseteq \{1, \ldots, n\}$, *such that* $\cup_i I_i = \{1, \ldots, n\}$, *and* $I_i \cap I_j = \emptyset$ *if* $i \neq j$, *it holds that*

$$H(Y|X_{I_i}) = H(Y)$$

for at least $s - 1$ *of* $I_1, \ldots, I_s$.

The definition means that *at most one subset* of shares can give information about $f(x)$.

It is easy to see that, when the ramp secret sharing scheme used in the first phase of the protocol is the trivial $(0, p, p)$-RS (shares/pieces of one bit), Definition 6 is equivalent to fully Zig-zag functions.

An Almost Optimal Reduction. Using a $(\frac{n}{2}, n, n)$-RS it is immediate to see that, to acquire information on x_i, the adversary needs at least $\frac{n}{2} + 1$ shares. Hence, recovering partial information on one secret *rules out* the possibility of recovering partial information on another secret. Notice that with such a scheme, if each secret has size p and ℓ divides p, the bound on the size of the shares (see Appendix D) implies $n \geq \frac{2p}{\ell}$ (number of invocations of the given $\binom{N}{1}$-OT$^\ell$). An implementation meeting the bound for several values of p and ℓ can be set up using, for example, the protocol described in [17]. In this case the function f used in the second phase can be simply the identity function!

6   Ramp Secret Sharing Schemes with Shares of One Bit

Fully Zig-zag, s-Zig-zag and Zig-zag functions give rise to ramp secret sharing schemes with shares of one bit. The idea is the following: the dealer, given one of these functions, say $f : X^n \to X^m$, chooses a secret $y \in X^m$ and computes a random pre-image $x \in f^{-1}(y)$. Then, he distributes the secret among the set of

n participants giving, as a share, a single bit of the pre-image x to each of them. It is immediate to see that

- some subsets of participants do not gain *any* information about the secret, even if they pool together their shares. These subsets are the subsets of $\{1, \ldots, n\}$ with respect to the function f is *unbiased*.
- some subsets of participants are able to recover *partial* information about the secret. These are the subsets of $\{1, \ldots, n\}$ with respect to f is biased
- *all* the participants are able to recover the *whole* secret.

The idea of such constructions was recently described in [8] (see Remark 9) as an application of ℓ-AONT transforms. In that construction, however, the dealer distributes among the participants the *bits of the image* of the secret while we distribute the bits of a pre-image of the secret.

7   Conclusions

In this paper we have shown how to achieve efficient unconditionally secure reductions of $\binom{N}{1}$-OT^L to $\binom{n}{1}$-OT^ℓ, proving that Zig-zag functions can be used to reduce $\binom{N}{1}$-OT^L to $\binom{n}{1}$-OT^ℓ *for each* $N \geq n$ and $L \geq \ell$. Finally, we have studied a generalization of these functions, identifying a combinatorial characterization and a relation with ramp schemes with shares of one bit. Some interesting questions arise from this study. To name a few:

- The constructions presented before are almost optimal but do not meet the bounds of Theorems 1 and 2 by equality. Hence, the question of how to reach (if it is possible) these bounds is still open.
- Do *cryptographic applications* of s-Zig-zag exist? We have pointed out the interesting relation with efficient ramp schemes, where each share is a single bit. Is it possible to say more?
- Linear Zig-zag are equivalent to *self-intersecting codes*. Is there any characterization in *terms of codes* for s-Zig-zag functions? And what about some *efficient constructions*? Is it possible, along the same line of [6], to set up any deterministic or probabilistic method?

Acknowledgements

This research was done while the first author was visiting the Department of Combinatorics and Optimization at the University of Waterloo. He would like to thank the Department for its hospitality. Moreover, he would like to thanks Yevgeniy Dodis for a helpful discussion.

D.R. Stinson's research is supported by NSERC grants IRC 216431-96 and RGPIN 203114-98.

References

1. G.R. Blakley, Security of Ramp Schemes, *Advances in Cryptology: Crypto '84*, pp. 547-559, LNCS Vol. 196, pp. 242-268, 1984.

2. M. Bellare and S. Micali, Non-interactive Oblivious Transfer and Applications, *Advances in Cryptology: Crypto '89*, Springer-Verlag, pp. 547-559, 1990.

3. M. Blum, How to Exchange (Secret) Keys, *ACM Transactions of Computer Systems*, Vol. 1, No. 2, pp. 175-193, 1993

4. G. Brassard, C. Crepéau, and J.-M. Roberts, Information Theoretic Reductions Among Disclosure Problems, *Proceedings of 27th IEEE Symposium on Foundations of Computer Science*, pp. 168-173, 1986.

5. G. Brassard, C. Crepéau, and J.-M. Roberts, All-or-Nothing Disclosure of Secrets, *Advances in Cryptology: Crypto '86*, Springer-Verlag, Vol. 263, pp. 234-238, 1987.

6. G. Brassard, C. Crepéau, and M. Sántha, Oblivious Transfer and Intersecting Codes, *IEEE Transaction on Information Theory*, special issue in coding and complexity, Vol. 42, No. 6, pp. 1769-1780, 1996.

7. C. Blundo, A. De Santis, and U. Vaccaro, Efficient Sharing of Many Secrets, *Proceedings of STACS*, LLNCS Vol. 665, pp.692-703, 1993.

8. R. Canetti, Y. Dodis, S. Halevi, E. Kushilevitz, and A. Sahai, Exposure-Resilient Functions and All-or-Nothing Transforms, *Advances in Cryptology: Proceedings of EuroCrypt 2000*, Springer-Verlag, LLNCS vol. 1807, pp. 453-469, 2000.

9. G. Cohen and A. Lempel, Linear Intersecting Codes, *Discrete Mathematics*, Vol. 56, pp. 35-43, 1985.

10. T. M. Cover and J. A. Thomas, Elements of Information Theory, *John Wiley & Sons*, 1991.

11. C. Crepéau, A Zero-Knowledge Poker Protocol that Achieves Confidentiality of the Players' Strategy or How to Achieve an Electronic Poker Face, *Advances in Cryptology: Proceedings of Crypto '86*, Springer-Verlag, pp. 239-247, 1987.

12. C. Crepéau, Equivalence between to flavors of oblivious transfers, *Advances in Cryptology: Proceedings of Crypto '87*, Springer-Verlag, vol. 293, pp. 350-354, 1988.

13. C. Crepéau and J. Kilian, Achieving Oblivious Transfer Using Weakened Security Assumptions, *Proceedings of 29th IEEE Symposium on Foundations of Computer Science*, pp. 42-52, 1988.

14. Y. Dodis and S. Micali, Lower Bounds for Oblivious Transfer Reduction, *Advances in Cryptology: Proceedings of Eurocrypt '99*, vol. 1592, pp. 42-54, Springer Verlag, 1999.

15. S. Even, O. Goldreich, and A. Lempel, A Randomized Protocol for Signing Contracts, *Advances in Cryptology: Proceedings of Crypto '83*, Plenum Press, New York, pp. 205-210, 1983.

16. M. Fisher, S. Micali, and C. Rackoff, A Secure Protocol for the Oblivious Transfer, *Journal of Cryptology*, vol. 9, No. 3, pp. 191-195, 1996.

17. M. Franklin and M. Yung, Communication Complexity of Secure Computation, *Proceedings of the 24th Annual Symposium on Theory of Computing*, pp. 699-710, 1992.

18. O. Goldreich, S. Micali, and A. Wigderson, How to play ANY mental game or: A Completeness Theorem for Protocols with Honest Majority, *Proceedings of 19th Annual Symposium on Theory of Computing*, pp. 20-31, 1987.

19. W.-A. Jackson and K. Martin, A Combinatorial Interpretation of Ramp Schemes, *Australasian Journal of Combinatorics*, vol. 14, pp. 51-60, 1996.

20. A. Hedayat, N.J.A. Sloane, and J. Stufken, Orthogonal Arrays : Theory and Applications, *Springer Verlag*, 1999.
21. J. Kilian, Founding Cryptography on Oblivious Transfer, *Proceedings of 20th Annual Symposium on Theory of Computing*, pp. 20-31, 1988.
22. M. Naor and B. Pinkas, Computationally Secure Oblivious Transfer, available at *http://www.wisdom.weizmann.ac.il/ naor/onpub.html*
23. M. Naor and B. Pinkas, Efficient Oblivious Transfer Protocols, available at *http://www.wisdom.weizmann.ac.il/ naor/onpub.html*
24. M. Rabin, How to Exchange Secrets by Oblivious Transfer, *Technical Memo TR-81, Aiken Computation Laboratory*, Harvard University, 1981.
25. D.R. Stinson, Some Results on Nonlinear Zig-zag Functions, *The Journal of Combinatorial Mathematics and Combinatorial Computing*, No. 29, pp. 127-138, 1999.
26. D.R. Stinson, Resilient Functions and Large Set of Orthogonal Arrays, *Congressus Numerantium*, Vol. 92, 105-110, 1993.
27. S. Wiesner, Conjugate Coding, *SIGACT News*, Vol. 15, pp. 78-88, 1983.

A   Information Theory Elements

This appendix briefly recalls some elements of information theory (the reader is referred to [10] for details).

Let $\mathbf{X}$ be a random variable taking values on a set X according to a probability distribution $\{P_{\mathbf{X}}(x)\}_{x \in X}$. The *entropy* of $\mathbf{X}$, denoted by $H(\mathbf{X})$, is defined as

$$H(\mathbf{X}) = -\sum_{x \in X} P_{\mathbf{X}}(x) \log P_{\mathbf{X}}(x),$$

where the logarithm is to the base 2. The entropy satisfies

$$0 \leq H(\mathbf{X}) \leq \log |X|,$$

where $H(\mathbf{X}) = 0$ if and only if there exists $x_0 \in X$ such that $Pr(\mathbf{X} = x_0) = 1$; whereas, $H(\mathbf{X}) = \log |X|$ if and only if $Pr(\mathbf{X} = x) = 1/|X|$, for all $x \in X$. The entropy of a random variable is usually interpreted as

– a measure of the equidistribution of the random variable
– a measure of the amount of information given on average by the random variable

Given two random variables $\mathbf{X}$ and $\mathbf{Y}$ taking values on sets X and Y, respectively, according to the joint probability distribution $\{P_{\mathbf{XY}}(x, y)\}_{x \in X, y \in Y}$ on their cartesian product, the *conditional entropy* $H(\mathbf{X}|\mathbf{Y})$ is defined as

$$H(\mathbf{X}|\mathbf{Y}) = -\sum_{y \in Y} \sum_{x \in X} P_{\mathbf{Y}}(y) P_{\mathbf{X}|\mathbf{Y}}(x|y) \log P_{\mathbf{X}|\mathbf{Y}}(x|y).$$

It is easy to see that

$$H(\mathbf{X}|\mathbf{Y}) \geq 0.$$

with equality if and only if X is a function of Y. The conditional entropy is a measure of the amount of information that $\mathbf{X}$ still has, once given $\mathbf{Y}$.

The *mutual information* between $\mathbf{X}$ and $\mathbf{Y}$ is given by

$$I(\mathbf{X};\mathbf{Y}) = H(\mathbf{X}) - H(\mathbf{X}|\mathbf{Y}),$$

and it enjoys the following properties,

$$I(\mathbf{X};\mathbf{Y}) = I(\mathbf{Y};\mathbf{X}), \text{ and } I(\mathbf{X};\mathbf{Y}) \geq 0.$$

The mutual information is a measure of the common information between $\mathbf{X}$ and $\mathbf{Y}$.

B   A Fully Zig-zag Function

In this section, we show an example of a fully Zig-zag function. Let $X = GF(2)$, and let $f : X^6 \to X^3$ be the function defined by $f(x) = xM^T$ where

$$M = \begin{bmatrix} 1\,0\,1\,1\,0\,0 \\ 1\,1\,0\,0\,1\,0 \\ 0\,1\,1\,0\,0\,1 \end{bmatrix}$$

To prove that f is fully Zig-zag it is necessary to show that, for any $1 < s \leq 6$, for each partition of $\{1, \ldots, 6\}$ into s parts, f is unbiased with respect to at least $s - 1$ of them. An easy proof can be obtained using the following theorem, which can be found in [25].

Theorem 8. *Let M be a generating matrix for an $[n, m]$ q-ary code, C, and let H be a parity-check matrix for C. The function $f(x) = xM^T$ is unbiased with respect to $I \subseteq \{1, \ldots, n\}$ if and only if the columns of H indexed by I are linearly independent.*

The parity-check matrix H for the generating matrix M is

$$H = \begin{bmatrix} 1\,0\,0\,1\,1\,0 \\ 0\,1\,0\,0\,1\,1 \\ 0\,0\,1\,1\,0\,1 \end{bmatrix}$$

Applying the above theorem, it is not difficult to see that f is unbiased with respect to

a) any subset of size 1.
b) any subset of size 2.
c) any subset of size 3, except $\{1, 2, 5\}, \{1, 3, 4\}, \{2, 3, 6\}$, and $\{4, 5, 6\}$.

Therefore, for any $2 \leq s \leq 6$, and for any s-partition, f is unbiased with respect to at least $s - 1$ subsets of the s subsets.

C   An Example of an s-Zig-zag

In this Appendix we show an example of a 3-Zig-zag function (where $n < 2m-1$). Let $X = GF(2)$, and let $f : X^4 \to X^3$ be the function defined by $f(x) = xM^T$ where

$$M = \begin{bmatrix} 1 & 0 & 0 & 1 \\ 0 & 1 & 0 & 1 \\ 0 & 0 & 1 & 1 \end{bmatrix}$$

In this case, the parity-check matrix H for the generating matrix M is simply

$$H = \begin{bmatrix} 1 & 1 & 1 & 1 \end{bmatrix}$$

Applying Theorem 8, it is easy to see that f is unbiased with respect to each subset of size 1. Since any 3-partition contains 2 subsets of size 1 and a subset of size 2, it follows that f is unbiased with respect to exactly 2 subsets.

Hence, s-Zig-zag functions can exist where Zig-zag functions and fully Zig-zag functions cannot exist.

D   Ramp Secret Sharing Schemes

A ramp secret sharing schemes $((t_1, t_2, n)$-RS, for short) is a protocol by means of which a dealer distributes a secret s among a set of n participants $\mathcal{P}$ in such a way that subsets of $\mathcal{P}$ of size greater than or equal to t_2 can reconstruct the value of s, any subset of $\mathcal{P}$ of size less than or equal to t_1 cannot determine anything about the value of the secret, while a subset of size $t_1 < t < t_2$ can recover *some* information about the secret [1]. Using information theory, the three properties of a (linear) (t_1, t_2, n)-RS can be stated as follows. Assuming that P denotes both a subset of participants and the set of shares these participants receive from the dealer to share a secret $s \in S$, and denoting the corresponding random variables in bold, it holds

– *Any subset of participants of size less than or equal to t_1 has no information on the secret value:* Formally, for each subset $P \in \mathcal{P}$ of size $|P| \leq t_1$, $H(\mathbf{S}|\mathbf{P}) = H(\mathbf{S})$.
– *Any subset of participants of size $t_1 < |P| < t_2$ has some information on the secret value:* Formally, for each subset $P \in \mathcal{P}$ of size $t_1 < |P| < t_2$, $H(\mathbf{S}|\mathbf{P}) = \frac{|P|-t_1}{t_2-t_1} H(\mathbf{S})$.
– *Any subset of participants of size greater than t_2 can compute the whole secret:* Formally, for each subset $P \in \mathcal{P}$ of size $|P| \geq t_2$, $H(\mathbf{S}|\mathbf{P}) = 0$.

In a (t_1, t_2, n)-RS, the size of each share must be greater than or equal to $\frac{H(\mathbf{S})}{t_2-t_1}$ (see [7,19]).

A Simple Algebraic Representation of Rijndael

Niels Ferguson[1], Richard Schroeppel[2], and Doug Whiting[3]

[1] Counterpane Internet Security,
niels@ferguson.net
[2] Sandia National Laboratory[†]
rschroe@sandia.gov
[3] Hi/fn, Inc.,
dwhiting@hifn.com

Abstract. We show that there is a very straightforward closed algebraic formula for the Rijndael block cipher. This formula is highly structured and far simpler then algebraic formulations of any other block cipher we know. The security of Rijndael depends on a new and untested hardness assumption: it is computationally infeasible to solve equations of this type. The lack of research on this new assumption raises concerns over the wisdom of using Rijndael for security-critical applications.

1 Introduction

Rijndael has been selected by NIST to become the AES. In this paper we look at the algebraic structure in Rijndael. After RC6, Rijndael is the most elegant of the AES finalists. It turns out that this elegant structure also results in an elegant algebraic representation of the Rijndael cipher.

We assume that the reader is familiar with Rijndael. We will concentrate on the version with 128-bit block size and 128-bit keys, and occasionally mention the versions with larger key sizes. Unless otherwise noted all formulae and equations will be in the $GF(2^8)$ field used by Rijndael.

2 Algebraic Formulae for Rijndael

In [DR98, section 8.5] the Rijndael designers note that the S-box can be written as an equation of the form

$$S(x) = w_8 + \sum_{d=0}^{7} w_d x^{255-2^d}$$

for certain constants $w_0, \ldots, w_8$.

[†] Sandia is a multiprogram laboratory operated by Sandia Corporation, a Lockheed Martin Company, for the United States Department of Energy under Contract DE-AC04-94AL85000.

S. Vaudenay and A. Youssef (Eds.): SAC 2001, LNCS 2259, pp. 103–111, 2001.

The first simplification that we make is to get rid of the constant w_8 in that formula. In a normal round the output of four S-boxes is multiplied by the MDS matrix and then four key bytes are added to the result. As the key addition and the MDS matrix are linear, we can replace the w_8 constant in the S-box by the addition of a suitable constant to the key bytes [MR00]. In the last round there is no MDS matrix but there is still a key addition, so the same trick works. This gives us the following formula for the S-box

$$S(x) = \sum_{d=0}^{7} w_d x^{255-2^d}$$

and we have to keep in mind that we now work with a modified key schedule where a suitable constant is added to each expanded key byte.

The next simplification is to rewrite the equation as

$$S(x) = \sum_{d=0}^{7} w_d x^{-2^d}$$

which is nearly equivalent as $x^{255} = 1$ for all x except $x = 0$. For the remainder of the paper we introduce the convention that $a/0 := 0$ for any value a in $GF(2^8)$. This makes the new equation equivalent to the previous one.[1]

The form of this equation can be explained from the structure of the S-box. The S-box consists of an inversion in $GF(2^8)$ with 0 mapped to 0, followed by a bit-linear function, followed by the addition of a constant. As noted earlier, we can move this last constant into the key schedule, so we will ignore it. Any bit-linear function can be expressed in $GF(2^8)$ by a polynomial where all exponents are powers of two. The easiest way to see this is to observe that squaring in $GF(2^8)$ is a bit-linear operation. After all, $(a+b)^2 = a^2 + b^2$ in $GF(2^8)$. Therefore any polynomial whose exponents are powers of two implements a bit-linear operation. None of these polynomials implements the same function, and the number of polynomials of this form equals the number of bit-linear functions. Therefore, any bit-linear function can be written as a polynomial with exponents that are powers of two.

2.1   One-Round Equation

We use the notation from [FKL+00] to discuss the internal values of a Rijndael encryption. Let $a_{i,j}^{(r)}$ be the byte at position (i,j) at the input of round r. As usual, state values in Rijndael are represented as a 4×4 square of bytes with the coordinates running from 0 to 3. For convenience we will assume that all coordinates are reduced modulo 4 so that for example $a_{8,4}^{(r)} = a_{0,0}^{(r)}$.

---

[1] Handling the case $x = 0$ correctly might not even be important, depending on the way we use these equations later. A single Rijndael encryption uses 160 S-box lookups. For random plaintexts there is a less than 50% chance that the case $x = 0$ will occur during the encryption. So even if we do not handle the case $x = 0$ well, the result still applies to more than half the plaintext/ciphertext pairs.

The first step in a normal round is to apply the S-box to each byte of the state in the ByteSub step. We get

$$s_{i,j}^{(r)} = S[a_{i,j}^{(r)}] = \sum_{d_r=0}^{7} w_{d_r}(a_{i,j}^{(r)})^{-2^{d_r}}$$

where $s_{i,j}^{(r)}$ is the state after ByteSub. The next step is the ShiftRow operation which we can write as

$$t_{i,j}^{(r)} = s_{i,i+j}^{(r)} = \sum_{d_r=0}^{7} w_{d_r}(a_{i,i+j}^{(r)})^{-2^{d_r}}$$

The third step in each round is the MixColumn. We can write this as

$$m_{i,j}^{(r)} = \sum_{e_r=0}^{3} v_{i,e_r} t_{e_r,j}^{(r)}$$

where the $v_{i,j}$ are the coefficients of the MDS matrix. Simple substitution now gives us

$$m_{i,j}^{(r)} = \sum_{e_r=0}^{3} v_{i,e_r} \sum_{d_r=0}^{7} w_{d_r}(a_{e_r,e_r+j}^{(r)})^{-2^{d_r}}$$

$$= \sum_{e_r=0}^{3} \sum_{d_r=0}^{7} w_{i,e_r,d_r}(a_{e_r,e_r+j}^{(r)})^{-2^{d_r}}$$

for some suitable constants $w_{i,j,k}$. The final step of the round is the key addition, and results in the input to the next round.

$$a_{i,j}^{(r+1)} = m_{i,j}^{(r)} + k_{i,j}^{(r)}$$

$$= k_{i,j}^{(r)} + \sum_{e_r=0}^{3} \sum_{d_r=0}^{7} w_{i,e_r,d_r}(a_{e_r,e_r+j}^{(r)})^{-2^{d_r}}$$

where $k_{i,j}^{(r)}$ is the round key of round r at position (i,j). We now have a fairly simple algebraic expression for a single round of Rijndael. We can write this formula in a couple of interesting ways.

$$a_{i,j}^{(r+1)} = k_{i,j}^{(r)} + \sum_{\substack{e_r \in \mathcal{E} \\ d_r \in \mathcal{D}}} w_{i,e_r,d_r}(a_{e_r,e_r+j}^{(r)})^{-2^{d_r}} \tag{1}$$

$$a_{i,j}^{(r+1)} = k_{i,j}^{(r)} + \sum_{f_r=0}^{31} w_{i,f_r}(a_{\lfloor f_r/8 \rfloor, \lfloor f_r/8 \rfloor + j}^{(r)})^{-2^{f_r}} \tag{2}$$

$$a_{i,j}^{(r+1)} = k_{i,j}^{(r)} + \sum_{f_r=0}^{31} w_{i,f_r}(a_{f_r,f_r+j}^{(r)})^{-2^{\lfloor f_r/4 \rfloor}} \tag{3}$$

Equation (1) is a more compact rewrite of the one we already had. We define $\mathcal{E} := \{0,\ldots,3\}$ and $\mathcal{D} := \{0,\ldots,7\}$ to get the same ranges. Equation (2) is derived by setting $f_r := 8e_r + d_r$. The $w_{i,j}$ are suitable constants. Note that we do not need to reduce f_r modulo 8 when using it as an exponent as this is done automatically. In $\mathrm{GF}(2^8)$ we have that for all k and all x, $x^k = x^{k \bmod 255}$. The exponent 2^{f_r} can thus be taken modulo 255, which makes the exponent 2^8 be equivalent to 2^0. In other words, only the $(f_r \bmod 8)$ part of f_r can affect the result and we do not need to take the modulo ourselves. Equation (3) is derived in a similar manner by setting $f_r = 4d_r + e_r$, and requires a suitable rearrangements of the constants $w_{i,j}$. We find (1) the most elegant and will use that in the rest of the paper. However, one of the other formulae could also have been used with similar results.

Finally there is one more interesting way to rewrite equation (1).

$$a_{i,j}^{(r+1)} = k_{i,j}^{(r)} + \sum_{\substack{e_r \in \mathcal{E} \\ d_r \in \mathcal{D}}} \frac{w_{i,e_r,d_r}}{\left(a_{e_r,e_r+j}^{(r)}\right)^{2^{d_r}}}$$

The reason that this is interesting becomes clear when we start to consider the formula for two or more rounds. Then $a_{e_r,e_r+j}^{(r)}$ is replaced by a formula with several terms, and it in turn is raised to an even power. As the field we are working in has characteristic 2 we can use the Freshman's Dream: $(a+b)^2 = a^2 + b^2$. This generalises to exponents that are powers of two, and thus it allows the exponent 2^{d_r} to be applied to each term individually instead of to the sum of terms. If we ever want to write out the full expression without the use of summation symbols this prevents the creation of many cross-product terms and thus keeps the size of the expression under control.

2.2   Multiple-Round Equations

Expressions for multiple rounds of Rijndael are easily derived by substitution. For simplicity we choose an actual value for r. For two rounds of Rijndael we get

$$a_{i,j}^{(3)} = k_{i,j}^{(2)} + \sum_{\substack{e_2 \in \mathcal{E} \\ d_2 \in \mathcal{D}}} \frac{w_{i,e_2,d_2}}{\left(k_{e_2,e_2+j}^{(1)} + \sum_{\substack{e_1 \in \mathcal{E} \\ d_1 \in \mathcal{D}}} \frac{w_{e_2,e_1,d_1}}{\left(a_{e_1,e_1+e_2+j}^{(1)}\right)^{2^{d_1}}}\right)^{2^{d_2}}} \tag{4}$$

and the three-round version is

$$a_{i,j}^{(4)} = k_{i,j}^{(3)} +$$
$$\sum_{\substack{e_3 \in \mathcal{E} \\ d_3 \in \mathcal{D}}} \frac{w_{i,e_3,d_3}}{\left(k_{e_3,e_3+j}^{(2)} + \sum_{\substack{e_2 \in \mathcal{E} \\ d_2 \in \mathcal{D}}} \frac{w_{e_3,e_2,d_2}}{\left(k_{e_2,e_2+e_3+j}^{(1)} + \sum_{\substack{e_1 \in \mathcal{E} \\ d_1 \in \mathcal{D}}} \frac{w_{e_2,e_1,d_1}}{\left(a_{e_1,e_1+e_2+e_3+j}^{(1)}\right)^{2^{d_1}}}\right)^{2^{d_2}}}\right)^{2^{d_3}}}$$

Applying the Freshman's Dream to equation 4 gives us

$$a_{i,j}^{(3)} = k_{i,j}^{(2)} + \sum_{\substack{e_2 \in \mathcal{E} \\ d_2 \in \mathcal{D}}} \frac{w_{i,e_2,d_2}}{(k_{e_2,e_2+j}^{(1)})^{2^{d_2}} + \sum_{\substack{e_1 \in \mathcal{E} \\ d_1 \in \mathcal{D}}} \frac{w_{e_2,e_1,d_1}^{2^{d_2}}}{(a_{e_1,e_1+e_2+j}^{(1)})^{2^{d_1+d_2}}}}$$

in which all exponentiations are on individual terms. This formula still looks rather complicated, but most of the complications are not essential to the structure of the formula. The subscripts get more complex the deeper into the recursion we go, but all subscripts are known and are independent of the key or plaintext. The same holds for the exponents, they are known and independent of the plaintext and key. We therefore introduce a somewhat sloppy notation which clarifies the structure. We write K for any expanded key byte, with the understanding that the exact position of that key byte in the key schedule is known to us. All constants are written as C even though they might not be all the same value. We replace the remaining subscripts and powers by a $*$. Again, each $*$ stands for a value that we can compute and that is independent of the plaintext and key. Finally, we use the fact that we can write the inputs to the first round by $a_{i,j}^{(1)} = p_{4j+i} + k_{i,j}^{(0)}$ where the p_i's are the plaintext bytes. All in all this gives us

$$a_{i,j}^{(3)} = K + \sum_{\substack{e_2 \in \mathcal{E} \\ d_2 \in \mathcal{D}}} \frac{C}{K^* + \sum_{\substack{e_1 \in \mathcal{E} \\ d_1 \in \mathcal{D}}} \frac{C}{p_*^* + K^*}}$$

We can now write the five-round formula

$$a_{i,j}^{(6)} = K + \sum_{\substack{e_5 \in \mathcal{E} \\ d_5 \in \mathcal{D}}} \frac{C}{K^* + \sum_{\substack{e_4 \in \mathcal{E} \\ d_4 \in \mathcal{D}}} \frac{C}{K^* + \sum_{\substack{e_3 \in \mathcal{E} \\ d_3 \in \mathcal{D}}} \frac{C}{K^* + \sum_{\substack{e_2 \in \mathcal{E} \\ d_2 \in \mathcal{D}}} \frac{C}{K^* + \sum_{\substack{e_1 \in \mathcal{E} \\ d_1 \in \mathcal{D}}} \frac{C}{K^* + p_*^*}}}}} \tag{5}$$

Keep in mind that every K is some expanded key byte, each C is a known constant, and each $*$ is a known exponent or subscript, but that these values depend on the summation variables that enclose the symbol.

Equation 5 gives us the intermediate values in an encryption after five rounds as a function of the plaintext and the expanded key. It is possible to write a similar formula for the value after five rounds as a function of the ciphertext and the expanded key. The S-box is constructed from an inversion in GF(2^8) followed by a bit-linear function. The inverse S-box is constructed from a bit-linear function followed by an inversion. The inverse of the MixColumn operation is another MixColumn operation with different constants. The inverse cipher is thus constructed from the same components, and leads to a formula similar to equation 5. (For simplicity we ignore the fact that there is no MixColumn operation in the last round which makes the last round simpler than all other rounds.)

As the results from equation 5 and the inverse equation must agree, we get a closed algebraic equation which consists of two formulae similar to equation 5. Alternatively we can write out the 10-round equation which will be about twice the size.

2.3   Fully Expanded Equations

If we try to write out equation 5 without summation symbols then we get a very large formula. Instead of a summation we simply write out 32 copies of the equation that we are summing, and substitute the appropriate summation variable values in each copy. As there are 5 summations, we end up with about 2^{25} individual terms of the form $C/(K^* + p_*^*)$. This formula would be too large to include here, but it would fit in the memory of a computer. Even the full 10-round formula would require only 2^{50} terms or so, which is certainly computable within the workload allowed for an attack on a 128-bit cipher. The 256-bit key version of Rijndael has 14 rounds. The expanded equation for half the cipher would have about 2^{35} terms, and the expanded formula for the full cipher about 2^{70} terms.

3   Other Ciphers

We know of no other 'serious' block cipher that has an algebraic description that is anywhere near as simple as the one for Rijndael. There are some general techniques that work for any block cipher, but these do not lead to practical attacks.

For example, any block cipher can be written as a boolean circuit, and then translated to a set of equations with one equation per boolean gate. However, this results in a system of equations and not a closed algebraic formula. It is equivalent to rewriting the problem of finding the cipher's key as an instance of SAT [Meh84], for which no efficient algorithms are known.

If one tries to rewrite the equations into a closed formula there is an explosion of terms. For example, in DES each output bit of the round function depends on 6 input bits. The boolean expressions for the S-boxes are fairly complicated [Kwa00], and each input bit will be used at least 16 times on average in the full boolean expressions for the output bits. A fully expanded boolean formula for DES therefore has at least around $16^{16} = 2^{64}$ terms, and due to the 'random' structure of the S-boxes this formula has no neat structures to take advantage of. Quite clearly this will not result in an attack that is faster than exhaustive search.

Another idea is to write the formula for the cipher in conjunctive normal form. This results in a simple formula: the entire function can be written using some constants and a few summation-type operators. Of course, the underlying problem formulation for an attack is still SAT. Furthermore, a direct evaluation of this formula is impossible. First of all, the constants cannot be determined without the entire plaintext/ciphertext mapping. Even if the constants were known, the direct evaluation would cost in the order of 2^{b+k} steps where b is the block size and k the key size. This is obviously slower than an exhaustive search which requires 2^k steps.

4 An Algebraic Attack?

The real question is of course whether we can turn these formulae for Rijndael into an attack. At this moment we do not have an attack on Rijndael that uses this algebraic representation.

If the formula was a simple polynomial it would be trivial to solve, but this is nothing new. To make the formula a polynomial we would have to eliminate the $1/x$ function in the S-box. If we simply throw the $1/x$ away it makes the entire cipher affine, and there are easier ways of attacking an affine cipher. Using the equivalence $1/x = x^{254}$ and converting the formula to a polynomial leads to a polynomial with too many terms to be useful.

Another idea would be to write the formula as the quotient of two polynomials. Again, the number of terms grows very rapidly which makes this approach unpromising.

We feel that an algebraic attack would have to handle equations of the form of equation 5 directly. The form of this equation is similar to that of continued fractions, and can be seen as a generalisation. There is quite a lot of knowledge about "solving" continued fraction, but it is unclear to us whether that can be applied to these formulae. This is outside the area of expertise of the authors. We therefore have to leave this as an open problem: is there a way of solving for the key bytes K in equation 5 given enough plaintext/ciphertext pairs?

We can give a few observations. A fully expanded version of equation 5 has 2^{25} terms. If we ignore the fact that some of the key bytes in the formula must be equal we can write it as a formula in about 2^{25} individual key bytes. Computing the same intermediate state from the ciphertext gives us another formula of similar size, and setting the two equal gives us an equation in about 2^{26} expanded key bytes. From a purely information-theoretical standpoint this would require at least 2^{22} known plaintext/ciphertext pairs, but this is not a problem. The attack can even afford an algorithm of order $O(n^4)$ in the number of terms of the equation before the workload exceeds the 128-bit key size limit. Larger key sizes are even more advantageous to the attacker in finding an attack with complexity less than that of exhaustive key search. The 256-bit key version uses 14 rounds, so each equation for half the cipher would have about 2^{35} terms. An algebraic equation solver with a workload in the order of $O(n^7)$ in the number of terms might very well lead to an attack.

If the attack were to use an expanded formula for the full cipher it would have about 2^{50} terms. Again, the required plaintext/ciphertext pairs are not a problem, and an algorithm that is quadratic in the number of terms is good enough. For 256-bit keys an $O(n^3)$ algorithm would even be good enough.

Any algorithm to solve these equations can also use the fact that many of the expanded key bytes in the formula must be equal. After all, there are only 176 expanded key bytes overall, and all the key bytes in the formula are chosen from that set. As we know exactly which key value in the formula corresponds to which key byte in the expanded key, we can derive these additional equations. The Rijndael key schedule also introduces many linear equations between the various expanded key bytes which might be used.

Note that adding more rounds to Rijndael does not help as much as one would think. Each extra round adds a factor of 2^5 to the size of the fully-expanded equation. Compare this to other attacks where attacking an extra round very often involves guessing a full round key, which corresponds to a factor of 2^{128}.

5   Conclusions

The Rijndael cipher can be expressed in a very neat and compact algebraic formula. We know of no other cipher for which it is possible to derive an algebraic formula that is anywhere near as elegant. This implies that the security of Rijndael relies on a new computational hardness assumption: it is computationally infeasible to solve algebraic equations of this form. As this problem has not been studied, we do not know whether this is a reasonable assumption to make.

This puts us in a difficult situation. We have no attack on Rijndael that uses these formulae, but there might very well exist techniques for handling this type of formula that we are unaware of, or somebody might develop them in the next 20 years or so. This is a somewhat disingenuous argument; any cipher could be attacked in the future. Yet our experience teaches us that in cryptography it is best to be cautious. A system that uses Rijndael automatically bases its security on a new hardness assumption, whereas this new assumption can be avoided by using a different block cipher. In that light we are concerned about the use of Rijndael in security-critical applications.

Acknowledgements

We would like to thank John Kelsey, Mike Stay, and Yoshi Kohno for their helpful comments.

References

[DR98]   Joan Daemen and Vincent Rijmen.   AES proposal: Rijndael.   In *AES Round 1 Technical Evaluation, CD-1: Documentation*. NIST, August 1998. See http://www.esat.kuleuven.ac.be/~rijmen/rijndael/ or http://www.nist.gov/aes.

[FKL+00] Niels Ferguson, John Kelsey, Stefan Lucks, Bruce Schneier, Mike Stay, David Wagner, and Doug Whiting. Improved cryptanalysis of Rijndael. In Bruce Schneier, editor, *Fast Software Encryption 2000*, volume 1978 of *Lecture Notes in Computer Science*. Springer-Verlag, 2000.

[Kwa00]   Matthew Kwan. Reducing the gate count of bitslice DES. Cryptology ePrint Archive, Report 2000/051, 2000. http://eprint.iacr.org/.

[Meh84]   Kurt Mehlhorn. *Data Structures and Algorithms 2: Graph Algorithms and NP-Completeness*. EATCS Monographs on Theoretical Computer Science. Springer-Verlag, 1984.

[MR00]   Sean Murphy and Matt Robshaw. New observations on Rijndael. Available from http://www.isg.rhul.ac.uk/mrobshaw/, August 2000. Preliminary Draft.

Improving the Upper Bound on the Maximum Average Linear Hull Probability for Rijndael

Liam Keliher[1], Henk Meijer[1], and Stafford Tavares[2]

[1] Department of Computing and Information Science,
Queen's University at Kingston, Ontario, Canada, K7L 3N6
{keliher,henk}@cs.queensu.ca
[2] Department of Electrical and Computer Engineering,
Queen's University at Kingston, Ontario, Canada, K7L 3N6
tavares@ee.queensu.ca

Abstract. In [15], Keliher et al. present a new method for upper bounding the maximum average linear hull probability (MALHP) for SPNs, a value which is required to make claims about provable security against linear cryptanalysis. Application of this method to Rijndael (AES) yields an upper bound of $UB = 2^{-75}$ when 7 or more rounds are approximated, corresponding to a lower bound on the data complexity of $\frac{32}{UB} = 2^{80}$ (for a 96.7% success rate). In the current paper, we improve this upper bound for Rijndael by taking into consideration the distribution of linear probability values for the (unique) Rijndael 8×8 s-box. Our new upper bound on the MALHP when 9 rounds are approximated is 2^{-92}, corresponding to a lower bound on the data complexity of 2^{97} (again for a 96.7% success rate). [This is after completing 43% of the computation; however, we believe that values have stabilized—see Section 7.]

Keywords: linear cryptanalysis, maximum average linear hull probability, provable security, Rijndael, AES

1 Introduction

The *substitution-permutation network* (SPN) [9,1,12] is a fundamental block cipher architecture based on Shannon's principles of *confusion* and *diffusion* [22]. These principles are implemented through substitution and linear transformation (LT), respectively. Recently, SPNs have been the focus of increased attention. This is due in part to the selection of the SPN Rijndael [6] as the U.S. Government Advanced Encryption Standard (AES).

Linear cryptanalysis (LC) [18] and differential cryptanalysis (DC) [4] are generally considered to be the two most powerful cryptanalytic attacks on block ciphers. In this paper we focus on the linear cryptanalysis of SPNs. As a first attempt to quantify the resistance of a block cipher to LC, the *expected linear characteristic probability* (ELCP) of the *best linear characteristic* often is evaluated. However, Nyberg [21] showed that the use of linear characteristics can underestimate the success of LC. To guarantee *provable security*, a block cipher

S. Vaudenay and A. Youssef (Eds.): SAC 2001, LNCS 2259, pp. 112–128, 2001.

designer needs to consider *linear hulls* instead of linear characteristics, and the *maximum average linear hull probability* (MALHP) instead of the ELCP of the best linear characteristic.

Since the MALHP is difficult, if not infeasible, to compute exactly, researchers have adopted the approach of upper bounding it [2,13,15]. In [15], Keliher et al. present a new general method for upper bounding the MALHP for SPNs. They apply their method to Rijndael, obtaining an upper bound on the MALHP of $UB = 2^{-75}$ when 7 or more rounds are approximated, corresponding to a lower bound on the data complexity of $\frac{32}{UB} = 2^{80}$ (for a 96.7% success rate—see Table 1).[1]

The current paper is based on the following observation: *the general method of Keliher et al. in [15] can potentially be improved by incorporating specific information about the distribution of linear probability (LP) values for the SPN s-boxes.* Due to the fact that Rijndael has only one (repeated) s-box, and because of the structure of this s-box, this observation applies readily to Rijndael, and allows us to improve the upper bound on the MALHP to $UB = 2^{-92}$ when 9 rounds are approximated, for a lower bound on the data complexity of 2^{97} (again for a 96.7% success rate). (This value is based on completion of 43% of the computation, although we believe that the values have stabilized—see Section 7.

Conventions

The Hamming weight of a binary vector $\mathbf{x}$ is written $wt(\mathbf{x})$. If $\mathbf{Z}$ is a random variable, $E[\mathbf{Z}]$ denotes the expected value of $\mathbf{Z}$. And we use $\#\mathcal{A}$ to indicate the number of elements in the set $\mathcal{A}$.

2   Substitution-Permutation Networks

A block cipher is a bijective mapping from N bits to N bits (N is the *block size*) parameterized by a bitstring called a *key*, denoted $\mathbf{k}$. Common block sizes are 64 and 128 bits (we consider Rijndael with a block size of 128 bits). The input to a block cipher is called a *plaintext*, and the output is called a *ciphertext*.

An SPN encrypts a plaintext through a series of R simpler encryption steps called *rounds*. (Rijndael with a key size of 128 bits consists of 10 rounds.) The input to round r ($1 \leq r \leq R$) is first bitwise XOR'd with an N-bit *subkey*, denoted $\mathbf{k}^r$, which is typically derived from the key, $\mathbf{k}$, via a separate *key-scheduling algorithm*. The *substitution stage* then partitions the resulting vector into M sub-blocks of size n ($N = Mn$), which become the inputs to a row of bijective $n \times n$ *substitution boxes* (*s-boxes*)—bijective mappings from $\{0,1\}^n$ to $\{0,1\}^n$. Finally, the *permutation stage* applies an invertible linear transformation (LT) to the output of the s-boxes (classically, a bitwise permutation). Often the permutation stage is omitted from the last round. A final subkey, $\mathbf{k}^{R+1}$, is XOR'd with

---

[1] In [15], the value 2^{80} was incorrectly given as 2^{78} due to an error in the table corresponding to Table 1. See Remark 2 for clarification.

the output of round R to form the ciphertext. Figure 1 depicts an example SPN with $N = 16$, $M = n = 4$, and $R = 3$.

We assume the most general situation for the key, namely, that $\mathbf{k}$ is an *independent key* [3], a concatenation of $(R + 1)$ subkeys chosen independently from the uniform distribution on $\{0, 1\}^N$—symbolically, $\mathbf{k} = \langle \mathbf{k}^1, \mathbf{k}^2, \ldots, \mathbf{k}^{R+1} \rangle$. We use $\mathcal{K}$ to denote the set of all independent keys.

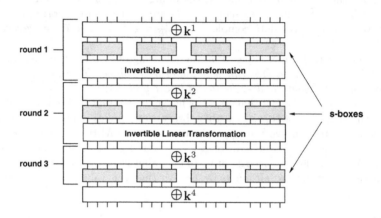

Fig. 1. SPN with $N = 16$, $M = n = 4$, $R = 3$

3   Linear Probability

In this section, and in Section 4, we make use of some of the treatment and notation from Vaudenay [23].

Definition 1. *Suppose $B : \{0, 1\}^d \to \{0, 1\}^d$ is a bijective mapping. Let $\mathbf{a}, \mathbf{b} \in \{0, 1\}^d$ be fixed, and let $\mathbf{X} \in \{0, 1\}^d$ be a uniformly distributed random variable. The linear probability $LP(\mathbf{a}, \mathbf{b})$ is defined as*

$$LP(\mathbf{a}, \mathbf{b}) \stackrel{\text{def}}{=} (2 \cdot \text{Prob}_{\mathbf{X}} \{\mathbf{a} \bullet \mathbf{X} = \mathbf{b} \bullet B(\mathbf{X})\} - 1)^2. \tag{1}$$

If B is parameterized by a key, $\mathbf{k}$, we write $LP(\mathbf{a}, \mathbf{b}; \mathbf{k})$, and the expected LP (ELP) is defined as

$$ELP(\mathbf{a}, \mathbf{b}) \stackrel{\text{def}}{=} E[LP(\mathbf{a}, \mathbf{b}; \mathbf{K})],$$

where $\mathbf{K}$ is a random variable uniformly distributed over the space of keys.

Note that LP values lie in the interval $[0, 1]$. A nonzero LP value indicates a correlation between the input and output of B, with a higher value indicating a stronger correlation (in fact, $LP(\mathbf{a}, \mathbf{b})$ is the square of entry $[\mathbf{a}, \mathbf{b}]$ in the *correlation matrix* for B [5]).

The values $\mathbf{a}/\mathbf{b}$ in Definition 1 are referred to as input/output *masks*. For our purposes, the bijective mapping B may be an s-box, a single encryption round, or a sequence of consecutive encryption rounds.

The following lemma derives immediately from Parseval's Theorem [20].

Lemma 1. *Let $B : \{0,1\}^d \to \{0,1\}^d$ be a bijective mapping parameterized by a key, $\mathbf{k}$, and let $\mathbf{a}, \mathbf{b} \in \{0,1\}^d$. Then*

$$\sum_{\mathbf{x} \in \{0,1\}^d} LP(\mathbf{a}, \mathbf{x}; \mathbf{k}) = \sum_{\mathbf{x} \in \{0,1\}^d} LP(\mathbf{x}, \mathbf{b}; \mathbf{k}) = 1$$

$$\sum_{\mathbf{x} \in \{0,1\}^d} ELP(\mathbf{a}, \mathbf{x}) = \sum_{\mathbf{x} \in \{0,1\}^d} ELP(\mathbf{x}, \mathbf{b}) = 1.$$

3.1   LP Values for the Rijndael S-box

Consider the (unique) Rijndael 8×8 s-box (see the Rijndael reference code [7]) as the bijective mapping B in Definition 1. A short computation yields the following interesting fact.

Lemma 2. *Let the bijective mapping under consideration be the 8×8 Rijndael s-box. If $\mathbf{a} \in \{0,1\}^8 \setminus \mathbf{0}$ is fixed, and $\mathbf{b}$ varies over $\{0,1\}^8$, then the distribution of values $LP(\mathbf{a}, \mathbf{b})$ is constant, and is given in the following table (ρ_i is the LP value, and ϕ_i is the number of times it occurs, for $1 \le i \le 9$). The same distribution is obtained if $\mathbf{b} \in \{0,1\}^8 \setminus \mathbf{0}$ is fixed, and $\mathbf{a}$ varies over $\{0,1\}^8$.*

i	1	2	3	4	5	6	7	8	9
ρ_i	$\left(\frac{8}{64}\right)^2$	$\left(\frac{7}{64}\right)^2$	$\left(\frac{6}{64}\right)^2$	$\left(\frac{5}{64}\right)^2$	$\left(\frac{4}{64}\right)^2$	$\left(\frac{3}{64}\right)^2$	$\left(\frac{2}{64}\right)^2$	$\left(\frac{1}{64}\right)^2$	0
ϕ_i	5	16	36	24	34	40	36	48	17

4   Linear Cryptanalysis of Markov Ciphers

It will be useful to consider linear cryptanalysis (LC) in the general context of Markov ciphers [17].

4.1   Markov Ciphers

Let $\mathcal{E} : \{0,1\}^N \to \{0,1\}^N$ be an R-round cipher, for which round r is given by the function $\mathbf{y} = \epsilon_r(\mathbf{x}; \mathbf{k}^r)$ ($\mathbf{x} \in \{0,1\}^N$ is the round input, and $\mathbf{k}^r \in \{0,1\}^N$ is the round-r subkey). Then $\mathcal{E}$ is a Markov cipher with respect to the XOR group operation ($\oplus$) on $\{0,1\}^N$ if, for $1 \le r \le R$, and any $\mathbf{x}, \Delta\mathbf{x}, \Delta\mathbf{y} \in \{0,1\}^N$,

$$\mathrm{Prob}_\mathbf{K} \{\epsilon_r(\mathbf{x}; \mathbf{K}) \oplus \epsilon_r(\mathbf{x} \oplus \Delta\mathbf{x}; \mathbf{K}) = \Delta\mathbf{y}\} =$$

$$\mathrm{Prob}_{\mathbf{K},\mathbf{X}} \{\epsilon_r(\mathbf{X}; \mathbf{K}) \oplus \epsilon_r(\mathbf{X} \oplus \Delta\mathbf{x}; \mathbf{K}) = \Delta\mathbf{y}\} \quad (2)$$

(where $\mathbf{X}$ and $\mathbf{K}$ are uniformly distributed and independent). That is, the probability over the key that a fixed input difference produces a fixed output difference is independent of the round input.

It is easy to show that the SPN model we are using is a Markov cipher, as are certain Feistel ciphers [10], such as DES [8].

Remark 1. The material in the remainder of Section 4 applies to any Markov cipher. Although we are dealing with LC, which ostensibly does not involve the $\oplus$ operation, the relevance of the Markov property given in (2) is via an interesting connection between linear probability and *differential probability* (see, for example, equations (3) and (4) in [23]).

4.2   Linear Cryptanalysis

Linear cryptanalysis (LC) is a known-plaintext attack (ciphertext-only in some cases) introduced by Matsui [18]. The more powerful version is known as Algorithm 2 (Algorithm 1 extracts only a single subkey bit). Algorithm 2 can be used to extract (pieces of) the round-1 subkey, $\mathbf{k}^1$. Once $\mathbf{k}^1$ is known, round 1 can be stripped off, and LC can be reapplied to obtain $\mathbf{k}^2$, and so on.

We do not give the details of LC here, as it is treated in many papers [18,3,14,15]. It suffices to say that the attacker wants to find input/output masks $\mathbf{a}, \mathbf{b} \in \{0,1\}^N$ for the bijective mapping consisting of rounds $2 \ldots R$, for which $LP(\mathbf{a}, \mathbf{b}; \mathbf{k})$ is maximal. Based on this value, the attacker can determine the number of known $\langle$plaintext, ciphertext$\rangle$ pairs, $\mathcal{N}_L$ (called the *data complexity*), required for a successful attack. Given an assumption about the behavior of round-1 output [18], Matsui shows that if

$$\mathcal{N}_L = \frac{c}{LP(\mathbf{a}, \mathbf{b}; \mathbf{k})},$$

then Algorithm 2 has the success rates in Table 1, for various values of the constant, c. Note that this is the same as Table 3 in [18], except that the constant values differ by a factor of 4, since Matsui uses *bias* values, not LP values.

Remark 2. The table in [15] corresponding to Table 1 has an error, in that the constants have *not* been multiplied by 4 to reflect the use of LP values.

Notational Issues. Above, we have discussed input and output masks and the associated LP values for rounds $2 \ldots R$ of an R-round cipher. It is useful to consider these and other related concepts as applying to any $T \geq 2$ consecutive

Table 1. Success rates for LC Algorithm 2

c	8	16	32	64
Success rate	48.6%	78.5%	96.7%	99.9%

"core" rounds (we say that these are the rounds being *approximated*). For Algorithm 2 as outlined above, $T = R - 1$, and the "first round," or "round 1," is actually round 2 of the cipher.

We use superscripts for individual rounds, so $LP^t(\mathbf{a}, \mathbf{b}; \mathbf{k}^t)$ and $ELP^t(\mathbf{a}, \mathbf{b})$ are LP and ELP values, respectively, for round t. On the other hand, we use t as a *subscript* to refer to values which apply to the first t rounds as a unit, so, for example, $ELP_t(\mathbf{a}, \mathbf{b})$ is an ELP value over rounds $1 \ldots t$.

4.3   Linear Characteristics

For fixed $\mathbf{a}, \mathbf{b} \in \{0, 1\}^N$, direct computation of $LP_T(\mathbf{a}, \mathbf{b}; \mathbf{k})$ for T core rounds is generally infeasible, first since it requires encrypting all N-bit vectors through rounds $1 \ldots T$, and second because of the dependence on an unknown key. The latter difficulty is usually handled by working instead with the expected value $ELP_T(\mathbf{a}, \mathbf{b})$. The data complexity of Algorithm 2 for masks $\mathbf{a}$ and $\mathbf{b}$ is now taken to be

$$\mathcal{N}_L = \frac{c}{ELP_T(\mathbf{a}, \mathbf{b})} . \tag{3}$$

The implicit assumption is that $LP_T(\mathbf{a}, \mathbf{b}; \mathbf{k})$ is approximately equal to $ELP_T(\mathbf{a}, \mathbf{b})$ for almost all values of $\mathbf{k}$ (this derives from the *Hypothesis of Stochastic Equivalence* in [17]).

The problem of computational complexity is usually treated by approximating $ELP_T(\mathbf{a}, \mathbf{b})$ through the use of *linear characteristics* (or simply *characteristics*). A T-round characteristic is a $(T + 1)$-tuple $\Omega = \langle \mathbf{a}^1, \mathbf{a}^2, \ldots, \mathbf{a}^T, \mathbf{a}^{T+1} \rangle$. We view $\mathbf{a}^t$ and $\mathbf{a}^{t+1}$ as input and output masks, respectively, for round t.

Definition 2. *Let* $\Omega = \langle \mathbf{a}^1, \mathbf{a}^2, \ldots, \mathbf{a}^T, \mathbf{a}^{T+1} \rangle$ *be a* T-round characteristic. The linear characteristic probability (LCP) and expected LCP (ELCP) of Ω are defined as

$$LCP(\Omega; \mathbf{k}) = \prod_{t=1}^{T} LP^t(\mathbf{a}^t, \mathbf{a}^{t+1}; \mathbf{k}^t)$$

$$ELCP(\Omega) = \prod_{t=1}^{T} ELP^t(\mathbf{a}^t, \mathbf{a}^{t+1}).$$

4.4   Choosing the Best Characteristic

In carrying out LC, the attacker typically runs an algorithm to find the T-round characteristic, Ω, for which $ELCP(\Omega)$ is maximal; such a characteristic (not necessarily unique) is called the *best characteristic* [19]. If $\Omega = \langle \mathbf{a}^1, \mathbf{a}^2, \ldots, \mathbf{a}^T, \mathbf{a}^{T+1} \rangle$, and if the input and output masks used in Algorithm 2 are taken to be $\mathbf{a} = \mathbf{a}^1$ and $\mathbf{b} = \mathbf{a}^{T+1}$, respectively, then $ELP_T(\mathbf{a}, \mathbf{b})$ (used to determine $\mathcal{N}_L$ in (3)) is approximated by

$$ELP_T(\mathbf{a}, \mathbf{b}) \approx ELCP(\Omega) . \tag{4}$$

The approximation in (4) has been widely used to evaluate the security of block ciphers against LC [12,14]. Knudsen calls a block cipher *practically secure* if the data complexity determined by this method is prohibitive [16]. However, by introducing the concept of *linear hulls*, Nyberg demonstrated that the above approach can underestimate the success of LC [21].

4.5   Linear Hulls

Definition 3 (Nyberg). *Given N-bit masks $\mathbf{a}, \mathbf{b}$, the corresponding linear hull, denoted $\mathrm{ALH}(\mathbf{a}, \mathbf{b})$,[2] is the set of all T-round characteristics (for the T rounds under consideration) having $\mathbf{a}$ as the input mask for round 1 and $\mathbf{b}$ as the output mask for round T, i.e., all characteristics of the form*

$$\Omega = \langle \mathbf{a}, \mathbf{a}^2, \mathbf{a}^3, \ldots, \mathbf{a}^T, \mathbf{b} \rangle.$$

Theorem 1 (Nyberg). *Let $\mathbf{a}, \mathbf{b} \in \{0,1\}^N$. Then*

$$ELP_T(\mathbf{a}, \mathbf{b}) = \sum_{\Omega \in \mathrm{ALH}(\mathbf{a}, \mathbf{b})} ELCP(\Omega).$$

It follows immediately from Theorem 1 that (4) does not hold in general, since $ELP_T(\mathbf{a}, \mathbf{b})$ is seen to be equal to a sum of terms $ELCP(\Omega)$ over a (large) set of characteristics, and therefore, in general, the ELCP of any characteristic will be strictly *less than* the corresponding ELP value. This is referred to as the *linear hull effect*. An important consequence is that an attacker may overestimate the number of ⟨plaintext, ciphertext⟩ pairs required for a given success rate.

Remark 3. It can be shown that the linear hull effect is significant for Rijndael, since, for example, the ELCP of any characteristic over $T = 8$ rounds is upper bounded by 2^{-300} [6],[3] but the largest ELP value has 2^{-128} as a trivial lower bound.[4]

The next lemma follows easily from Theorem 1 and Definition 2 (recall the conventions for superscripts and subscripts).

Lemma 3. *Let $T \geq 2$, and let $\mathbf{a}, \mathbf{b} \in \{0,1\}^N$. Then*

$$ELP_T(\mathbf{a}, \mathbf{b}) = \sum_{\mathbf{x} \in \{0,1\}^N} ELP_{T-1}(\mathbf{a}, \mathbf{x}) \cdot ELP^T(\mathbf{x}, \mathbf{b}).$$

---

[2] Nyberg [21] originally used the term *approximate linear hull*, hence the abbreviation ALH, which we retain for consistency with [15].

[3] Any 8-round characteristic, Ω, has a minimum of 50 active s-boxes, and the maximum LP value for the Rijndael s-box is 2^{-6}, so $ELCP(\Omega) \leq \left(2^{-6}\right)^{50} = 2^{-300}$.

[4] This follows by observing that Lemma 1 is contradicted if the maximum ELP value is less than 2^{-d}.

4.6 Maximum Average Linear Hull Probability

An SPN is considered to be *provably secure* against LC if the maximum ELP,

$$\max_{\mathbf{a},\mathbf{b}\in\{0,1\}^N\setminus\mathbf{0}} ELP_T(\mathbf{a},\mathbf{b}), \tag{5}$$

is sufficiently small that the resulting data complexity is prohibitive for any conceivable attacker.[5] The value in (5) is also called the *maximum average linear hull probability* (MALHP). We retain this terminology for consistency with [15].

Since evaluation of the MALHP appears to be infeasible in general, researchers have adopted the approach of upper bounding this value [2,13,15]. If such an upper bound is sufficiently small, provable security can be claimed.

5   SPN-Specific Considerations

In the current section, we adapt certain results from Section 4 to the SPN model. Note that where matrix multiplication is involved, we view all vectors as column vectors. Also, if $\mathcal{M}$ is a matrix, $\mathcal{M}'$ denotes the transpose of $\mathcal{M}$.

Lemma 4. *Consider T core SPN rounds. Let $1 \le t \le T$, and $\mathbf{a},\mathbf{b},\mathbf{k}^t \in \{0,1\}^N$. Then $LP^t(\mathbf{a},\mathbf{b};\mathbf{k}^t)$ is independent of $\mathbf{k}^t$, and therefore*

$$LP^t(\mathbf{a},\mathbf{b};\mathbf{k}^t) = ELP^t(\mathbf{a},\mathbf{b}).$$

Proof. Follows by observing the interchangeable roles of the round input, $\mathbf{x}$, and $\mathbf{k}^t$, and from a simple change of variables $\hat{\mathbf{x}} = \mathbf{x} \oplus \mathbf{k}^t$ when evaluating (1).

Corollary 1. *Let Ω be a T-round characteristic for an SPN. Then $LCP(\Omega) = ELCP(\Omega)$.*

Definition 4. *Let $\mathbf{L}$ denote the N-bit LT of the SPN represented as a binary $N \times N$ matrix, i.e., if $\mathbf{x},\mathbf{y} \in \{0,1\}^N$ are the input and output, respectively, for the LT, then $\mathbf{y} = \mathbf{Lx}$.*

Lemma 5 ([5]). *If $\mathbf{b} \in \{0,1\}^N$ and $\mathbf{a} = \mathbf{L}'\mathbf{b}$, then $\mathbf{a} \bullet \mathbf{x} = \mathbf{b} \bullet \mathbf{y}$ for all N-bit inputs to the LT, $\mathbf{x}$, and corresponding outputs, $\mathbf{y}$ (i.e., if $\mathbf{b}$ is an output mask for the LT, then $\mathbf{a} = \mathbf{L}'\mathbf{b}$ is the (unique) corresponding input mask).*

It follows from Lemma 5 that if $\mathbf{a}^t$ and $\mathbf{a}^{t+1}$ are input and output masks for round t, respectively, then the resulting input and output masks for the *substitution stage* of round t are $\mathbf{a}^t$ and $\mathbf{b}^t = \mathbf{L}'\mathbf{a}^{t+1}$. Further, $\mathbf{a}^t$ and $\mathbf{b}^t$ determine input and output masks for each s-box in round t. Let the masks for S_i^t be

---

[5] For Algorithm 2 as described above, this must hold for $T = R - 1$. Since variations of LC can be used to attack the first and last SPN rounds simultaneously, it may also be important that the data complexity remain prohibitive for $T = R - 2$.

denoted $\mathbf{a}_i^t$ and $\mathbf{b}_i^t$, for $1 \leq i \leq M$ (we number s-boxes from left to right). Then from Matsui's Piling-up Lemma [18] and Lemma 4,

$$ELP^t(\mathbf{a}^t, \mathbf{a}^{t+1}) = \prod_{i=1}^{M} LP^{S_i^t}(\mathbf{a}_i^t, \mathbf{b}_i^t). \tag{6}$$

From the above, any characteristic $\Omega \in \mathrm{ALH}(\mathbf{a}, \mathbf{b})$ determines an input and an output mask for each s-box in rounds $1 \ldots T$. If this yields at least one s-box for which the input mask is zero and the output mask is nonzero, or vice versa, the linear probability associated with that s-box will trivially be 0, and therefore $ELCP(\Omega) = 0$ by (6) and Definition 2. We exclude such characteristics from consideration via the following definition.

Definition 5. *For $\mathbf{a}, \mathbf{b} \in \{0,1\}^N$, let $\mathrm{ALH}(\mathbf{a}, \mathbf{b})^*$ consist of the elements $\Omega \in \mathrm{ALH}(\mathbf{a}, \mathbf{b})$ such that for each s-box in rounds $1 \ldots T$, the input and output masks determined by Ω for that s-box are either both zero or both nonzero.*

Remark 4. In [23], the characteristics in $\mathrm{ALH}(\mathbf{a}, \mathbf{b})^*$ are called *consistent*.

Definition 6 ([3]). *Any T-round characteristic, Ω, determines an input and an output mask for each s-box in rounds $1 \ldots T$. Those s-boxes having nonzero input and output masks are called* active.

Definition 7. *Let $\mathbf{v}$ be an input or an output mask for the substitution stage of round t. Then the active s-boxes in round t can be determined from $\mathbf{v}$ (without knowing the corresponding output/input mask). We define $\gamma_\mathbf{v}$ to be the M-bit vector which encodes the pattern of active s-boxes: $\gamma_\mathbf{v} = \gamma_1 \gamma_2 \ldots \gamma_M$, where $\gamma_i = 1$ if the ith s-box is active, and $\gamma_i = 0$ otherwise, for $1 \leq i \leq M$.*

Definition 8 ([15]). *Let $\gamma, \hat{\gamma} \in \{0,1\}^M$. Then*

$$W[\gamma, \hat{\gamma}] \stackrel{\mathrm{def}}{=} \#\left\{\mathbf{y} \in \{0,1\}^N : \gamma_\mathbf{x} = \gamma, \gamma_\mathbf{y} = \hat{\gamma}, \quad \text{where } \mathbf{x} = \mathbf{L}'\mathbf{y}\right\} .$$

Remark 5. Informally, the value $W[\gamma, \hat{\gamma}]$ represents the number of ways the LT can "connect" a pattern of active s-boxes in one round (γ) to a pattern of active s-boxes in the next round ($\hat{\gamma}$).

We now proceed to our improved method for upper bounding the MALHP for Rijndael.

6   Improved Upper Bound on MALHP for Rijndael

6.1   Technical Lemmas

Lemma 6 ([15]). *Let $m \geq 2$, and suppose $\{c_i\}_{i=1}^m$, $\{d_i\}_{i=1}^m$ are sequences of nonnegative values. Let $\{\dot{c}_i\}_{i=1}^m$, $\left\{\dot{d}_i\right\}_{i=1}^m$ be the sequences obtained by sorting $\{c_i\}$ and $\{d_i\}$, respectively, in nonincreasing order. Then $\sum_{i=1}^m c_i d_i \leq \sum_{i=1}^m \dot{c}_i \dot{d}_i$.*

Lemma 7 ([15]). *Suppose* $\{\dot{c}_i\}_{i=1}^m$, $\{\ddot{c}_i\}_{i=1}^m$, *and* $\left\{\dot{d}_i\right\}_{i=1}^m$ *are sequences of non-negative values, with* $\left\{\dot{d}_i\right\}$ *sorted in nonincreasing order. Suppose there exists* $\tilde{m}$, $1 \leq \tilde{m} \leq m$, *such that*

(a) $\ddot{c}_i \geq \dot{c}_i$, *for* $1 \leq i \leq \tilde{m}$
(b) $\ddot{c}_i \leq \dot{c}_i$, *for* $(\tilde{m}+1) \leq i \leq m$
(c) $\sum_{i=1}^m \dot{c}_i \leq \sum_{i=1}^m \ddot{c}_i$

Then $\sum_{i=1}^m \dot{c}_i \dot{d}_i \leq \sum_{i=1}^m \ddot{c}_i \dot{d}_i$.

6.2   Distribution of LP Values for Multiple Active S-boxes

Definition 9. *Let* $\mathbf{a} \in \{0,1\}^{128} \setminus \mathbf{0}$ *be a fixed input mask for the substitution stage of Rijndael, and let* $\mathbf{b}$ *be an output mask which varies over* $\{0,1\}^{128}$, *with the restriction that* $\gamma_{\mathbf{a}} = \gamma_{\mathbf{b}}$. *If* A *is the number of s-boxes made active* ($A = wt(\gamma_{\mathbf{a}})$), *define* $\mathcal{D}_A$ *to be the set of distinct LP values produced as* $\mathbf{b}$ *varies, and let* $D_A = \#\mathcal{D}_A$. *Define* $\langle \rho_1^A, \rho_2^A, \ldots, \rho_{D_A}^A \rangle$ *to be the sequence obtained by sorting* $\mathcal{D}_A$ *in decreasing order, and let* ϕ_j^A *be the number of occurrences of the value* ρ_j^A, *for* $1 \leq j \leq D_A$.

Note that if $A = 1$, then $D_A = 9$, and ρ_j^1 and ϕ_j^1 are as given in Lemma 2.

Lemma 8. *For* $A \geq 2$,

$$\mathcal{D}_A = \left\{ \rho_s^1 \cdot \rho_t^{A-1} : 1 \leq s \leq D_1, \ 1 \leq t \leq D_{A-1} \right\},$$

and for each j, $1 \leq j \leq D_A = \#\mathcal{D}_A$,

$$\phi_j^A = \sum \left\{ \phi_s^1 \cdot \phi_t^{A-1} \ : \ \rho_s^1 \cdot \rho_t^{A-1} = \rho_j^A, \ 1 \leq s \leq D_1, \ 1 \leq t \leq D_{A-1} \right\}.$$

Proof. Follows easily from Lemma 4 and (6).

Definition 10. *For* $A \geq 1$ *and* $1 \leq J \leq D_A$, *we define the partial sums*

$$\Phi_J^A = \sum_{j=1}^J \phi_j^A$$

$$\Lambda_J^A = \sum_{j=1}^J \rho_j^A \cdot \phi_j^A.$$

Also, we define $\mathcal{S}_A$ *to be the sequence*

$$\underbrace{\rho_1^A, \ldots, \rho_1^A}_{\phi_1^A \ terms}, \ \underbrace{\rho_2^A, \ldots, \rho_2^A}_{\phi_2^A \ terms}, \ \ldots, \underbrace{\rho_{D_A}^A, \ldots, \rho_{D_A}^A}_{\phi_{D_A}^A \ terms}.$$

Remark 6. For $1 \leq A \leq M$, $\Lambda_{D_A}^A = 1$ by Lemma 1.

6.3 Derivation of Improved Upper Bound

Convention: In this subsection, whenever we deal with values of the form $ELP_t(\mathbf{a}, \mathbf{b})$ or $ELP^t(\mathbf{a}, \mathbf{b})$ ($1 \leq t \leq T$), we omit the LT from round t. This is simply a technical matter that simplifies the proofs which follow.

Let $T \geq 2$. As in [15], our approach is to compute an upper bound for each nonzero pattern of active s-boxes in round 1 and round T—that is, we compute $UB_T[\gamma, \hat{\gamma}]$, for $\gamma, \hat{\gamma} \in \{0,1\}^M \setminus \mathbf{0}$, such that the following holds:

UB Property for T.   For all $\mathbf{a}, \mathbf{b} \in \{0,1\}^N \setminus \mathbf{0}$, $ELP_T(\mathbf{a}, \mathbf{b}) \leq UB_T[\gamma_{\mathbf{a}}, \gamma_{\mathbf{b}}]$.

If the *UB Property for T* holds, then the MALHP is upper bounded by

$$\max_{\gamma, \hat{\gamma} \in \{0,1\}^M \setminus \mathbf{0}} UB_T[\gamma, \hat{\gamma}].$$

The case $T = 2$ is handled in Theorem 2, and the case $T \geq 3$ in Theorem 3.

Theorem 2. *Let the values $UB_2[\gamma, \hat{\gamma}]$ be computed using the algorithm in Figure 2. Then the* UB Property for 2 *holds.*

Proof. In this proof, "Line X" refers to the Xth line in Figure 2. Let $\gamma, \hat{\gamma} \in \{0,1\}^M \setminus \mathbf{0}$ be fixed, and let $\mathbf{a}, \mathbf{b} \in \{0,1\}^N \setminus \mathbf{0}$ such that $\gamma_{\mathbf{a}} = \gamma$ and $\gamma_{\mathbf{b}} = \hat{\gamma}$. We want to show that $ELP_2(\mathbf{a}, \mathbf{b}) \leq UB_2[\gamma, \hat{\gamma}]$. There are $W = W[\gamma, \hat{\gamma}]$ ways that the LT can "connect" the f active s-boxes in round 1 to the ℓ active s-boxes in round 2. Let $\mathbf{x}_1, \mathbf{x}_2, \cdots, \mathbf{x}_W$ be the corresponding output masks for the substitution stage of round 1 (and therefore the input masks for the round-1 LT), and let $\mathbf{y}_1, \mathbf{y}_2, \cdots, \mathbf{y}_W$ be the respective output masks for the round-1 LT (and therefore the input masks for the substitution stage of round 2). So $\gamma_{\mathbf{x}_i} = \gamma$ and $\gamma_{\mathbf{y}_i} = \hat{\gamma}$, for $1 \leq i \leq W$. Let $c_i = ELP^1(\mathbf{a}, \mathbf{x}_i)$ and $d_i = ELP^2(\mathbf{y}_i, \mathbf{b})$, for $1 \leq i \leq W$. It follows from Lemma 3 that $ELP_2(\mathbf{a}, \mathbf{b}) = \sum_{i=1}^{W} c_i d_i$.

Without loss of generality, $f \leq \ell$, so $A_{\min} = f$ and $A_{\max} = \ell$. Let $\{\dot{c}_i\}$ ($\{\dot{d}_i\}$) be the sequence obtained by sorting $\{c_i\}$ ($\{d_i\}$) in nonincreasing order. Then $\sum_{i=1}^{W} c_i d_i \leq \sum_{i=1}^{W} \dot{c}_i \dot{d}_i$ by Lemma 6. Let $\{\ddot{c}_i\}$ ($\{\ddot{d}_i\}$) consist of the first W terms of $\mathcal{S}_f$ ($\mathcal{S}_\ell$). Since the terms $\dot{c}_i$ ($\dot{d}_i$) are elements of $\mathcal{S}_f$ ($\mathcal{S}_\ell$), it follows that $\dot{c}_i \leq \ddot{c}_i$ ($\dot{d}_i \leq \ddot{d}_i$), for $1 \leq i \leq W$, so

$$ELP_2(\mathbf{a}, \mathbf{b}) = \sum_{i=1}^{W} c_i d_i \leq \sum_{i=1}^{W} \dot{c}_i \dot{d}_i \leq \sum_{i=1}^{W} \ddot{c}_i \ddot{d}_i .$$

It is not hard to see that the value $UB_2[\gamma, \hat{\gamma}]$ computed in Figure 2 is exactly $\sum_{i=1}^{W} \ddot{c}_i \ddot{d}_i$. For computational efficiency, we do not sum "element-by-element" (i.e., for each i), but instead take advantage of the fact that $\{\ddot{c}_i\}$ has the form

$$\underbrace{\rho_1^f, \ldots, \rho_1^f}_{\phi_1^f \text{ terms}}, \underbrace{\rho_2^f, \ldots, \rho_2^f}_{\phi_2^f \text{ terms}}, \underbrace{\rho_3^f, \ldots, \rho_3^f}_{\phi_3^f \text{ terms}}, \cdots,$$

1.	For each $\gamma \in \{0,1\}^M \setminus \mathbf{0}$
2.	For each $\hat{\gamma} \in \{0,1\}^M \setminus \mathbf{0}$
3.	$W \leftarrow W[\gamma, \hat{\gamma}]$
4.	$f \leftarrow wt(\gamma), \ \ell \leftarrow wt(\hat{\gamma})$
5.	$A_{\min} \leftarrow \min\{f, \ell\}, \ A_{\max} \leftarrow \max\{f, \ell\}$
6.	$\lambda \leftarrow 0, \ \text{Sum} \leftarrow 0$
7.	$h \leftarrow 1$
8.	While $(h \leq D_{A_{\min}})$ and $(\Phi_h^{A_{\min}} \leq W)$
9.	$\text{Sum} \leftarrow \text{Sum} + \text{NextTerm2}\,(\Phi_h^{A_{\min}})$
10.	$h \leftarrow h + 1$
11.	If $(h \leq D_{A_{\min}})$ and $(\Phi_h^{A_{\min}} > W)$
12.	$\text{Sum} \leftarrow \text{Sum} + \text{NextTerm2}\,(W)$
13.	$UB_2[\gamma, \hat{\gamma}] \leftarrow \text{Sum}$
14.	Function NextTerm2 (Z)
15.	$J \leftarrow \min\left\{ j \ : \ 1 \leq j \leq D_{A_{\max}}, \ \ \Phi_j^{A_{\max}} \geq Z \right\}$
16.	$\Delta\lambda \leftarrow \left(\Lambda_J^{A_{\max}} - \lambda \right) - \left[\left(\Phi_J^{A_{\max}} - Z \right) * \rho_J^{A_{\max}} \right]$
17.	$\lambda \leftarrow \lambda + \Delta\lambda$
18.	return $\left(\rho_h^{A_{\min}} * \Delta\lambda \right)$

Fig. 2. Algorithm to compute $UB_2[\ ]$

and similarly for $\{\ddot{d}_i\}$ (replace f with ℓ). Viewing these sequences as "groups" of consecutive identical elements, the algorithm in Figure 2 proceeds "group-by-group." The variable h is the index of the current group in $\{\ddot{c}_i\}$. The function NextTerm2() identifies the corresponding elements in $\{\ddot{d}_i\}$, and computes the equivalent of the element-by-element product, which is added to the growing sum in Line 9. The situation in which $\{\ddot{c}_i\}_{i=1}^W$ is a truncated version of $\mathcal{S}_f$ is handled by the conditional statement in Lines 11–12.

Theorem 3. Let $T \geq 3$. Assume that the values $UB_{T-1}[\gamma, \hat{\gamma}]$ have been computed for all $\gamma, \hat{\gamma} \in \{0,1\}^M \setminus \mathbf{0}$ such that the UB Property for $(T-1)$ holds. Let the values $UB_T[\gamma, \hat{\gamma}]$ be computed using the algorithm in Figure 3. Then the UB Property for T holds.

Proof. Throughout this proof, "Line X" refers to the Xth line in Figure 3. Let $\mathbf{a}, \mathbf{b} \in \{0,1\}^N \setminus \mathbf{0}$. It suffices to show that if $\gamma = \gamma_\mathbf{a}$ in Line 1 and $\hat{\gamma} = \gamma_\mathbf{b}$ in

1. For each $\gamma \in \{0,1\}^M \setminus \mathbf{0}$
2. 　For each $\hat{\gamma} \in \{0,1\}^M \setminus \mathbf{0}$
3. 　　$\ell \leftarrow wt(\hat{\gamma})$
4. 　　$\Gamma \leftarrow \{\xi \in \{0,1\}^M \setminus \mathbf{0} : W[\xi, \hat{\gamma}] \neq 0\}$
5. 　　Order the $H = \#\Gamma$ elements of Γ as $\gamma_1, \gamma_2, \ldots, \gamma_H$ such that
6. 　　　$UB_{T-1}[\gamma, \gamma_1] \geq UB_{T-1}[\gamma, \gamma_2] \geq \cdots \geq UB_{T-1}[\gamma, \gamma_H]$
7. 　　$U_h \leftarrow UB_{T-1}[\gamma, \gamma_h]$, for $1 \leq h \leq H$
8. 　　$W_h \leftarrow W[\gamma_h, \hat{\gamma}]$, for $1 \leq h \leq H$
9. 　　$\Psi \leftarrow 0, \quad \lambda \leftarrow 0, \quad W_{\text{total}} \leftarrow 0, \quad \text{Sum} \leftarrow 0$
10. 　　$h \leftarrow 1$
11. 　　While $(h \leq H)$ and $(U_h > 0)$ and $(\Psi + (U_h * W_h) \leq 1)$ and $(\lambda < 1)$
12. 　　　$W_{\text{total}} \leftarrow W_{\text{total}} + W_h$
13. 　　　$\text{Sum} \leftarrow \text{Sum} + \text{NextTermT}(W_{\text{total}})$
14. 　　　$h \leftarrow h + 1$
15. 　　If $(h \leq H)$ and $(U_h > 0)$ and $(\Psi + (U_h * W_h) > 1)$ and $(\lambda < 1)$
16. 　　　$W_{\text{total}} \leftarrow W_{\text{total}} + (1 - \Psi)/U_h$
17. 　　　$\text{Sum} \leftarrow \text{Sum} + \text{NextTermT}(W_{\text{total}})$
18. 　　$UB_T[\gamma, \hat{\gamma}] \leftarrow \text{Sum}$

---

19. Function NextTermT (Z)
20. 　$J \leftarrow \min\left\{j : 1 \leq j \leq D_\ell, \ \Phi_j^\ell \geq Z\right\}$
21. 　$\Delta\lambda \leftarrow (\Lambda_J^\ell - \lambda) - [(\Phi_J^\ell - Z) * \rho_J^\ell]$
22. 　$\Psi \leftarrow \Psi + (U_h * W_h)$
23. 　$\lambda \leftarrow \lambda + \Delta\lambda$
24. 　return $\left(\rho_h^{A_{\min}} * \Delta\lambda\right)$

Fig. 3. Algorithm to compute $UB_T[\ ]$ for $T \geq 3$

Line 2, then the value $UB_T[\gamma, \hat{\gamma}]$ computed in Figure 3 satisfies $ELP_T(\mathbf{a}, \mathbf{b}) \leq UB_T[\gamma, \hat{\gamma}]$. Enumerate the elements of $\{0,1\}^N \setminus \mathbf{0}$ as $\mathbf{y}_1, \mathbf{y}_2, \ldots, \mathbf{y}_{2^N-1}$. We view these as input masks for round T, and hence as *output* masks for the LT of round $(T-1)$. For each $\mathbf{y}_i$, let $\mathbf{x}_i$ be the corresponding input mask for the LT. It follows from Lemma 3 that $ELP_T(\mathbf{a}, \mathbf{b}) = \sum_{i=1}^{2^N-1} ELP_{T-1}(\mathbf{a}, \mathbf{x}_i) \cdot ELP^T(\mathbf{y}_i, \mathbf{b})$. If $\gamma_{\mathbf{y}_i} \neq \gamma_{\mathbf{b}} \ (= \hat{\gamma})$, then $ELP^T(\mathbf{y}_i, \mathbf{b}) = 0$ (this follows from (6)), so we remove these $\mathbf{y}_i$ from consideration, leaving $\bar{\mathbf{y}}_1, \bar{\mathbf{y}}_2, \ldots, \bar{\mathbf{y}}_L$ (for some L), and corresponding input masks, $\bar{\mathbf{x}}_1, \bar{\mathbf{x}}_2, \ldots, \bar{\mathbf{x}}_L$, respectively.

Let $c_i = ELP_{T-1}(\mathbf{a}, \bar{\mathbf{x}}_i)$ and $d_i = ELP^T(\bar{\mathbf{y}}_i, \mathbf{b})$, for $1 \leq i \leq L$. Then $ELP_T(\mathbf{a}, \mathbf{b}) = \sum_{i=1}^{L} c_i d_i$. Note that $\sum c_i \leq 1$, $\sum d_i \leq 1$ by Lemma 1. Let

$\{\dot{c}_i\}$ ($\{\dot{d}_i\}$) be the sequence obtained by sorting $\{c_i\}$ ($\{d_i\}$) in nonincreasing order. Then $\sum_{i=1}^{L} c_i d_i \leq \sum_{i=1}^{L} \dot{c}_i \dot{d}_i$ by Lemma 6. If $\{\ddot{d}_i\}$ consists of the first L terms of $\mathcal{S}_\ell$ ($\ell = wt(\hat{\gamma})$ as in Line 3), then $\dot{d}_i \leq \ddot{d}_i$, for $1 \leq i \leq L$ (since the $\dot{d}_i$ are elements of $\mathcal{S}_\ell$), so $\sum_{i=1}^{L} \dot{c}_i \dot{d}_i \leq \sum_{i=1}^{L} \dot{c}_i \ddot{d}_i$.

Let $u_i = UB_{T-1}[\mathbf{a}, \bar{\mathbf{x}}_i]$, for $1 \leq i \leq L$, and let $\{\dot{u}_i\}$ be obtained by sorting $\{u_i\}$ in nonincreasing order. Clearly $\dot{c}_i \leq \dot{u}_i$, for $1 \leq i \leq L$. Using notation from Lines 4–8, $\{\dot{u}_i\}$ has the form

$$\underbrace{U_1, \ldots, U_1}_{W_1 \text{ terms}}, \underbrace{U_2, \ldots, U_2}_{W_2 \text{ terms}}, \underbrace{U_3, \ldots, U_3}_{W_3 \text{ terms}}, \ldots \tag{7}$$

If $\sum_{i=1}^{L} \dot{u}_i \leq 1$, let $\{\ddot{c}_i\}$ be identical to the sequence $\{\dot{u}_i\}$. If $\sum_{i=1}^{L} \dot{u}_i > 1$, let L_u ($1 \leq L_u \leq L$) be minimum such that $\sum_{i=1}^{L_u} \dot{u}_i > 1$, and let $\{\ddot{c}_i\}$ consist of the first L terms of

$$\dot{u}_1, \dot{u}_2, \ldots, \dot{u}_{L_u-1}, \left(1 - \sum_{i=1}^{L_u-1} \dot{u}_i\right), 0, 0, 0, \ldots \tag{8}$$

It follows that $\sum_{i=1}^{L} \dot{c}_i \ddot{d}_i \leq \sum_{i=1}^{L} \ddot{c}_i \ddot{d}_i$ by Lemma 7 (with $\{\ddot{d}_i\}$ playing the role of $\{d_i\}$ in the statement of the lemma). Combining inequalities gives

$$ELP_T(\mathbf{a}, \mathbf{b}) \leq \sum_{i=1}^{L} \ddot{c}_i \ddot{d}_i . \tag{9}$$

The value $\sum_{i=1}^{L} \ddot{c}_i \ddot{d}_i$ in (9) is exactly the upper bound computed in Figure 3. We argue similarly to the $T = 2$ case. Since $\{\ddot{c}_i\}$ and $\{\ddot{d}_i\}$ are derived from sequences which consist of groups of consecutive identical elements (the sequence in (7) and $\mathcal{S}_\ell$, respectively), the algorithm operates group-by-group, not element-by-element. Beginning at Line 10, the variable h is the index of the current group in $\{\ddot{c}_i\}$ (having element value U_h and size W_h). Function NextTermT() identifies the corresponding elements in $\{\ddot{d}_i\}$, and computes the equivalent of the element-by-element product.

If the terms in $\{\ddot{c}_i\}$ (resp. $\{\ddot{d}_i\}$) shrink to 0 because the corresponding terms in (7) (resp. $\mathcal{S}_\ell$) become 0, the check ($U_h > 0$) (resp. ($\lambda < 1$)) in Line 11 or Line 15 will fail, and the algorithm will exit. The check ($\Psi + (U_h * W_h) > 1$) in Line 15 detects the case that in the derivation of $\{\ddot{c}_i\}$ from $\{\dot{u}_i\}$ above, $\sum_{i=1}^{L} \dot{u}_i > 1$, and therefore $\{\ddot{c}_i\}$ is based on the truncated sequence in (8).

7   Computational Results

We estimate that running the above algorithm to completion will take up to 200,000 hours on a single Sun Ultra 5. We are currently running on about 50 CPUs, and have completed 43% of the computation for $2 \leq T \leq 10$.

It is worth noting that in progressing from 11% to 43% of the computation, there was no change in the upper bound for $2 \leq T \leq 10$. Combined with our experience in running the algorithm of [15], for which the numbers also stabilized quickly, we expect that the final results will be the same as those presented below.

In Figure 4, we plot our improved upper bound against that of [15] for $2 \leq T \leq 10$. Note that the new bound is noticeably superior to that of [15] for $T \geq 4$. When $T = 9$ rounds are being approximated, the upper bound value is $UB = 2^{-92}$. For a success rate of 96.7%, this corresponds to a data complexity of $\frac{32}{UB} = 2^{97}$ (Table 1). The corresponding upper bound value from [15] is 2^{-75}, for a data complexity of 2^{80}. This represents a significant improvement in the calculation of the provable security of Rijndael against linear cryptanalysis.

We also plot *very* preliminary results for $11 \leq T \leq 15$, in order to gain a sense of the behavior of the upper bound (for these values of T, we have completed only 1.5% of the necessary computation, hence the label "Extrapolation"). Unlike the upper bound in [15], the new upper bound does not appear to flatten out, but continues a downward progression as T increases.

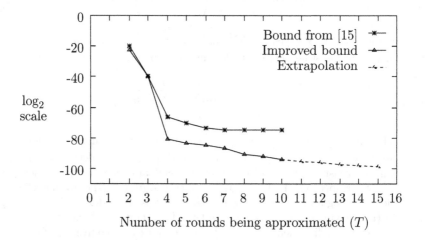

Fig. 4. Improved upper bound on MALHP for Rijndael

7.1   Presentation of Final Results

Upon completion of computation, we will post our final results in the IACR Cryptology ePrint Archive (eprint.iacr.org) under the title *Completion of Computation of Improved Upper Bound on the Maximum Average Linear Hull Probability for Rijndael*.

8 Conclusion

We have presented an improved version of the algorithm given in [15] (which computes an upper bound on the maximum average linear hull probability (MALHP) for SPNs) in the case of Rijndael. The improvement is achieved by taking into account the distribution of linear probability values for the (unique) Rijndael s-box. When 9 rounds of Rijndael are approximated, the new upper bound is 2^{-92}, which corresponds to a lower bound on the data complexity of 2^{97}, for a 96.7% success rate. (This is based on completion of 43% of the computation. However, we expect that the values obtained so far for $2 \le T \le 10$ core rounds will remain unchanged—see Section 7.) This is a significant improvement over the corresponding upper bound from [15], namely 2^{-75}, for a data complexity of 2^{80} (also for a 96.7% success rate). The new result strengthens our confidence in the provable security of Rijndael against linear cryptanalysis.

Acknowledgments

We are grateful to the reviewers for comments which improved the content and presentation of this paper. We are also grateful to the following for help in obtaining access to significant computational resources: the High Performance Computing Virtual Laboratory (Canada), the San Diego Supercomputer Center, Tom Bradshaw, Randy Ellis, Peter Hellekalek, Alex MacPherson, and Gerhard Wesp.

References

1. C.M. Adams, *A formal and practical design procedure for substitution-permutation network cryptosystems,* Ph.D. Thesis, Queen's University, Kingston, Canada, 1990.
2. K. Aoki and K. Ohta, *Strict evaluation of the maximum average of differential probability and the maximum average of linear probability,* IEICE Trans. Fundamentals, Vol. E80-A, No. 1, January 1997.
3. E. Biham, *On Matsui's linear cryptanalysis,* Advances in Cryptology—EUROCRYPT'94, LNCS 950, Springer-Verlag, pp. 341–355, 1995.
4. E. Biham and A. Shamir, *Differential cryptanalysis of DES-like cryptosystems,* Journal of Cryptology, Vol. 4, No. 1, pp. 3–72, 1991.
5. J. Daemen, R. Govaerts, and J. Vandewalle, *Correlation matrices,* Fast Software Encryption : Second International Workshop, LNCS 1008, Springer-Verlag, pp. 275–285, 1995.
6. J. Daemen and V. Rijmen, *AES proposal: Rijndael,* http://csrc.nist.gov/encryption/aes/rijndael/Rijndael.pdf.
7. J. Daemen and V. Rijmen, *AES (Rijndael) reference code in ANSI C,* http://csrc.nist.gov/encryption/aes/rijndael/.
8. *Data Encryption Standard (DES),* National Bureau of Standards FIPS Publication 46, 1977.
9. H. Feistel, *Cryptography and computer privacy,* Scientific American, Vol. 228, No. 5, pp. 15–23, May 1973.

10. H. Feistel, W.A. Notz, and J.L. Smith, *Some cryptographic techniques for machine to machine data communications,* Proceedings of the IEEE, Vol. 63, No. 11, pp. 1545–1554, November 1975.
11. C. Harpes, G. Kramer, and J. Massey, *A generalization of linear cryptanalysis and the applicability of Matsui's piling-up lemma,* Advances in Cryptology—EUROCRYPT'95, LNCS 921, Springer-Verlag, pp. 24–38, 1995.
12. H.M. Heys and S.E. Tavares, *Substitution-permutation networks resistant to differential and linear cryptanalysis,* Journal of Cryptology, Vol. 9, No. 1, pp. 1–19, 1996.
13. S. Hong, S. Lee, J. Lim, J. Sung, and D. Cheon, *Provable security against differential and linear cryptanalysis for the SPN structure,* Fast Software Encryption (FSE 2000), LNCS 1978, Springer-Verlag, pp. 273–283, 2001.
14. L. Keliher, H. Meijer, and S. Tavares, *Modeling linear characteristics of substitution-permutation networks,* Sixth Annual International Workshop on Selected Areas in Cryptography (SAC'99), LNCS 1758, Springer-Verlag, pp. 78–91, 2000.
15. L. Keliher, H. Meijer, and S. Tavares, *New method for upper bounding the maximum average linear hull probability for SPNs,* Advances in Cryptology—EUROCRYPT 2001, LNCS 2045, Springer-Verlag, pp. 420–436, 2001.
16. L.R. Knudsen, *Practically secure Feistel ciphers,* Fast Software Encryption, LNCS 809, Springer-Verlag, pp. 211–221, 1994.
17. X. Lai, J. Massey, and S. Murphy, *Markov ciphers and differential cryptanalysis,* Advances in Cryptology—EUROCRYPT'91, LNCS 547, Springer-Verlag, pp. 17–38, 1991.
18. M. Matsui, *Linear cryptanalysis method for DES cipher,* Advances in Cryptology—EUROCRYPT'93, LNCS 765, Springer-Verlag, pp. 386–397, 1994.
19. M. Matsui, *On correlation between the order of s-boxes and the strength of DES,* Advances in Cryptology—EUROCRYPT'94, LNCS 950, Springer-Verlag, pp. 366–375, 1995.
20. W. Meier and O. Staffelbach, *Nonlinearity criteria for cryptographic functions,* Advances in Cryptology—EUROCRYPT'89, LNCS 434, Springer-Verlag, pp. 549–562, 1990.
21. K. Nyberg, *Linear approximation of block ciphers,* Advances in Cryptology—EUROCRYPT'94, LNCS 950, LNCS 950, Springer-Verlag, pp. 439–444, 1995.
22. C.E. Shannon, *Communication theory of secrecy systems,* Bell System Technical Journal, Vol. 28, no. 4, pp. 656–715, 1949.
23. S. Vaudenay, *On the security of CS-Cipher,* Fast Software Encryption (FSE'99), LNCS 1636, Springer-Verlag, pp. 260–274, 1999.

Polynomial Reconstruction Based Cryptography
(A Short Survey)

Aggelos Kiayias[1] and Moti Yung[2]

[1] Graduate Center, CUNY, NY USA,
akiayias@gc.cuny.edu
[2] CertCo, NY USA
moti@cs.columbia.edu

Abstract. Cryptography and Coding Theory are closely knitted in many respects. Recently, the problem of Decoding Reed Solomon Codes (aka Polynomial Reconstruction) was suggested as an intractability assumption upon which the security of cryptographic protocols can be based. This has initiated a line of research that exploited the rich algebraic structure of the problem and related subproblems of which in the cryptographic setting. Here we give a short overview of recent works on the subject and the novel applications that were enabled due to this development.

1 Background

The polynomial reconstruction (PR) problem with parameters n, k, t is a natural way of expressing the problem of (list-)decoding Reed Solomon Codes: given a set of n points over a finite field, it asks for all polynomials of degree less than k that "fit" into at least t of the points. Translated to the coding theoretic context, PR asks for all messages that agree with at least t positions of the received codeword, for a Reed Solomon code of rate k/n.

Naturally, PR received a lot of attention from a "positive" side, i.e. how to solve it efficiently. When $t \geq \frac{n+k}{2}$ then PR has only one solution and it can be found with the algorithm of Berlekamp and Welch [BW86] ($\frac{n+k}{2}$ is the error-correction bound of Reed-Solomon codes). The problem has been investigated further for smaller values of t ([Sud97,GS98,GSR95]). These works have pointed to a certain threshold for the solvability of PR. Specifically, the problem appears to be hard if t is smaller than $\sqrt{kn}$, (the best algorithm known, by Guruswami and Sudan [GS98], finds all solutions when $t \geq \sqrt{kn}$).

We note here that apart from any direct implications of efficient list-decoding methods in the context of coding theory, these algorithms have proved instrumental in a number of computational complexity results such as the celebrated PCP theorem. There are numerous other works in computational complexity that utilize (list-)decoding techniques such as: the average-case hardness of the permanent [FL92,CPS99], hardness amplification [STV99], hardness of predicting witnesses for NP-predicates [KS99] etc.

S. Vaudenay and A. Youssef (Eds.): SAC 2001, LNCS 2259, pp. 129–133, 2001.

Perhaps the most notable work which applies the "negative" side of error-correction decoding (i.e., its inherent hardness for certain parameters) is the McEliece's cryptosystem [McE78]. Recently, in the work of Naor and Pinkas [NP99], the above well-studied Reed-Solomon list-decoding problem (namely: PR) has been looked at from a "negative" perspective, i.e. as a hard problem which cryptographic applications can base their security on.

It is important to stress that from a cryptographic perspective we are not interested in the worst-case hardness of PR but rather on the hardness of PR on the average. It is easy to see that PR on the average (termed also noisy PR) has only one solution with very high probability (note that we consider PR in a *large* prime finite field). It is believed that the noisy PR is not easier than the PR. This is because given an instance of the PR it is possible to randomize the solution polynomial (but it is not known how to randomize the noise, only to k-wise randomize it). This justification was presented by [NP99] who gave the basic suggestion to exploit the cryptographic intractability of this problem.

2   The Work of [NP99]

In [NP99] the PR problem is first exploited in concrete and efficient cryptographic protocol design. They presented a useful cryptographic primitive termed "Oblivious Polynomial Evaluation" (OPE) that allowed the secure evaluation of the value of a univariate polynomial between two parties. In their protocol the security of the receiving party was based on closely related problem to PR (later, due to the investigation of [BN00], the protocol of OPE was easily modified to be based directly on the PR problem [BN00,NP01]). We note here that a related intractability assumption appeared independently in [MRW99].

Various useful cryptographic applications based on OPE, were presented in [NP99] such as password authentication and secure list intersection computation. OPE proved to be a useful primitive in other settings, see e.g. [Gil99,KWHI01].

The assumption of [NP01] is essentially the following: given a (random) instance of PR, the value of the (unique with high probability) solution polynomial over 0 is pseudorandom for every polynomially bounded observer. Under this "pseudorandomness" assumption it can be easily shown that the receiving party in the OPE protocol is secure. Note that this assumption appears to be stronger than merely assuming hardness on the average.

3   Structural Investigation of PR

In [KY01b] we investigate cryptographic hardness properties of PR. The main theme of this work is outlined below.

Given a supposedly computationally hard problem, it is important to identify reasonable related (sub)problems upon which the security of advanced cryptographic primitives such as semantically-secure encryption and pseudorandom functions can be based. This practice is ubiquitous in cryptography, e.g. the

Decision-Diffie-Hellman problem is a subproblem related to the discrete-logarithm problem upon which the semantic security of ElGamal encryption is based; the Quadratic Residuosity Problem is a subproblem related to Factoring (and modular square roots) upon which the semantic security of [GM84] is based, etc. In [KY01b] a similar route is followed: first a suitable related subproblem of PR is identified and then advanced cryptographic primitives based on this problem are extracted. The problem is related to distinguishing one of the indices that correspond to the polynomial points in a PR-instance. Distinguishing between the points of the polynomial solution and the random points in a PR-instance appears to be naturally related to the supposed hardness of PR. The corresponding assumption is called Index-PR-Assumption (IPR). Subsequently under this assumption, we show

1. A PR instance conceals its solution in a semantic level: any algorithm that computes a function on a new value of the polynomial-solution (which is not given in the input) that is distributed according to an adversarially chosen probability distribution has negligible advantage.
2. The PR-Instances are pseudorandom.

Regarding the Polynomial Reconstruction Problem itself as the assumption, we show that it has interesting robustness properties under the assumption of almost everywhere hardness. In particular, solving PR with overwhelming probability of success is shown to be equivalent to:

1. Computing a value of the solution-polynomial at a new point with non-negligible success for almost all PR-instances.
2. Computing the least-significant-bit of a new value with non-negligible advantage for almost all PR-instances.

These results suggest that PR and its related subproblem are very robust in the cryptographic sense and seem to be suitable problems for further cryptographic exploitation. A direct application of our work is that the OPE protocol of [NP99,NP01] can be shown semantically secure (based on the IPR assumption instead).

4   Multisample Polynomial Reconstruction

A straightforward way to generalize PR so that additional cryptographic applications are allowed is the following: we can associate with any PR instance a set of indices (called the index-set) that includes the indices of the "good" points that correspond to the graph of the (with high probability unique) polynomial that "fits into" the instance. In the Multisample Polynomial Reconstruction (MPR) Problem, the given instance contains a set of r (random) PR-instances with the same index-set. The challenge is to solve all PR-instances.

MPR was defined in [KY01a] and further investigated in [BKY01]. This latter work points to a hardness threshold for the parameter r. Specifically MPR appears to be hard when r is smaller than n/t. MPR has similar robustness properties as PR and is likewise sensitive to partial information extraction. These properties are investigated in [KY01c] under the corresponding Index-MPR-Assumption.

5   Cryptographic Applications

In [KY01a] a general family of two-player games was introduced together with an efficient protocol construction that allowed a variety of novel applications, such as a deterministically correct, polylogarithmic Private Information Retrieval (PIR) protocol. The security of these games, that involved the composition of many multivariate polynomials, bilaterally contributed by the two parties, was based on the hardness of MPR. Other applications of this work include: secure computation of the *Lists' Intersection Predicate* (a stringent version of the List Intersection Problem [NP99] where the two parties want to securely check if the two private lists have a non-empty intersection without revealing any items) and *Settlement Escrows* and *Oblivious Bargaining/Negotiations*, which are protocol techniques that are useful in the e-commerce setting.

In [KY01b,KY01c] PR and MPR are employed in the setting of symmetric encryption to produce stream/block ciphers with novel attributes including:

- Semantic Security.
- Error-Correcting Decryption.
- The capability of sending messages that are superpolynomial in the security parameter (namely, a cryptosystem with a very short (sublinear) key size).
- Double homomorphic encryption over the underlying finite field operations (with bounded number of multiplications).

6   Conclusion

The rich algebraic structure of Polynomial Reconstruction (PR), its related problem (IPR) and its multisample version (MPR), has proved valuable in the cryptographic setting. On the one hand, PR and its variants appear to be robust in the cryptographic sense and can be used as a basis for advanced cryptographic primitives (as exemplified in [KY01b,KY01c]). On the other hand, several interesting cryptographic protocols that take advantage of the algebraic properties of the problem have been introduced together with their applications in secure computing and e-commerce (as seen in [NP99,KY01a]).

References

[BW86] Elwyn R. Berlekamp and L. Welch, *Error Correction of Algebraic Block Codes.* U.S. Patent, Number 4,633,470 1986.
[BKY01] Daniel Bleichenbacher, Aggelos Kiayias and Moti Yung, *Batched Decoding of Reed-Solomon Codes with Correlated Errors*, work in progress, 2001.

[BN00] Daniel Bleichenbacher and Phong Nguyen, *Noisy Polynomial Interpolation and Noisy Chinese Remaindering*. In the Proceedings of EUROCRYPT2000, Lecture Notes in Computer Science, Springer, 2000.

[CPS99] Jin-Yi Cai, A. Pavan, and D. Sivakumar, *On the Hardness of the Permanent*, In the Proceedings of the 16th International Symposium on Theoretical Aspects of Computer Science, 1999.

[FL92] Uriel Feige and Carsten Lund, *On the Hardness of Computing the Permanent of Random Matrices*, In the Proceedings of the 24th ACM Symposium on the Theory of Computing, 1992.

[Gil99] Niv Gilboa, *Two Party RSA Key Generation*, CRYPTO 1999.

[GSR95] Oded Goldreich, Madhu Sudan and Ronitt Rubinfeld, *Learning Polynomials with Queries: The Highly Noisy Case.* In the Proceedings of the 36th Annual Symposium on Foundations of Computer Science, 1995.

[GM84] Shafi Goldwasser and Silvio Micali, *Probabilistic Encryption*, JCSS 28(2): 270-299, 1984.

[GS98] Venkatesan Guruswami and Madhu Sudan, *Improved Decoding of Reed-Solomon and Algebraic-Geometric Codes.* In the Proceedings of the 39th Annual Symposium on Foundations of Computer Science, 1998.

[KWHI01] Hirotaka Komaki, Yuji Watanabe, Goichiro Hanaoka, and Hideki Imai, *Efficient Asymmetric Self-Enforcement Scheme with Public Traceability*, International Workshop on Practice and Theory in Public Key Cryptography, 2001.

[KY01a] Aggelos Kiayias and Moti Yung, *Secure Games with Polynomial Expressions*, In the Proceedings of the 28th International Colloquium in Algorithms, Languages and Programming, 2001, pp. 939-950.

[KY01b] Aggelos Kiayias and Moti Yung, *Cryptographic Hardness based on the Decoding of Reed-Solomon Codes*, manuscript, 2001.

[KY01c] Aggelos Kiayias and Moti Yung, *Symmetric Encryption based on Polynomial Reconstruction*, manuscript, 2001.

[KS99] S. Ravi Kumar and D. Sivakumar, *Proofs, Codes and Polynomial-time Reducibilities*, In the Proceedings of the 14th IEEE Conference on Computational Complexity, 1999.

[McE78] Richard J. McEliece, *A Public-Key Cryptosystem Based on Algebraic Coding Theory*, JPL Deep Space Network Progress Report 42-44, pp. 114-116, 1978.

[MRW99] Fabian Monrose, Michael K. Reiter, and Suzanne Wetzel, *Password Hardening based on Keystroke Dynamics.* In the Proceedings of the 6th ACM Computer and Communications Security Conference, Singapore, November, 1999.

[NP99] Moni Naor and Benny Pinkas, *Oblivious Transfer and Polynomial Evaluation.* In the Proceedings of the 31th ACM Symposium on the Theory of Computing, 1999.

[NP01] Moni Naor and Benny Pinkas, *Oblivious Polynomial Evaluation*, manuscript 2001, available at http://www.wisdom.weizmann.ac.il/ naor/onpub.html.

[Sud97] Madhu Sudan, *Decoding of Reed Solomon Codes beyond the Error-Correction Bound.* Journal of Complexity 13(1), pp. 180–193, 1997.

[STV99] Madhu Sudan, Luca Trevisan and Salil Vadhan, *Pseudorandom Generators without the XOR Lemma*, In the Proceedings of the 31th ACM Symposium on the Theory of Computing, 1999.

An Improved Implementation
of Elliptic Curves over $GF(2^n)$
when Using Projective Point Arithmetic

Brian King

Motorola Labs, Schaumburg IL, USA,
Indiana University Purdue University at Indianapolis
briking@iupui.edu

Abstract. Here we provide a comparison of several projective point transformations of an elliptic curve defined over $GF(2^n)$ and rank their performance. We provide strategies to achieve improved implementations of each. Our work shows that under certain conditions, these strategies can alter the ranking of these projective point arithmetic methods.

1 Introduction

In [9,17], Koblitz and Miller independently proposed to use elliptic curves over a finite field to implement cryptographic primitives. One important primitive is the Diffie-Hellman key exchange (the elliptic curve version of the protocol is called the elliptic curve Diffie-Hellman key exchange, abbreviated as ECDH). The underlying task is computing the scalar multiple kP of a point P, where k is the user's private key.

The focus of this paper will be with elliptic curves (EC) defined over fields $GF(2^n)$. For the finite field $GF(2^n)$, the standard equation or Weierstrass equation for a non supersingular elliptic curve is:

$$y^2 + xy = x^3 + ax^2 + b$$

where $a, b \in GF(2^n)$. The points $P = (x, y)$, where $x, y \in GF(2^n)$, that satisfy the equation, together with the point $\mathcal{O}$, called the point of infinity, form an additive abelian group G. Here addition in G is defined by: for all $P \in G$

- $P + \mathcal{O} = P$,
- for $P = (x, y) \neq \mathcal{O}$, $-P = (x, x + y)$
- and for all $P_1 = (x_1, y_1)$, $P_2 = (x_2, y_2)$, both not equal to the identity and $P_1 \neq -P_2$, $P_1 + P_2 = P_3 = (x_3, y_3)$ where $x_3, y_3 \in GF(2^n)$ and satisfy:

$$x_3 = \begin{cases} \left(\dfrac{y_1 + y_2}{x_1 + x_2}\right)^2 + \dfrac{y_1 + y_2}{x_1 + x_2} + x_1 + x_2 + a & \text{if } P_1 \neq P_2 \\ x_1^2 + \dfrac{b}{x_1^2} & \text{if } P_1 = P_2 \end{cases} \quad (1)$$

S. Vaudenay and A. Youssef (Eds.): SAC 2001, LNCS 2259, pp. 134–150, 2001.

$$y_3 = \begin{cases} \left(\dfrac{y_1 + y_2}{x_1 + x_2}\right)(x_1 + x_3) + x_3 + y_1 & \text{if } P_1 \neq P_2 \\ x_1^2 + \left(x_1 + \dfrac{y_1}{x_1}\right)x_3 + x_3 & \text{if } P_1 = P_2 \end{cases} \tag{2}$$

The computation of a scalar multiple of a point kP can be performed by expressing k in binary form $k = k_r k_{r-1} \ldots k_1 k_0$ and applying the "double and add" method. That is,

$$kP = 2(\cdots 2((2k_r P) + k_{r-1} P) + \cdots) + k_0 P.$$

The "add" operation requires 2 field multiplications, 1 square, and 1 inverse. The "double" operation requires 2 field multiplications, 1 square, and 1 inverse[1].

An alternate method to computing kP is to use projective point coordinates. The use of projective point arithmetic on an elliptic curve, rather than the standard affine arithmetic is such that in projective point arithmetic one delays the computation of an inverse until the very end of the process of computing kP. By doing so one will naturally see a rise in the number of required field multiplications. That is, one inverse will take place during the computation of the key kP, whereas in the affine method, one inverse takes place for each "add" function invoked (and as well, for each "double" function invoked).

The decision "affine arithmetic" vs. "projective point arithmetic" should be decided based on the ratio

$$\frac{\text{time to compute an inverse}}{\text{time to multiply}}.$$

The larger this ratio, the more attractive it is to implement projective point arithmetic. Although there exists improved methods to compute inverses [21,7], the computation of an inverse will take significantly more time than a multiplication. For example, our implementation of field operations in $GF(2^{163})$, using generating polynomial $x^{163} + x^7 + x^6 + x^3 + 1$, was such that the performance of an inverse was over 10 times the time it took to perform a multiplication. This would be equivalent to "add" and " double" functions which requires approximately 13 multiplications each. This led us to investigate projective point arithmetic.

In the following, we discuss alternate methods to develop projective point arithmetic, in an effort to determine the optimal method. We include benchmarks to illustrate which is the optimal method, ranking the methods by performance. We propose strategies which can lead to further improvement in performance, and in some cases lead to different conclusions with regard to such rankings.

2   Some Implementation Strategies That Improve Efficiency

There are a number of resources that discuss efficiency improvements for elliptic curve implementations. Some examples of efficiency improvements include: im-

---

[1] These requirements reflect the most efficient "add" and the most efficient "double".

proved key representations [2,24], improved field multiplication algorithms [18,6], the use of projective point coordinates [1,4,14], the use of a halving a point algorithm [22,13], and using the Frobenius map to improve efficiency [8,19,23]. Two resources that provide excellent overviews are [5,3]. For a review of ECC implementations in literature see [15]. In this section we provide a limited list of strategies. This list incorporates strategies considered in our implementations.

2.1   Field Multiplication Using a Lookup Table

In [6], Hasan described a method which uses lookup tables to improve performance of the field multiplication. The idea is to precompute all 2^g possible g-bit multiples of a multiplicand, and place them in a lookup table. Then you compute the multiplication by sliding over all $\lceil \frac{n}{g} \rceil$ many non overlapping g-bit windows of the other multiplicand. In all of our implementations, we use a four-bit window. This precomputation is placed into a lookup table. The product is then computed by sliding over the $\lceil \frac{n}{4} \rceil$ many non overlapping 4 bit windows of the other multiplicand. Therefore each call to a field multiplication creates a lookup table for one of the multiplicands.

2.2   Alternate Key Representations When Computing kP

Express k in NAF form

To reduce the number of "additions", one may express k in NAF (non adjacent form) form (see [3,23,2,6,5]). The NAF representation of an integer is a "signed" binary representation such that no two consecutive bits are nonzero. For example, $30 = 16 + 8 + 4 + 2 = 11110_2$, a NAF form for $30 = 100\bar{1}0_2$, we use $\bar{1}$ to represent -1. In [2], it was shown that the expected weight of a NAF of length l is approximately $l/3$.

Use a windowing technique on k

Rather than computing kP using the binary representation of $k = k_r \ldots k_1 k_0$. One could precompute the first $b - 1$ multiples of P, then express k in base b (see [18,12,3]).

2.3   Koblitz Curves

In [11], Koblitz suggested using anomalous binary curves (or *Koblitz curves*), which possess properties that can lead to improvements in efficiency. A curve described by the equation $y^2 + xy = x^3 + ax^2 + b$, where $b = 1$ and either $a = 0$ or $a = 1$, describes a Koblitz curve. In this setting, the Frobenius map, denoted by τ, is such that $\tau : (x, y) \mapsto (x^2, y^2)$ and satisfies the equation $2 = \tau - \tau^2$. Using this equation, we express the key k in τ-adic form. That is, since $k \in \mathbf{Z}[\tau]$, k which is $(k_i)_2$ can be expressed as $k = (t_j)_\tau$. For example, $010_2 = \bar{1}10_\tau$ describes the equation $2 = \tau - \tau^2$ (we use $\bar{1}$ to represent -1). This allows one to compute kP, as $\sum t_{i_\tau} P$, allowing us to use a "τ and add" method rather than a "double and add" method [11,19,8]. The efficiency improvement is accomplished because we are replacing the required 2 multiplications, 1 inverse and 1 square in the "double" by 2 squares (the "τ" of a point). Observe that in the "τ and add" method, the time-consuming operation is the "add".

2.4   Other Implementation Issues

The goal is to make relevant comparisons between different projective point representations. In particular, we would like to determine the most efficient representations and strategies for both Koblitz curves and "Random" curves. To make these comparisons, we will be using two curves contained in the WAP/WTLS list of curves [26]. One, a Koblitz curve with Weierstrass equation $y^2 + xy = x^3 + x^2 + 1$, where the field is $GF(2^{163})$ defined by generating polynomial $x^{163} + x^7 + x^6 + x^3 + 1$ (this curve has been included in many standards, and is identified in the WTLS standard as Curve 3). The other curve is defined by $y^2 + xy = x^3 + ax^2 + b$ [2] and the underlying field is $GF(2^{163})$ defined by the generating polynomial $x^{163} + x^8 + x^2 + x + 1$ (this is identified in WTLS standard by Curve 5). Further assumptions, we will express the key in a NAF form (τ-NAF when the curve is a Koblitz curve). We will not apply any windowing techniques to the key. Our implementation of the field multiplication is such that it will create a four bit lookup table of one of the multiplicands (all possible four bit products of this multiplicand). Remember, the intent of this work is two-fold. To provide an overview of projective point methods, and secondly, to discuss efficiency improvements when using a projective coordinates with a field multiplication which uses lookup tables. The intention is to use benchmarking as a comparison between methods, and the effect of different strategies on these methods.

All benchmarks were created on a HP 9000/782 with a 236 MHz Risc processor (32 bit). The table belows illustrates the performance of the basic field operations in $GF(2^{163})$ with generating polynomial $x^{163} + x^7 + x^6 + x^3 + 1$ on this platform.

Operation	Time to compute
inverse	6.493 microsec
multiply	0.540 microsec
square	0.046 microsec

3   Projective Point Arithmetic

A projective plane over a field $\mathcal{F}$, can be defined by fixing positive integers α, β and creating an equivalence relation where $(x, y, z) \sim (x', y', z')$ if $(x', y', z') =$

---

[2] where in hexadecimal EC parameter
a is 072546B5435234A422E0789675F432C89435DE5242 and
b is 00C9517D06D5240D3CFF38C74B20B6CD4D6F9DD4D9

$(\lambda^\alpha x, \lambda^\beta y, \lambda z)$ where $\lambda \in \mathcal{F}$, $\lambda \neq 0$. An equivalence class is called a projective point in $\mathcal{F}$. Each affine point $(x', y') \in \mathcal{F} \times \mathcal{F}$ can be associated with the equivalence class $(x', y', 1)$. All ordered triples (x, y, z) within this class satisfy $x' = \frac{x}{z^\alpha}$ and $y' = \frac{y}{z^\beta}$. The projective plane can be thought of as the union of the affine plane together with all equivalence classes for which $z = 0$.

Each affine point (x, y) can be mapped into the projective plane by ϕ : $(x, y) \rightarrow (x, y, 1)$. This map allows us to make a natural transformation from the affine plane to the projective plane. Then the $Image(\phi)$ consist of all equivalence classes such that $z \neq 0$. Each equivalence class (x, y, z) in the $Image(\phi)$ can be mapped to the affine plane by $x' = \frac{x}{z^\alpha}$ and $y' = \frac{y}{z^\beta}$, let us call this map ψ. Observe then that $(\psi \circ \phi)(x, y) = (x, y)$.

Recall that we are considering an elliptic curve E given by $y^2 + xy = x^3 + ax^2 + b$ defined over $GF(2^n)$. The set of affine points satisfying this equation form an additive abelian group G, and we have denoted this addition by $+$. Together Equations (1) and Equations (2) describe the addition operation in G. (Altogether we have four equations, two to describe x_3' and two to describe y_3'). The projective point addition will be denoted by $+_*$. That is, for all $P, Q \in EC$, there exists a $\zeta \in \phi(E)$, such that $\zeta = \phi(P) +_* \phi(Q)$, and $\psi(\zeta) = P + Q$. The goal is to describe $+_*$ using the projective point coordinates $\phi(P)$ and $\phi(Q)$ so that a field inverse operation in $GF(2^n)$ is not required. For a brief discussion on how to generate alternate projective point representations see the Appendix.

In practice the concern is to compute kP, where P is a fixed affine point (x_2, y_2). (In fact if the intention is to compute the ECDH key, then P represents the other user's public key.) This point P is transformed to the projective point $(x_2, y_2, 1)$. Q will represent a projective point (x_1, y_1, z_1), and will represent the partial computation of kP as we parse the key k. Thus we assume that P is of the form $P = (x_2, y_2, 1)$, that is, $z_2 = 1$. This assumption is adopted for all projective point formulas described here (this assumption is referred to as using mixed coordinates). From now on we will always make this assumption, and we introduce all projective point arithmetic operations under this assumption. Do note that when adding two EC points one needs to test the cases: are the two points equal, is one point equal to the point at infinity, and is one point the negative of the other. However, our intention is to consider the computation of kP where P is a point in a subgroup of prime order. Thus except for the case when $k = subgroup\ order$ -1, these cases will never arise. So when discussing the "add", we omit testing for these cases.

3.1   The Homogeneous Projective Point Representation

The Homogeneous projective point transformation [1,16,10] is such that the relationship between affine points (x', y') and projective point (x, y, z), where $z \neq 0$, is given by $x' = \frac{x}{z}$ and $y' = \frac{y}{z}$. For this setting, the projective point (x, y, z) which belongs to $\phi(E)$ must satisfy $zy^2 + zxy = x^3 + azx^2 + bz^3$.

Assume Q is a projective point (x_1, y_1, z_1) and P is $(x_2, y_2, 1)$, further we assume both P and $Q \neq \mathcal{O}$ and that $P \neq -Q$. There are two cases two consider: the "add" (when $P \neq Q$)and the "double" (when $Q = P$). If $P \neq \pm Q$, then

$P +_* Q = (x_3, y_3, z_3)$

$$x_3 = AD$$
$$y_3 = CD + A^2(Bx_1 + Ay_1) \tag{3}$$
$$z_3 = A^3 z_1$$

$A = x_2 z_1 + x_1$, $B = y_2 z_1 + y_1$, $C = A + B$ and $D = A^2(A + az_1) + z_1 BC$. In the case where $P = Q$, $2Q = (x_3, y_3, z_3)$

$$x_3 = AB$$
$$y_3 = x_1^4 A + B(x_1^2 + y_1 z_1 + A) \tag{4}$$
$$z_3 = A^3$$

where $A = x_1 z_1$ and $B = bz_1^4 + x_1^4$.

3.2   The Jacobian Projective Point Representation

In [4], G. Chudnovsky and D. Chudnovsky described the Jacobian projective point representation. The IEEE working group P1363 [25] is developing a public-key cryptography standard and has incorporated the Jacobian transformation as their recommended manner to perform elliptic curve arithmetic when using projective point representation. The Jacobian transformation is given by $x' = \frac{x}{z^2}$ and $y' = \frac{y}{z^3}$ to perform elliptic curve addition when using projective point coordinates. For this projective to affine transformation, the projective point (x, y, z) which belongs to $\phi(E)$ must satisfy $y^2 + zxy = x^3 + az^2 x^2 + bz^6$.

To compute $P +_* Q = (x_3, y_3, z_3)$, with $Q = (x_1, y_1, z_1)$ and $P = (x_2, y_2, 1)$ where both P and $Q \neq \mathcal{O}$: if $P \neq \pm Q$

$$
\begin{array}{llll}
U_1 = x_1 & S_2 = y_2 z_1^3 & z_3 = Lz_1 & \\
S_1 = y_1 & R = S_1 + S_2 & T = R + z_3 & \\
U_2 = x_2 z_1^2 & L = z_1 W & x_3 = az_3^2 + TR + W^3 & \\
W = U_1 + U_2 & V = Rx_2 + Ly_1 & y_3 = Tx_3 + VL^2. &
\end{array} \tag{5}
$$

In the case where $P = Q$, $2Q = (x_3, y_3, z_3)$

$$
\begin{array}{ll}
c = b^{2^{n-2}} & U = z_3 + x_1^2 + y_1 z_1 \\
z_3 = x_1 z_1^2 & y_3 = x_1^4 z_3 + U x_1. \\
x_3 = (x_1 + cz_1^2)^4 &
\end{array} \tag{6}
$$

Here $c = \sqrt{b}$ in $GF(2^n)$. The equations collectively numbered (5) are performed columnwise left to right, top to bottom.

3.3   Lopez & Dahab

In [14], Lopez and Dahab described an alternate projective point representation. Here the relationship between affine point (x', y') and projective point (x, y, z)

is $x' = \frac{x}{z^2}$ and $y' = \frac{y}{z}$. Then all projective points (x, y, z) satisfy $y^2 + xyz = zx^3 + az^2x^2 + bz^4$. To compute $P +_* Q = (x_3, y_3, z_3)$, with $Q = (x_1, y_1, z_1)$ and $P = (x_2, y_2, 1)$ where both P and $Q \neq \mathcal{O}$: if $P \neq \pm Q$

$$
\begin{aligned}
A &= y_2 z_1^2 + y_1 & z_3 &= C^2 & G &= x_3 + y_2 Z_3 \\
B &= x_2 z_1 + x_1 & E &= AC & y_3 &= EF + z_3 G \\
C &= z_1 B & x_3 &= A^2 + D + E \\
D &= B^2(C + az_1^2) & F &= x_3 + x_2 z_3
\end{aligned}
\tag{7}
$$

In the case where $P = Q$, $2Q = (x_3, y_3, z_3)$

$$
\begin{aligned}
z_3 &= z_1^2 x_1^2 \\
x_3 &= x_1^4 + bz_1^4 \\
y_3 &= bz_1^4 z_3 + x_3(az_3 + y_1^2 + bz_1^4)
\end{aligned}
\tag{8}
$$

3.4   A Comparison of the Projective Point Representations

Table 1 describes the computational requirements for each of the projective point methods. Note, if the EC parameter a is "sparse" and if it has to be multiplied to the field element t, then the time to compute at is not equivalent to a field multiplication of two arbitrary elements. In many sources, for example [14], they disregard the number of field multiplications performed with the EC parameters a and b, assuming that one can choose an elliptic curve with sparse parameters. However, in practice one may have to implement curves which have been defined in standards (although it maybe possible to make a transformation so that one or more of the transformed EC parameters is sparse). For example as the WAP/WTLS standard developed, there were initially only two strong elliptic curves defined over $GF(2^n)$, one a Koblitz curve (Curve 3) which has parameters $a = b = 1$, and the other a random curve (Curve 5) where both a and b are not sparse. And so to disregard field multiplications with EC parameters is not realistic. In Table 1 we have counted all field multiplications.

Table 1.

		no. of mult.	no. of squares
Homogeneous	Add	13	1
	Double	7	5
Jacobian	Add	11	4
	Double	5	5
Lopez& Dahab	Add	10	5
	Double	5	5

3.5 Are There More Efficient Projective Point Representations?

Of the three primary field operations in $GF(2^n)$ needed to perform projective point arithmetic: add, multiply and square; the field multiply is the time consuming operation. Although algebraically, a square is a field multiplication, as an implementation, the square can be performed very efficiently in $GF(2^n)$. For example, if you are using a normal basis, a square is a cyclic shift. If you are using a polynomial basis, then the square can be implemented by inserting 0's between terms (see [3,21]), then reducing. That is, if $\zeta = (\zeta_0, \ldots, \zeta_{n-1}) \in GF(2^n)$ then $\zeta^2 = (\zeta_0, 0, \zeta_1, 0, \ldots, \zeta_{n-1}, 0, \zeta_n)$. The representation of ζ^2 uses $n+n-1 = 2n-1$ terms, so a significant reduction needs to take place. But this can be achieved very efficiently.

Our interest in efficient projective point representations, led us to pursue alternate equations in an effort to reduce the time-consuming field multiplication. Consider the set of equations described by the Jacobian transformation, although they are algebraically efficient, the equations force an increase in the number of multiplications by the choice of an odd power. That is, to compute ζ raised to an odd power ensures the need of a multiplication, whereas, if one needs to compute ζ raised to 2,4, 8, ... all that is required is a series of squares.

Consider the projective point to affine point transformation $x' = \frac{x}{z^\alpha}$ and $y' = \frac{y}{z^\beta}$. In the Jacobian, $\alpha = 2$ and $\beta = 3$. The odd power in the denominator for y' forces additional field multiplications. In the Homogeneous transformation, both α and β are 1. This is inefficient in that both of the affine equations that generate y', Equation (2), include the term x'. Hence β should be greater than α. (This is why the Homogeneous representation has so many more field multiplications than the Jacobian.) It would appear that the "best" implementation of ECC using projective point arithmetic with fields of the form $GF(2^n)$ would satisfy α, β and $\beta - \alpha$ are powers of 2. A solution is to have $\alpha = 2^j$ and $\beta = 2^{j+1}$ for non-negative integer j. This incorporates the suggestion that $\beta > \alpha$ and the observation that the computation of z^{2^j} and $z^{2^{j+1}}$ can be done by performing a series of squares. Of course this selection criteria determines a family of projective point representations. It is trivial to show that the most efficient of this family of projective point representations is the case when $j = 0$, which is the Lopez & Dahab representation. Also note that the number of field multiplications[3] for the case $j = 1$ is equal to the number of field multiplications for the case $j = 0$ (the case $j = 1$ will require one more square operation than the $j = 0$ case). This observation could have been generated solely from manipulating the Lopez & Dahab representation.

4 Efficiency Improvements

Clearly Table 1 shows that the most efficient projective point representation is the one developed by Lopez & Dahab, next the Jacobian, and last the Homogeneous projective point representation. However we have found that there can

---

[3] Our count refers to the number of field operations required to perform an "add" and a "double" when you use this projective point representations.

be other factors that may influence performance, in addition to the number of field operations. Further, one must consider the type of curve that one is implementing. For example, if the curve is a Koblitz curve, then only the "add" must be gauged. Whereas, if the curve is a random curve then both the "double" and the "add" must be gauged. In this case, further factors that may play a role are whether the EC parameters a and/or b are sparse?

4.1   Create a Pipeline Effect–Reusing Constructed Lookup Table

A technique used to speed up computations is to pipeline common instructions together. Recall that our field multiplication is utilizing lookup tables to generate the product. To achieve a pipeline effect we examined all products and looked for similar multiplicands, allowing us to share lookup table for more than one multiply. We applied this technique to all three projective point methods. There are various strategies, the first strategy we employed was to allow at most one lookup table to be created in RAM at a time.

We illustrate the multiplications that would take place in the "add" of a Koblitz curve using the Lopez & Dahab method. Each rectangle represents a lookup table to be generated.

$z_1^2 \cdot y_2$	$z_1 \cdot x_2$	$\sqrt{z_3} \cdot (x_1 + z_1 x_2)^2$
	$z_1 \cdot (x_1 + z_1 x_2)$	$\sqrt{z_3} \cdot (y_1 + z_1^2 y_2)$
$z_3 \cdot x_2$	$(z_3 \cdot (y_1 + z_1^2 y_2)) \cdot (z_3 \cdot x_2 + x_3)$	
$z_3 \cdot y_2$		
$z_3 \cdot (x_3 + z_3 y_2)$		

The result is that the nine required field multiplications within the "add" can be arranged so that only 5 lookup tables are required. A word of warning, the existence of common multiplicands does not imply that a lookup table reduction can take place, for there may be a dependency between two multiplications. We have implemented a pipe which allowed us to add an output with a field element and take it as the next input (which is why we used the same lookup table for $z_3 \cdot y_2$ and $z_3(x_3 + z_3 \cdot y_2)$). All our piping was restricted to allowing at most one field addition.

In our implementation, we saw a dramatic improvement (in proportion to the previous benchmark) when we incorporated a pipelined multiplication (which minimizes the number of multiplication lookup tables created within the "add"). The following table highlights our benchmarks for the pipelined version of the Jacobian, Homogeneous, and Lopez & Dahab. We point out that the Homogeneous transformation performed slightly better than the Jacobian transformation (a reverse of what one may infer from Table 1). Within the Jacobian transformation we found many of the multiplications were dependent upon each other which limited our reduction of lookup tables.

Table 2. For a Koblitz curve

VERSION	Number of mult.	Compute kP nonpipeline	Lookup tables pipeline-add	Compute kP pipeline	Improvement
Lopez & Dahab	9	3.550ms	5	2.982ms	16%
Jacobian	10	3.926ms	7	3.525ms	10%
Homogeneous	12	4.240ms	6	3.486ms	18%

4.2   Applying the Pipelining to a Random Curve

We then applied the technique of pipelining multiplications with a common multiplicand in our implementation when using a random curve. To achieve a minimal amount of lookup tables, the piping choices made for a random curve will differ with the choices made for a Koblitz curve. For example, pipelines for lookup tables for the "add" and "double", respectively, using the Lopez and Dahab implementation is such that the "add" required 5 lookup tables and the "double" required 4 lookup tables. (See the appendix for an illustration of how to construct a minimum number of lookup tables for the add and the double of a Random curve).

Table 3. For a Random curve

VERSION	Number of mult. add	Number of mult. double	Compute kP nonpipeline	Lookup tables pipeline add	Lookup tables pipeline double	Compute kP pipeline
L & D	10	5	7.987ms	5	4	6.908ms
Jacobian	11	5	8.276ms	8	5	7.910ms
Homog.	13	7	10.445 ms	6	4	8.259ms

As expected the Lopez & Dahab method performed the best. Again we see a dramatic improvement in the Homogeneous method when we pipeline like multiplicands together. Although it doesn't in this case, out-perform the Jacobian method. When implementing all three methods on a Koblitz curve, the time to compute "τ" of a point is the same for all three methods. Thus the comparison of the three methods for a Koblitz curve is really a comparison of the "add" for each. Whereas in the implementation for the Random curve, the "double" is computed for each bit of the key, and so this comparison illustrates how "double" affects performance.

5   Other Strategies

Although we have achieved an implementation which requires a minimum number of lookup tables for both the "double" and "add", there are further improvements that can be achieved, however some may require more memory. That is, presently we have created a elliptic curve implementation over binary fields which will utilize lookup tables for field multiplications such that only one lookup table will exist in RAM at a time. We can improve on this implementation slightly at a cost of memory by allowing more than one lookup table to exist.

Choices we made earlier were based on the requirement that only one lookup table existed at a given time. Thus the RAM requirement for this implementation is the same as the RAM requirement for a nonpipelined implementation of the computation of kP. The pipeline was developed to meet this requirement and minimize the number of lookup tables generated. Suppose we relax this requirement and allow multiple lookup tables to exist at a given time. Observe that a number of field multiplications will contain an EC parameter as a multiplicand (i.e. a and/or b). Consequently, we can generate lookup tables for both a and b during the initial stage and save them throughout the computation. Normally a lookup table represent all 4 bit multiples of the field element. A field element in $GF(2^{163})$ requires at most 21 bytes. The size of a lookup table is at least 336 bytes. (In practice our lookup tables were such that each multiple was expressed using 6 words requiring 384 bytes). Because a table for a and/or b is such that it is used throughout the entire kP computation, the expected number of times a 4-bit multiple of a is used (say from the "add") is heuristically $160 \cdot \frac{1}{2} \cdot \frac{1}{2^4}$ for each occurrence of a multiply of an a in an "add" and $160 \cdot \frac{1}{2^4}$ for each occurrence of a multiply of an a in the "double". A similar analysis is true for multiples of b. Thus one finds an argument for increasing the window size from 4-bit to 6-bit or 8-bit, consequently we generated larger lookup tables for a and b, saved them and used these tables throughout the computation of kP.

Note that in the "add", there exist products which have a multiplicand x_2 and/or y_2, where P of the kP computation is such that $P = (x_2, y_2)$. Fortunately, we were not implementing any windowing. However, because we are working with a NAF form, we will utilize both P and $-P$. Consequently, there are two possible y_2 and one possible x_2. Again we can generate lookup tables for x_2 and both y_2 of P and the y_2 of $-P$ during the initial stage, save them and use these tables throughout the computation of kP. (Do note that if one uses a b-bit window in their implementation, and opts for the strategy of generating a precomputed lookup table for y_2 and saving it throughout the computation, then it is required to compute and save $2^b - 1$ lookup tables.)

Lastly, observe that in the Lopez & Dahab method when we compute the "add", a required multiplication is $z_3 x_3$ (this is required to compute y_3). Further note that a required computation in the "double" is $x_1 z_1$. Of course in every case, but the case when we are processing the last bit of the key, a "double" will follow an "add". Also, in every such case the product $x_3 z_3$ in the "add" is the same as $x_1 z_1$ in the following "double". Consequently, if we save the product $x_3 z_3$ from the "add" workspace and use it in the following "double", then we reduce the number of field multiplications for the "double" by one. Note we are suggesting saving a computation from a workspace, thus if we are computing the "double" after a "double" we will not have this product saved and so we must compute $x_1 z_1$. Consequently when we discuss the number of field multiplications for the "double", we must use "expected number of field multiplications". This "expected number" is of course correlated to the Hamming weight of the key. Perhaps a better way to incorporate this reduction of multiplications is to report this reduction for a multiplication as a reduction in the "add" (even though it occurs in the "double," the following table reports this reduction as a reduction within the "add" function).

In the table that follows, we have included the before and after benchmark performance of our Lopez & Dahab implementation (here the after refers to the use of the above strategies).

One can opt for any or all of the RAM increasing pipelining strategies. However, it is redundant to try to utilize all three strategies to improve performance. In fact for a Koblitz curve, only strategy 2 is relevant (precomputing lookup tables for x_2 and/or y_2). For a Random curve, we suggest strategies 1 and 3 and to increase the lookup table window for parameters a and b. In the appendix we describe an implementation for an "add" and a "double" which utilizes only strategy 1. The following table describes an implementation of a Random curve utilizes strategies 1 and 3, and uses an increased window for the lookup tables for a and b.

Random Curve using RAM increasing pipelining strategies

VERSION	Number of mult. add/double	Lookup tables add/double	Compute kP pipeline	Compute kP improved pipeline
Lopez & Dahab	9 / 5	4 / 3	6.908 ms	6.342 ms

Thus by using the Lopez & Dahab method, pipelining, and using the RAM increasing pipeline strategies that we have outlined, we have reduced a scalar multiplication from the original time of 7.983 ms to 6.342 ms. (Note that in table above we have reduced the number of multiplies in an "add" by one due to the strategy of saving the computation of $x_3 z_3$ in an "add" and using it as the $x_1 z_1$ in the following "double".)

6   A Suggestion on Squaring in $GF(2^m)$

Let $\mu \in GF(2^m)$, then $\mu = \mu_0 + \mu_1 x + \mu_2 x^2 + \cdots + \mu_{m-1} x^{m-1}$. We will also use $(\mu_0, \ldots, \mu_{m-1})$ to represent μ. Further, if $\mu_i = 0$ for all $i > j$ then we may represent μ by $(\mu_0, \ldots, \mu_j)$. Let $x^m + p(x)$ represent the generating polynomial of the field.

Several sources [21,3], have observed that a square can be computed by inserting 0's between terms and reducing. That is, $\mu^2 = \mu_0 + \mu_1 x^2 + \mu_2 x^4 + \cdots +$

$\mu_{m-1}x^{2(m-1)} = (\mu_0, 0, \mu_1, 0, \mu_2, \ldots, 0, \mu_{m-1}, 0)$. which needs to be reduced to compute the square. The required reduction can be inefficient, for it involves performing a series of shifts and adds. The shifts are determined by the terms of the generating polynomial, and they are performed on the terms of μ^2. This inefficiency exists because every other term of μ^2 is zero, thus performing an xor (an add in $GF(2^m)$) with a shift can be a waste. For example if the shift was an even length, we will see that we are wastefully computing an xor of a 0 with a 0. If the shift is an odd length then we are performing an xor with a value and a zero, rather than simply resetting a term.

Our suggestion involves using the odd and even "parts" of a polynomial and performing the shifting and adds in place. That is, performing shifts and adds on either the odd or even "parts" of a polynomial.

Consider the calculation of μ^2. Let $A = (\mu_0, 0, \mu_1, \ldots, \mu_{\frac{m-1}{2}})$ and $B = (0, \mu_{\frac{m-1}{2}+1}, 0, \mu_{\frac{m-1}{2}+2}, 0, \ldots, 0, \mu_{m-1})$, then both A and B are of degree $m-1$. Here A is an even polynomial and B is an odd polynomial. Note that $\mu^2 = A + x^m B$. So what remains is to reduce $x^m B$.

We will explain our algorithm using an example. Consider the field $GF(2^m)$ with generating polynomial $x^{163} + x^7 + x^6 + x^3 + 1$ (i.e. $m = 163$). So

$$x^m B = p(x)B = (x^7 + x^6 + x^3 + 1)B = x^7 B + x^6 B + x^3 B + 1B.$$

Since odd $\cdot$ odd is even and odd $\cdot$ even is odd, we see that $x^6 B$ and $1B$ are odd polynomials, and $x^7 B$ and $x^3 B$ are even polynomials. However, $x^6 B$, $x^7 B$, and $x^3 B$ are of degree greater than 162 (so they will require a multiplication with $p(x)$ but only for a few coefficients).

Next observe that A and B have alternating zeros, as well $x^6 B$, $x^7 B$, and $x^3 B$. We can efficiently encode this representation: encode A by $\mathcal{A} = (\mu_0, \mu_1, \ldots, \mu_{\frac{m-1}{2}})_{EVEN} = (\mu_0, \mu_1, \ldots, \mu_{81})_{EVEN}$ and encode B by $\mathcal{B} = (\mu_{\frac{n+1}{2}}, \ldots, \mu_{n-1})_{ODD} = (\mu_{82}, \ldots, \mu_{162})_{ODD}$. The subscript $EVEN$ and ODD refer to whether the polynomials are even or odd. Also note that $\mathcal{A}$ and $\mathcal{B}$ have different lengths (although we may view them as the same length but that the coefficient of the leading term of $\mathcal{B}$ is 0). Multiplication of $\mathcal{B}$ by x^i is a right shift, introducing zeros on the left and possibly generating a polynomial of degree $> m - 1$ (i.e. in this case $m - 1 = 162$). If this latter case occurs then we will call it an "overflow", the result is that these terms will need to be multiplied by $p(x)$. We will use SH_i to represent this right shift caused by multiplying by x^i. Then $xB = SH(\mathcal{B}) = (0, \mu_{82}, \mu_{83}, \ldots, \mu_{162})_{EVEN}$. $x^2 B = (0, \mu_{82}, \mu_{83}, \ldots, \mu_{161})_{ODD} + \mu_{162}p(x) = SH_2(\mathcal{B}) + \mu_{162}p(x)$. $x^3 B = (0, 0, \mu_{82}, \mu_{83}, \ldots, \mu_{161})_{EVEN} + \mu_{162}xp(x) = SH_3(\mathcal{B}) + \mu_{162}xp(x)$, and so forth. Note $x^6 B = (0,0,0,0,\mu_{82}, \mu_{83}, \ldots, \mu_{159})_{ODD} + (\mu_{162}x^4 + \mu_{161}x^2 + \mu_{160})p(x) = SH_6(\mathcal{B}) + (\mu_{162}x^4 + \mu_{161}x^2 + \mu_{160})p(x)$, and $x^7 B = (0, 0, 0, 0, \mu_{82}, \mu_{83}, \ldots, \mu_{159})_{EVEN} + (\mu_{162}x^4 + \mu_{161}x^2 + \mu_{160})p(x) = SH_7(\mathcal{B}) + (\mu_{162}x^5 + \mu_{161}x^3 + \mu_{160}x)p(x)$.

We can compute μ^2 by computing the even part, the odd part and the overflow part.

$$even\ part = \mathcal{A} + SH_3(\mathcal{B}) + SH_7(\mathcal{B})$$
$$odd\ part = \mathcal{B} + SH_6(\mathcal{B})$$

We write the overflow as:

$$overflow = p(x)\left(\mu_{162}(x^5 + x^4 + x + 1) + \mu_{161}(x^3 + x^2) + \mu_{160}(x + 1)\right)$$

The *overflow* is not decomposed in odd and even form.

What remains is to transfer even part and odd part to correct form. If f represents some algorithm which takes as input even part and odd part polynomial and outputs the polynomial, then $\mu^2 = overflow + f(even\ part, odd\ part)$. This algorithm f is very similar to the algorithm which inserts zeros between terms, except it will insert odd coefficients between even coefficients.

Consider the algorithm to determine a square by the method of inserting zeros between terms and reducing. The reduction consists of a multiplication of the generating polynomial to the higher degree terms, and creates a series of xor (addition) operations (1 less than the number of terms in the generating polynomial). When performing the xor operation, one would perform the operation on the entire multi precision integer, however every other bit is zero. Thus one does not need to perform the xor operation on the entire multi precision integer, and if one is, then they are performing twice the number of bitwise-xor operations then what is really needed. Our algorithm computes the square without performing the wasted bitwise-xor operations (essentially it will halve the number of bitwise-xor operations performed). However, in our algorithm there still is a need to compute the polynomial given its even and odd parts (this is comparable to inserting zeros between terms).

In practice one should find that this algorithm will perform better than the method that inserts zeros and reduces. Although the improvement will be marginal for small field sizes.

7   Conclusion

Our work has provided a survey and comparison of several projective point representations. In addition we have provided strategies that lead to efficiency improvements, and these stratgies may alter the comparison rankings of these projective point representations. We have provided further strategies that one can implement at a cost of memory. Lastly we have illustrated how to utilize the odd and even "parts" of a polynomial, to compute a square in $GF(2^n)$. The author wishes to thank the reviewers for their valuable and constructive suggestions.

References

1. Agnew, G., R. Mullin, S. Vanstone, "On the development of a fast elliptic curve processor chip", In *Advances in Cryptology - Crypto '91*, Springer-Verlag, 1991, pp 482-487.
2. Steven Arno and Ferrell S. Wheeler. "Signed Digit Representations of Minimal Hamming Weight". In *IEEE Transactions on Computers*. 42(8), 1993, p. 1007-1009.

3. I.F. Blake, Nigel Smart, and G. Seroussi, *Elliptic Curves in Cryptography*. London Mathematical Society Lecture Note Series. Cambridge University Press, 1999.
4. D. V. Chudnovsky and G. V. Chudnovsky. "Sequences of numbers generated by addition in formal groups and new primality and factorization tests." In *Adv. in Appl. Math.* 7 (1986) 385–434.
5. Darrel Hankerson, Julio Lopez Hernandez and Alfred Menezes. "Software Implementation of Elliptic Curve Cryptography over Binary Fields". In *CHES 2000*. p. 1-24.
6. M. A. Hasan, "Look-up Table Based Large Finite Field Multiplication in Memory Constrained Cryptosystems," In *Proceeding of the 7th IMA Conference on Cryptography and Coding*, Cirencester, UK, December 1999, LNCS, Springer Verlag, pp. 213-221.
7. M.A. Hasan. "Efficient computation of Multiplicative Inverses for Cryptographic Applications". In *CORR 2001-03 http://cacr.math.uwaterloo.ca*.
8. T. Kobayashi, H. Morita, K. Kobayashi, and F. Hoshino. "Fast elliptic curve algorithm combining Frobenius map and table reference to adapt to higher characteristic", *Advances in Cryptology - Eurocrypt '99*, Springer-Verlag, 1992, pp 176-189.
9. Neal Koblitz, *Elliptic curve cryptosystems*, Mathematics of Computation, Vol. 48, No. 177, 1987, 203-209.
10. Neal Koblitz, *Algebraic Aspects of Cryptography, Algorithms and Computation in Mathematics* ,Vol. 3, Springer-Verlag, New York, 1998.
11. Neal Koblitz, "CM-curves with good cryptographic properties", In *Advances in Cryptology - Crypto '91*, Springer-Verlag, 1992, 279-287.
12. Kenji Koyama and Yukio Tsuruoka. "Speeding up elliptic curve cryptosystems by using a signed binary window method", *Advances in Cryptology–Crypto'92*, Springer-Verlag, 1992, pp 345-357.
13. Erik Woodward Knudsen. "Elliptic Scalar Multiplication Using Point Halving". In *Advances in Cryptology - ASIACRYPT '99*. LNCS Vol. 1716, Springer, 1999, p. 135-149
14. Julio Lopez and Ricardo Dahab. "Improved Algorithms for Elliptic Curve Arithmetic in GF(2n)". In *Selected Areas in Cryptography '98, SAC'98*. LNCS 1556, Springer, 1999, p. 201-212
15. J. Lopez and R. Dahab. "Performance of elliptic curve cryptosystems". In *Technical report, IC-00-08*. May 2000. Available at http://www.dcc.unicamp.br/ic-main/publications-e.html
16. Alfred Menezes,*Elliptic Curve Public Key Cryptosystems*, Kluwer Academic Publishers, 1993.
17. Victor S. Miller, "Use of Elliptic Curves in Cryptography", In *Advances in Cryptology CRYPTO 1985*,Springer-Verlag, New York, 1985, pp 417-42
18. Volker Muller. "Efficient Algorithms for Multiplication on Elliptic Curves", Proceedings *GI–Arbeitskonferenz Chipkarten 1998*, TU Munchen, 1998.
19. Volker Muller. "Fast Multiplication on Elliptic curves over small fields of characteristic two", *Journal of Cryptology*, 11(4), 219-234, 1998.
20. NIST, *Recommended elliptic curves for federal use*, http://www.nist.gov
21. Richard Schroeppel, Hilarie Orman, Sean W. O'Malley, Oliver Spatscheck: "Fast Key Exchange with Elliptic Curve Systems", In *Advances in Cryptology - CRYPTO '95*, Lecture Notes in Computer Science, Vol. 963, Springer, 1995, pp 43-56.
22. Rich Schroeppel. "Elliptic Curves: Twice as Fast!". In*Rump session of CRYPTO 2000*.

23. Jerome A. Solinas, "An Improved Algorithm for Arithmetic on a Family of Elliptic Curves" In *Advances in Cryptology - CRYPTO 1997*, 357-371 1997, pp 357-371.
24. H. Wu and M. A. Hasan. "Closed-Form Expression for the Average Weight of Signed-Digit Representation". In *IEEE Trans. Computers*. vol. 48, August 1999, p. 848-851.
25. *IEEE P1363 Appendix A*. http://www.grouper.org/groups/1363
26. *WTLS Specification*, http://www.wapforum.org

8   Appendix

Using the strategies discussed concerning precomputing and saving lookup tables, one has great flexibility in deriving the "add" and "double" formula. In the following, we provide an example of a pipelined "add" and a pipelined "double". We utilize strategy 1, and require precomputed lookup tables for EC parameter a and EC parameter b.

Random curve: Lookup tables – add

$z_1 x_2$	$z_1^2 y_2$	$z_1(x_1 + z_1 x_2) \cdot (y_1 + z_1^2 y_2)$
$z_1(x_1 + z_1 x_2)$		$z_1(x_1 + z_1 x_2) \cdot (x_1 + z_1 x_2)^2$

$z_3 x_2$	$[z_1(x_1 + z_1 x_2) \cdot (y_1 + z_1^2 y_2)](z_3 x_2 + x_3)$
$z_3(y_2 + a)$	
$z_3 \cdot (x_3 + z_3 y_2))$	
$z_3 a$	

Random curve: Lookup tables – double

$z_1 x_1$	$z_1^4 b$	$(y_1 + x_1^2)x_3$
$(z_1 x_1) \cdot (b z_1^4)$		
$(z_1 x_1) \cdot (z_1 x_1 (b z_1^4 + (y_1 + x_1^2)x_3$		

Example of a pipelined add for a Random curve

(1)	$D \longleftarrow z_1^2$	(12) $z_3 \longleftarrow z_3^2$
(2)	make LT(z_1)	(13) $x_3 \longleftarrow a z_3$ NO LT
(3)	$A \longleftarrow z_1 x_2$	(14) $x_3 \longleftarrow x_3 + D$
(4)	$z_3 \longleftarrow z_1(A + x_1)$	(15) $x_3 \longleftarrow x_3 + A + B$
(5)	$B \longleftarrow a \cdot D$ NO LT	(16) make LT(z_3)
(6)	$B \longleftarrow B + y_1$	(17) $x_3 \longleftarrow z_3 x_2$
(7)	$A \longleftarrow A^2$	(18) $B \longleftarrow z_3 y_2$
(8)	make LT(z_3)	(19) $B \longleftarrow S(B + x_3)$
(9)	$D \longleftarrow z_3 B$	(20) $y_3 \longleftarrow x_3 + y_3$
(10)	$A \longleftarrow z_3 A$	(21) $y_3 \longleftarrow y_3 \cdot D$
(11)	$B \longleftarrow B^2$	(22) $y_3 \longleftarrow y_3 + B$

Example of a pipelined double for a random curve

(1)	$B \longleftarrow z_1^2$	(8)	$C \longleftarrow y_1^2$
(2)	$A \longleftarrow x_1^2$	(9)	$C \longleftarrow C + B$
(3)	$z_3 \longleftarrow A \cdot B$	(10)	$y_3 \longleftarrow a z_3$ NO LT
(4)	$B \longleftarrow B^2$	(11)	$y_3 \longleftarrow y_3 + C$
(5)	$B \longleftarrow bB$ NO LT	(12)	$y_3 \longleftarrow y_3 \cdot x_3 B$
(6)	$x_3 \longleftarrow A^2$	(13)	$C \longleftarrow z_3 \cdot B$
(7)	$x_3 \longleftarrow x_3 + b$	(14)	$y_3 \longleftarrow y_3 + C$

As stated earlier each multiplication requires a lookup table. For those cases where we share lookup table we have indicated the creation of a lookup table (by Make LT( )). Otherwise, the creation of the lookup table is generated within the multiplication algorithm. However, in line (5) we refer to NO LT. The implication is that no lookup table needs to be generated here. That is, b is a coefficient of the Weierstrass equation. One can create the lookup table for b at the start of the computation of kP, and save it.

8.1   How to Generate Alternate Projective Point Representations

We continue with the notation described in section 3.

Often there are more than one set of equations which describe $+_*$. Recall that $(x', y', 1) \sim (x, y, z)$ implies $x' = \frac{x}{z^\alpha}$ and $y' = \frac{y}{z^\beta}$. Fix positive integers α and β. Let $P = (x_1', y_1')$ and $Q = (x_2', y_2')$. Step one, replace x_i and y_i found in Equations (1) by the ratios $\frac{x_i}{z_i^\alpha}$ and $\frac{y_i}{z_i^\beta}$, respectively for i=1,2 (these equation determine x_3'). Step two, in Equations (2) replace x_i and y_i by the ratios $\frac{x_i}{z_i^\alpha}$ and $\frac{y_i}{z_i^\beta}$, respectively (i=1,2) and replace x_3 by the correct case of the two equations described in step one. The result will be, after simplification, four rational equations in the variables $x_1, y_1, z_1, x_2, y_2, z_2$, one set of equations for x_3' (the case where $P \neq Q$ and the other case for $P = Q$) and another set of equations for y_3' (one case for $P \neq Q$ and the other case for $P = Q$). To provide an example, choose z_3 to be the least common multiple of all four denominators (which is a polynomial in $x_1, y_1, z_1, x_2, y_2, z_2$). Set $x_3 = z_3^\alpha x_3'$, as α is a positive integer, the denominator will cancel out (in both cases $P \neq Q$ and $P = Q$), and that x_3 is equal to a polynomial in $x_1, y_1, z_1, x_2, y_2, z_2$. Set $y_3 = z_3^\beta y_3'$ and for a similar reason y_3 is equal to a polynomial in $x_1, y_1, z_1, x_2, y_2, z_2$. The result is four equations, two which describe x_3 and two which describe y_3. The pair of equations that describe z_3 are identical (the least common multiple) . For two projective points (x_1, y_1, z_1) and (x_2, y_2, z_2), both not equal to the identity, we describe the equations that define $+_*$ as: $(x_1, y_1, z_1) +_* (x_2, y_2, z_2) = (x_3, y_3, z_3)$, where z_3 was chosen to be least common multiple of all four denominators, and where $x_3 = z_3^\alpha x_3'$ and $y_3 = z_3^\beta y_3'$. If one of the points was the origin, then the sum would be the other point. The goal was to describe the equations so that the inverse did not need to be employed. The result $+_*$ is a binary operation on the $Image(\phi(E))$ such that $\psi(\phi(P) +_* \phi(Q)) = P + Q$. This argument illustrates how to define $+_*$ on $\phi(E)$. Most important, we are able to achieve the addition without the use of a field inverse.

Fast Generation of Pairs $(k, [k]P)$
for Koblitz Elliptic Curves

Jean-Sébastien Coron[1], David M'Raïhi[2], and Christophe Tymen[1]

[1] École Normale Supérieure, 45 rue d'Ulm, 75230 Paris, France
`christophe.tymen@gemplus.com,coron@clipper.ens.fr`
[2] Gemplus Card International, 3 Lagoon Drive - Suite 300,
Redwood City, CA 94065-1566, USA
`david.mraihi@gemplus.com`

Abstract. We propose a method for increasing the speed of scalar multiplication on binary anomalous (Koblitz) elliptic curves. By introducing a generator which produces random pairs $(k, [k]P)$ of special shape, we exhibit a specific setting where the number of elliptic curve operations is reduced by 25% to 50% compared with the general case when k is chosen uniformly. This generator can be used when an ephemeral pair $(k, [k]P)$ is needed by a cryptographic algorithm, and especially for Elliptic Curve Diffie-Hellman key exchange, ECDSA signature and El-Gamal encryption. The presented algorithm combines normal and polynomial basis operations to achieve optimal performance. We prove that a probabilistic signature scheme using our generator remains secure against chosen message attacks.

Key words: Elliptic curve, binary anomalous curve, scalar multiplication, accelerated signature schemes, pseudo-random generators.

1   Introduction

The use of the elliptic curves (EC) in cryptography was first proposed by Miller [8] and Koblitz [4] in 1985. Elliptic curves provide a group structure, which can be used to translate existing discrete logarithm-based cryptosystems. The discrete logarithm problem in a cyclic group G of order n with generator g refers to the problem of finding x given some element $y = g^x$ of G. The discrete logarithm problem over an EC seems to be much harder than in other groups such as the multiplicative group of a finite field, and no subexponential-time algorithm is known for the discrete logarithm problem in the class of *non-supersingular* EC which trace is different from zero and one. Consequently, keys can be much smaller in the EC context, typically about 160 bits.

Koblitz described in [5] a family of elliptic curves featuring several attractive implementation properties. In particular, these curves allow very fast scalar multiplication, *i.e.* fast computation of $[k]P$ from any point P belonging to the curve. The original algorithm proposed by Koblitz introduced an expansion method based on the Frobenius map to multiply points on elliptic curves defined over

S. Vaudenay and A. Youssef (Eds.): SAC 2001, LNCS 2259, pp. 151–164, 2001.

$\mathbb{F}_2$, $\mathbb{F}_4$, $\mathbb{F}_8$ and $\mathbb{F}_{16}$. An improvement due to Meier and Staffelbach was proposed in [6] and later on, Solinas introduced in [19] an even faster algorithm.

Many EC cryptographic protocols such as the Elliptic Curve Diffie-Hellman for key exchange [13], and the ECDSA for signature [13] require the production of fresh pairs $(k, [k]P)$ consisting of a random integer k and the point $[k]P$. A straightforward way of producing such pairs is to first generate k at random and then compute $[k]P$ using an efficient scalar multiplication algorithm. Another possiblity, introduced and analysed in [16,18,17,14,15], consists in randomly generating k and $[k]P$ at the same time, so that fewer elliptic curve operations are performed.

In this paper we focus on Koblitz (or anomalous) elliptic curves in $\mathbb{F}_{2^n}$. By introducing a generator producing random pairs $(k, [k]P)$, we are able to exhibit a specific setting where the number of elliptic curve additions is significantly reduced compared to the general case when k is chosen uniformly. The new algorithm combines normal and polynomial basis operations to achieve optimal performance. We provide a security proof for probabilistic signature schemes based on this generator.

The paper is organized as follows: in section 2 we briefly recall the basic definitions of elliptic curves and operations over a finite field of characteristic two. In section 3 we recall the definition of binary anomalous (Koblitz) curves for which faster scalar multiplication algorithms are available. We also recall the specific exponentiation techniques used on this type of curves. In section 4 we introduce the new generator of pairs $(k, [k]P)$. Section 6 provides a security proof for $(k, [k]P)$-based probabilistic signature schemes, through a fine-grained analysis of the distribution of probability of the generator (theorem 2), and using a new result on the security of probabilistic signature schemes (theorem 1). Finally, we propose in section 7 a choice of parameters resulting in a significant increase of speed compared to existing algorithms, with a proven security level.

2   Elliptic Curves on $\mathbb{F}_{2^n}$

2.1   Definition of an Elliptic Curve

An elliptic curve is the set of points (x, y) which are solutions of a bivariate cubic equation over a field K [7]. An equation of the form:

$$y^2 + a_1 xy + a_3 y = x^3 + a_2 x^2 + a_4 x + a_6 \ , \tag{1}$$

where $a_i \in K$, defines an elliptic curve over K.

In the field $\mathbb{F}_{2^n}$ of characteristic 2, equation (1) can be reduced to the form:

$$y^2 + xy = x^3 + ax^2 + b \text{ with } a, b \in \mathbb{F}_{2^n} \ .$$

The set of points on an elliptic curve, together with a special point $\mathcal{O}$ called the *point at infinity*, has an abelian group structure and therefore an addition operation. The formula for this addition is provided in [13].

2.2   Computing a Multiple of a Point

The operation of adding a point P to itself d times is called *scalar multiplication* by d and denoted $[d]P$. Scalar multiplication is the basic operation for EC protocols. Scalar multiplication in the group of points of an elliptic curve is analog to the exponentiation in the multiplicative group of integers modulo a fixed integer p.

Computing $[d]P$ is usually done with the *addition-subtraction method* based on the *nonadjacent form* (NAF) of the integer d, which is a signed binary expansion without two consecutive nonzero coefficients:

$$d = \sum_{i=0}^{\ell-1} c_i 2^i \ ,$$

with $c_i \in \{-1, 0, 1\}$ and $c_i \cdot c_{i+1} = 0$ for all $i \geq 0$. The NAF is said to be optimal because each positive integer has a unique NAF, and the NAF of d has the fewest nonzero coefficients of any signed binary expansion of d [2]. An algorithm for generating the NAF of any integer in described in [9].

3   Anomalous Binary Curves

3.1   Definition and Frobenius Map

The *anomalous binary curves* or Koblitz curves [5] are two curves E_0 and E_1 defined over $\mathbb{F}_2$ by

$$E_a : y^2 + xy = x^3 + ax^2 + 1 \text{ with } a \in \{0, 1\} \ . \tag{2}$$

We define $E_a(\mathbb{F}_{2^n})$ as the set of points (x, y) which are solutions of (2) over $\mathbb{F}_{2^n}$.

Since the anomalous curves are defined over $\mathbb{F}_2$, if $P = (x, y)$ is in $E_a(\mathbb{F}_{2^n})$, then the point (x^2, y^2) is also in $E_a(\mathbb{F}_{2^n})$. In addition, it can be checked that:

$$(x^4, y^4) + 2(x, y) = (-1)^{1-a}(x^2, y^2) \ , \tag{3}$$

where $+$ holds for the addition of points in the curve. Let τ be the Frobenius map over $\mathbb{F}_{2^n} \times \mathbb{F}_{2^n}$

$$\tau(x, y) = (x^2, y^2) \ .$$

Equation (3) can be rewritten for all $P \in E_a(\mathbb{F}_{2^n})$ as

$$\tau^2 P + [2]P = (-1)^{1-a} \tau P \ .$$

This shows that the squaring map is equivalent to a multiplication by the complex number τ satisfying

$$\tau^2 + 2 = (-1)^{1-a} \tau \ ,$$

and we say that E_a has a *complex multiplication* by τ [5]. Consequently, a point on E_a can be multiplied by any element of the ring $\mathbb{Z}[\tau] = \{x + y \cdot \tau \,|\, x, y \in \mathbb{Z}\}$.

3.2   Faster Scalar Multiplication

The advantage of using the multiplication by τ is that squaring is very fast in $\mathbb{F}_{2^n}$. Consequently, it is advantageous to rewrite the exponent d as a signed τ-adic NAF

$$d = \sum_{i=0}^{n+1} e_i \tau^i \quad \mathrm{mod} \ (\tau^n - 1) \ ,$$

with $e_i \in \{-1, 0, 1\}$ and $e_i \cdot e_{i+1} = 0$. This representation is based on the fact that $\mathbb{Z}[\tau]$ is an euclidian ring. An algorithm for computing the τ-adic NAF is given in [19]. This encoding yields the following scalar multiplication algorithm:

Algorithm 1 : Addition-substraction method with τ-adic NAF

Input:P
Output:Q
$Q \leftarrow [e_{n+1}]P$
for $i \leftarrow n$ to 0 do
    $Q \leftarrow \tau Q$
    if $e_i = 1$ then $Q \leftarrow Q + P$
    if $e_i = -1$ then $Q \leftarrow Q - P$
return Q

The algorithm requires approximately $n/3$ point additions instead of n doubles and $n/3$ additions for the general case [19]. If we neglect the cost of squarings, this is four times faster.

As in the general case, it is possible to reduce the number of point additions by precomputing and storing some "small" τ-adic multiples of P. [19] describes an algorithm which requires the storage of

$$C(\omega) = \frac{2^\omega - (-1)^\omega}{3} \ \mathrm{points} \ ,$$

where ω is a trade-off parameter. Precomputation requires $C(\omega) - 1$ elliptic additions, and the scalar multiplication itself requires approximately

$$\frac{n}{\omega + 1} \ \mathrm{elliptic \ additions} \ ,$$

which gives a total workload of

$$\simeq \frac{2^\omega}{3} + \frac{n}{\omega + 1} \ \mathrm{elliptic \ additions} \ .$$

For example, for the 163-bit curve $E_1(\mathbb{F}_{2^{163}})$ and $\omega = 4$, a scalar multiplication can be performed in approximately 35 additions, instead of 52 without precomputation.

When P is known in advance, as is the case for protocols such as Elliptic Curve Diffie-Hellman or ECDSA, it is possible to precompute and store the "small" τ-adic multiples of P once for all. The real time computation that remains is the scalar multiplication itself, which requires around $n/(\omega+1)$ operations when $C(w)$ points are stored. For example, for the 163-bit curve $E_1(\mathbb{F}_{2^{163}})$, a scalar multiplication can be performed with $\omega = 7$ in about 19 additions if 43 points are stored.

In the next section we describe an algorithm for producing random pairs $(k, [k]P)$ which requires even fewer additions for approximately the same number of points stored in memory. This algorithm appears to be well-suited for constrained environments such as smart-cards.

4   Fast Generation of $(k, [k]P)$

4.1   A Simple Generator

Many EC cryptographic protocols such as Elliptic Curve Diffie-Hellman for key exchange [13] and ECDSA for signature [13] require to produce pairs $(k, [k]P)$ consisting of a random integer k in the interval $[0, q-1]$ and the point $[k]P$, where q is a large prime divisor of the order of the curve, and P is a fixed point of order q.

For ECDSA this is the initial step of signature generation. The x coordinate of $[k]P$ is then converted into an integer c modulo q and the signature of m is (c, s) where $s = (H(m) + d \cdot c)/k \mod q$ and d is the private key associated to the public key $Q = d.P$.

[1] describes a simple method for generating random pairs of the form (x, g^x). This method can be easily adapted to the elliptic curve setting for computing pairs $(k, [k]P)$, where P is a point of order q.

Preprocessing:
  Generate t integers $k_1, \ldots, k_t \in \mathbb{Z}_q$.
  Compute $P_j = k_j.P$ for each j and store the k_j's and the P_j's in a table.
Pair generation:
  Randomly generate $S \subset [1, t]$ such that $|S| = \kappa$.
  Let $k = \sum_{i \in S} k_j \mod q$.
  Let $Q = \sum_{i \in S} P_j$ and return (k, Q).

The algorithm requires $\kappa - 1$ elliptic curve additions. Of course, the generated k is not uniformly distributed and the parameters have to be chosen with great care so that the distribution of the generated k is close to the uniform random distribution.

4.2   The New Generator

We consider the generator of figure 1 which produces random pairs of the form $(k, [k]P)$ on a Koblitz curve defined over $\mathbb{F}_{2^n}$.

Preprocessing:

        Generate t integers $k_1, \ldots, k_t \in \mathbb{Z}_q$.

        Compute $P_j = k_j.P$ for each j

        Store the k_j's and the P_j's in a table.

Pair generation:

        Generate κ random values $s_i = \pm 1$

        Generate κ random integers $e_i \in [0, n-1]$.

        Generate κ random indices $r_i \in [1, t]$.

        Let $k = \displaystyle\sum_{i=1}^{\kappa} s_i \cdot \tau^{e_i} \cdot k_{r_i} \mod q$.

        Let $Q = \displaystyle\sum_{i=1}^{\kappa} s_i \cdot \tau^{e_i} \cdot P_{r_i}$.

        Return (k, Q).

Fig. 1. Generation of $(k, [k]P)$ pairs on Koblitz curves

The difference with the previous generator is the use of the Frobenius map τ, which increases the entropy of the generated k. The new generator requires $\kappa - 1$ elliptic curve additions and t points stored in memory. In the next section we describe an efficient implementation of the new generator.

4.3   Implementing the Generator

The new generator uses the Frobenius map τ extensively, as on average $\kappa \cdot n/2$ applications of τ are performed for each generated pair, which represents $\kappa \cdot n$ squarings.

Squaring comes essentially for free when $\mathbb{F}_{2^n}$ is represented in terms of a *normal basis*: a basis over $\mathbb{F}_{2^n}$ of the form

$$\{\theta, \theta^2, \theta^{2^2}, \ldots, \theta^{2^{n-1}}\} \ .$$

Namely, in this representation, squaring a field element is accomplished by a one-bit cyclic rotation of the bitstring representing the element.

Elliptic curve additions will be performed using a *polynomial basis* representation of the elements, for which efficient algorithms for field multiplication and inversion are available. A polynomial basis is a basis over $\mathbb{F}_{2^n}$ of the form

$$\{1, x, x^2, \ldots, x^{n-1}\} \ .$$

The points P_j are stored using a normal basis representation. When a new pair is generated, the point $\tau^{e_i} \cdot P_{r_i}$ is computed by successive rotations of the coordinates of P_{r_i}. Then $\tau^{e_i} \cdot P_{r_i}$ is converted into a polynomial basis representation and it is added to the accumulator Q. To convert from normal to polynomial

Preprocessing:

Generate t random integers $k_1, \ldots, k_t \in \mathbb{Z}_q$.
Compute $P_j = k_j.P$ for each j
Store the k_j's and the P_j's in normal basis.

Pair generation:

Generate κ random integers $e_i \in [0, n-1]$
Sort the e_i: $e_1 \geq e_2 \geq \ldots \geq e_\kappa$
Set $e_{\kappa+1} \leftarrow 0$
Set $Q \leftarrow \mathcal{O}$ and $k \leftarrow 0$.
For $i \leftarrow 1$ to κ do:
    Generate a random integer $r \in [1, t]$
    Generate a random $s \leftarrow \pm 1$
    Compute $R \leftarrow s \cdot \tau^{e_i} \cdot P_r$ in normal basis.
    Convert R into polynomial basis.
    Compute $Q \leftarrow Q + R$
    Compute $k \leftarrow \tau^{e_i - e_{i+1}} \cdot (s \cdot k_j + k)$ in $\mathbb{Z}[\tau]$.
Convert k into an integer.
Return (k, Q).

Fig. 2. Algorithm for implementing the generator of $(k, [k]P)$ pairs for Koblitz curves

basis, we simply store the change-of-base matrix. The conversion's time is then approximately equivalent to one field multiplication, and this method requires to store $O(n^2)$ bits.

Before a new pair $(k, [k]P)$ is computed, the integers e_i's are sorted: $e_1 \geq e_2 \geq \ldots \geq e_\kappa$, so that k can be rewritten as

$$k = \tau^{e_\kappa} \left(s_\kappa \cdot k_{r_\kappa} + \tau^{e_{\kappa-1} - e_\kappa} \left(s_{\kappa-1} \cdot k_{r_{\kappa-1}} + \ldots \right) \right) \ .$$

The integer k is computed in the ring $\mathbb{Z}[\tau]$ as $k = k' + k'' \cdot \tau$ where $k', k'' \in \mathbb{Z}$. The element $k \in \mathbb{Z}[\tau]$ is finally converted into an integer by replacing τ by an integer T in $\mathbb{Z}_q$ solution of the equation

$$T^2 + 2 = (-1)^{1-a} T \mod q \ ,$$

so that for any point Q, we have $\tau(Q) = [T]Q$.

The implementation of the generator is summarized in figure 2.

5 Lattice Reduction Attacks and Hidden Subsets

When the generator is used in ECDSA, each signature (c, s) of a message m yields a linear equation

$$k \cdot s = H(m) + d \cdot c \mod q \ ,$$

where d is the unknown secret key and k is a sum of the hidden terms $\pm \tau^i \cdot k_i$.

The generator of [1] described in section 4.1 for which k is a sum of the hidden terms k_i has been attacked by Nguyen and Stern [11]. However, the attack requires the number of hidden k_i to be small (around 45 for a 160-bit integer k). The security of the generator relies on the difficulty of the hidden-subset sum problem studied in [11]: given a positive integer M and $b_1, \ldots, b_m \in \mathbb{Z}_M$, find $\alpha_1, \ldots, \alpha_n \in \mathbb{Z}_M$ such that each b_i is some subset sum of $\alpha_1, \ldots, \alpha_n$ modulo M.

For the new generator, if one simply considers all $\tau^i \cdot k_i$ to be hidden, this yields a large number of hidden terms ($n \cdot t$, where n is the field size and t the number of stored points) which can not be handled by [11]. We did not find any way of adapting [11] to our new generator.

6   Security Proof for Signature Schemes Using the New Generator

Since the generated integers k are not to be uniformly distributed, the security might be considerably weakened when the generator is used in conjunction with a signature scheme, a key-exchange scheme or an encryption scheme. In this section, we provide a security proof in the case of probabilistic signature schemes.

In the following, we relate the security of a signature scheme using a truly random generator with the security of the same signature scheme using our generator. Resistance against adaptive chosen message attacks is considered. This question has initially been raised by [12], and we improve the result of [12, p. 9].

Let $\mathcal{S}$ be a probabilistic signature scheme. Denote by $\mathcal{R}$ the set of random elements used to generate the signature. In our case of interest, $\mathcal{R}$ will be $\{0, \ldots, q-1\}$. Let G be a random variable on $\mathcal{R}$. Define $\mathcal{S}_G$ as the signature scheme identical to $\mathcal{S}$, except that its generation algorithm uses G as random source instead of a truly random number generator.

The following theorem shows that if a signature scheme using a truly random number generator is secure, the corresponding signature scheme using G will be secure if the distribution of G is sufficiently close to the uniform distribution. The proof is given in appendix.

If X is a random variable on a set Ω, we denote by $\delta_2(X)$ the statistical distance defined by $\delta_2(X) := \left(\sum_{\omega \in \Omega} \left| \Pr(X = \omega) - \frac{1}{|\Omega|} \right|^2 \right)^{1/2}$. In the same way, we define $\delta_1(X) := \sum_{\omega \in \Omega} \left| \Pr(X = \omega) - \frac{1}{|\Omega|} \right|$.

Theorem 1. *Let A_G be an adaptive chosen message attack against the signature scheme $\mathcal{S}_G$, during which at most m signature queries are performed. Let A be the corresponding attack on the signature scheme $\mathcal{S}$. The probabilities of existential* *forgery satisfy*

$$| \Pr (A \ succeeds) - \Pr (A_G \ succeeds) | \leq$$

$$\frac{(1 + |\mathcal{R}|\delta_2(G)^2)^{m/2} - 1}{(1 + |\mathcal{R}|\delta_2(G)^2)^{1/2} - 1} \sqrt{|\mathcal{R}| \Pr (A \ succeeds)} \delta_2(G) \ .$$

Note that asymptotically for $|\mathcal{R}|\delta_2(G)^2 \ll 1$, the bound of theorem 1 yields the inequality

$$| \Pr (A \ succeeds) - \Pr (A_G \ succeeds) | \leq m \sqrt{|\mathcal{R}| \Pr (A \ succeeds)} \delta_2(G) \ , \quad (4)$$

which has to be compared to the inequality of [12],

$$| \Pr (A \ succeeds) - \Pr (A_G \ succeeds) | \leq m \delta_1(G) \ .$$

In the following, we consider our generator of pairs $(k, [k]P)$ of section 4, which we denote by k, and compute its statistical distance $\delta_2(k)$ to the uniform distribution. Using the previous theorem with $G = k$ and $\mathcal{R} = \{0, \ldots, q-1\}$, this will provide a security proof for a signature scheme using our generator.

The following theorem is a direct application of a result exposed in [12]. It gives a bound on the expectation of $\delta_2(k)^2$, this expectation being considered on a uniform choice of $k_1, \ldots, k_t$.

Theorem 2. *If the k_i are independent random variables uniformly distributed in $\{0, \ldots, q-1\}$, then the average of $\delta_2(k)^2$ over the choice of $k_1, \ldots, k_t$ satisfies*

$$E[\delta_2(k)^2] \leq \frac{1}{(2n)^\kappa \binom{t}{\kappa}} \ .$$

In order to use this inequality, we have to link $\delta_2(k)$ to $E[\delta_2(k)^2]$; a simple application of Markov's inequality yields:

Theorem 3. *Let $\epsilon > 0$. With probability at least $1 - \epsilon$ (this probability being related to a uniform choice of $k_1, \ldots, k_t$), we have*

$$\delta_2(k) \leq \sqrt{\frac{E[\delta_2(k)^2]}{\epsilon}} \ .$$

Theorem 1 shows that the parameter which measures the security of the signature scheme using our generator is $\sqrt{|\mathcal{R}|}\delta_2(G) = \sqrt{q} \cdot \delta_2(k)$. In table 1 we summarize several values of the bound on $\sqrt{q \cdot E[\delta_2(k)^2]}$ of theorem 2, which using theorem 3 provides an upper bound for $\sqrt{q} \cdot \delta_2(k)$. We stress that the number κ of points to be stored has to be corrected by the amount of data that are required to convert from normal to polynomial basis. Roughly, one must add to κ the equivalent amount of $n/2$ points of the curve, to obtain the total amount of storage needed.

For example, consider the ECDSA signature scheme using our generator with a field size $n = 163$, $\kappa - 1 = 15$ point additions and $t = 100$ precomputed points.

Table 1. $\log_2 \sqrt{q \cdot E[\delta_2(k)^2]}$ for various values of κ and t for $n = 163$

κ/t	25	50	100	150	200
10	31	26	21	18	16
14	15	7	0	-4	-6
16	6	-2	-10	-15	-18
18	-1	-10	-19	-25	-28
20	-9	-19	-29	-35	-39
25	-27	-40	-53	-60	-65

Assume that up to $m = 2^{16}$ messages can be signed by the signer. Using table 1, we have $\sqrt{q \cdot E[\delta_2(k)^2]} \approx 2^{-10}$. Using the inequality of theorem 3, we know that, except with probability 2^{-10}, we have $\sqrt{q}\delta_2(k) \leq 2^{-10}/2^{-5} = 2^{-5}$. Assume that for a given time bound, the probability of any attack $\mathcal{A}$ breaking the ECDSA signature scheme with a truly random generator after $m = 2^{13}$ signature queries, is smaller than 2^{-60} for $n = 163$. Then the probability of breaking the ECDSA signature scheme with our generator in the same time bound is smaller than

$$\Pr\left(\mathcal{A}_G \text{ succeeds}\right) \leq 2^{13} \cdot \sqrt{2^{-60}} \cdot 2^{-5} = 2^{-19} \ .$$

This shows that the ECDSA signature scheme remains secure against chosen message attacks when using our generator for this set of parameters.

7   Parameters and Performances

We propose two sets of parameters for the field size $n = 163$. The first one is $\kappa = 16$ and $t = 100$ (which corresponds to 15 additions of points), the second is $\kappa = 11$ and $t = 50$ (which corresponds to 10 additions of points). The first set of parameters provides a provable security level according to the previous section, whereas the second set of parameters lies in a grey area where the existing attacks by lattice reduction do not apply, but security is not proven.

Recall that the scalar multiplication algorithm described in section 3.2 requires 19 elliptic curve additions with 43 points stored. Thus, the two proposed parameter sets induce a 21% and a 47% speed-up factor, respectively[1].

8   Conclusion

We have introduced a new generator of pairs $(k, [k]P)$ for anomalous binary curves. This pairs generator can be used for key exchange (ECDH), signature (ECDSA) and encryption (El-Gamal schemes). We have shown that for an appropriate choice of parameters, a probabilistic signature scheme using our generator

---

[1] If we neglect the cost of squaring the P_j's, converting from normal to polynomial basis and computing k.

remains secure against chosen message attacks. This result can be extended to key exchange schemes and encryption schemes.

We have provided a first set of parameters which provides a speed-up factor of 21% over existing techniques, with a proven security level. The second set of parameters provides a speed-up factor of 47%, but no security proof is available. However, since security is proven for slightly larger parameters, this provides a convincing argument to show that the generator has a sound design and should be secure even for smaller parameters.

Acknowledgements

We would like to thank Richard Schroeppel for a careful reading of the preliminary version of this paper, and for many useful comments. We are also grateful to Jacques Stern for his valuable input.

References

1. V. Boyko, M. Peinado, and R. Venkatesan. Speeding up discrete log and factoring based schemes via precomputations. In *Advances in Cryptology - Eurocrypt '98*, pages 221–235. Springer Verlag, 1998.
2. D.M. Gordon. A survey of fast exponentiation methods. *Journal of Algorithms*, 27:129–146, 1998.
3. B.S. Kaliski Jr. and T.L. Yin. Storage-efficient finite field basis conversion. In *Selected areas in Cryptography - SAC'98*, volume 1556, 1998.
4. N. Koblitz. Elliptic curve cryptosystems. *Mathematics of Computation*, 48:203–209, 1987.
5. N. Koblitz. CM-curves with good cryptographic properties. In Joan Feigenbaum, editor, *Advances in Cryptology - Crypto '91*, pages 279–287, Berlin, 1991. Springer-Verlag. Lecture Notes in Computer Science Volume 576.
6. W. Meier and O. Staffelbach. Efficient multiplication on certain non-supersingular elliptic curves. In *Advances in Cryptology - Crypto '92*, volume LNCS 740, pages 333–344. Springer Verlag, 1993.
7. A.J. Menezes. *Elliptic Curve Public Key Cryptosystems*. Kluwer Academic Publishers, 1993.
8. V.S. Miller. Use of elliptic curves in cryptography. In Springer Verlag, editor, *Proceedings of Crypto 85*, volume LNCS 218, pages 417–426. Springer Verlag, 1986.
9. F. Morain and J. Olivos. Speeding up the computation of an elliptic curve using addition-subtraction chains. *Inform. Theory Appl.*, 24:531–543, 1990.
10. P. Nguyen. *La géométrie des nombres en cryptologie*. PhD thesis, Université de Paris 7, 1999.
11. P. Nguyen and J. Stern. The hardness of the hidden subset sum problem and its cryptographic implications. In Michael Wiener, editor, *Advances in Cryptology - Crypto'99*, pages 31–46, Berlin, 1999. Springer-Verlag. Lecture Notes in Computer Science.
12. Phong Nguyen, Igor Shparlinsky, and Jacques Stern. Distribution of Modular Subset Sums and the Security of the Server Aided Exponentiation. In *Workshop on Cryptography and Computational Number Theory*, 1999.
13. IEEE P1363. *Standard Specifications for Public Key Cryptography*. August 1998.

14. P. de Rooij. On the security of the Schnorr scheme using preprocessing. In Donald W. Davies, editor, *Advances in Cryptology - EuroCrypt '91*, pages 71–80, Berlin, 1991. Springer-Verlag. Lecture Notes in Computer Science Volume 547.

15. P. de Rooij. On Schnorr's preprocessing for digital signature schemes. In Tor Helleseth, editor, *Advances in Cryptology - EuroCrypt '93*, pages 435–439, Berlin, 1993. Springer-Verlag. Lecture Notes in Computer Science Volume 765.

16. P. de Rooij. Efficient exponentiation using precomputation and vector addition chains. In Alfredo De Santis, editor, *Advances in Cryptology - EuroCrypt '94*, pages 389–399, Berlin, 1995. Springer-Verlag. Lecture Notes in Computer Science Volume 950.

17. C. P. Schnorr. Efficient identification and signatures for smart cards. In Jean-Jacques Quisquater and Joos Vandewalle, editors, *Advances in Cryptology - EuroCrypt'89*, pages 688–689, Berlin, 1989. Springer-Verlag. Lecture Notes in Computer Science Volume 434.

18. C. P. Schnorr. Efficient identification and signatures for smart cards. *Journal of Cryptology*, 4:161–174, 1991.

19. J.A. Solinas. An improved algorithm for arithmetic on a family of elliptic curves. In Burt Kaliski, editor, *Advances in Cryptology - Crypto '97*, pages 357–371, Berlin, 1997. Springer-Verlag. Lecture Notes in Computer Science Volume 1294.

A   Proof of Theorem 1

Theorem 3. *Let S be a probabilistic signature scheme. Let $\mathcal{R}$ be the set from which the signature generation algorithm chooses a random element when generating a signature. Let G be a random variable in $\mathcal{R}$, and S_G the scheme derived from S which uses G as random source instead of a random oracle for the signature generation. Let A_G be an adaptative attack with m chosen messages on S_G. If A is the corresponding attack on S, then the probabilities of existential forgery satisfy*

$$\left| \Pr\left(A \text{ succeeds}\right) - \Pr\left(A_G \text{ succeeds}\right) \right| \leq$$

$$\frac{(1 + |\mathcal{R}|\delta_2(G)^2)^{m/2} - 1}{(1 + |\mathcal{R}|\delta_2(G)^2)^{1/2} - 1} \sqrt{|\mathcal{R}| \Pr\left(A \text{ succeeds}\right)} \delta_2(G) \ .$$

Proof. An adaptative attack with m chosen messages makes m queries to a signature oracle. At each call, this oracle picks a random r in $\mathcal{R}$, and uses this r to produce a signature. If the signature scheme is S, r is chosen uniformly in $\mathcal{R}$, and is thus equal to the value of a random variable U uniformly distributed in $\mathcal{R}$. If the signature scheme is S_G, r is the value of the random variable G. Consequently, an attack with m chosen messages depends on a random variable defined over the probability space $\mathcal{R}^m$. This variable is either $U = (U_1, \ldots, U_m)$ in the case of an attack against S, or $G = (G_1, \ldots, G_m)$ in the case of an attack against S_G, where the U_i are pairwise independent and follow the same distribution as U, and the G_i are pairwise independent and follow the same distribution as G.

The following proof is a refinement of the result that can be found in [12] concerning accelerated signatures schemes. First note that as A and A_G are the same attacks (that is, are the same Turing machines making calls to the same signature oracle except that they use different random sources), for all $\boldsymbol{r} = (r_1, \ldots, r_m) \in \mathcal{R}^m$,

$$\Pr(A \text{ succeeds}|\boldsymbol{U} = \boldsymbol{r}) = \Pr(A_G \text{ succeeds}|\boldsymbol{G} = \boldsymbol{r}) \ .$$

Thus, using Bayes formula, we get

$$|\Pr(A \text{ succeeds}) - \Pr(A_G \text{ succeeds})| \leq$$

$$\sum_{\boldsymbol{r}=(r_1,\ldots,r_m)\in\mathcal{R}^m} |\Pr(\boldsymbol{G} = \boldsymbol{r}) - \Pr(\boldsymbol{U} = \boldsymbol{r})| \Pr(A \text{ succeeds}|\boldsymbol{U} = \boldsymbol{r}) \ . \tag{5}$$

Using the triangular inequality, the independence of the U_i and of the G_i, and the equidistribution property, we get also that

$$|\Pr(\boldsymbol{U} = \boldsymbol{r}) - \Pr(\boldsymbol{G} = \boldsymbol{r})| \leq$$

$$\sum_{k=1}^{m} \left(\prod_{1 \leq i < k} \Pr(G = r_i) \right) |\Pr(U = r_k) - \Pr(G = r_k)| \left(\prod_{m \geq i > k} \Pr(U = r_i) \right) , \tag{6}$$

with the convention that the product of zero terms is equal to 1.

Consequently, if we denote, for $k = 1, \ldots, m$, by $a_k(\boldsymbol{r})$ the quantity

$$\left(\prod_{1 \leq i < k} \Pr(G = r_i) \right) |\Pr(U = r_k) - \Pr(G = r_k)| \left(\prod_{m \geq i > k} \Pr(U = r_i) \right) ,$$

equation (5) can be rewritten as

$$|\Pr(A \text{ succeeds}) - \Pr(A_G \text{ succeeds})| \leq$$

$$\sum_{k=1}^{m} \sum_{\boldsymbol{r} \in \mathcal{R}^m} a_k(\boldsymbol{r}) \Pr(A \text{ succeeds}|\boldsymbol{U} = \boldsymbol{r}) \ . \tag{7}$$

Using Cauchy's inequality,

$$\sum_{\boldsymbol{r} \in \mathcal{R}^m} a_k(\boldsymbol{r}) \Pr(A \text{ succeeds}|\boldsymbol{U} = \boldsymbol{r}) =$$

$$\sum_{\boldsymbol{r} \in \mathcal{R}^m} \left(|\mathcal{R}|^{m/2} a_k(\boldsymbol{r}) \right) \left(|\mathcal{R}|^{-m/2} \Pr(A \text{ succeeds}|\boldsymbol{U} = \boldsymbol{r}) \right) \leq \tag{8}$$

$$\left(\sum_{\boldsymbol{r} \in \mathcal{R}^m} |\mathcal{R}|^m a_k(\boldsymbol{r})^2 \right)^{1/2} \left(\sum_{\boldsymbol{r} \in \mathcal{R}^m} |\mathcal{R}|^{-m} \Pr(A \text{ succeeds}|\boldsymbol{U} = \boldsymbol{r})^2 \right)^{1/2} \ .$$

And as $\Pr(A \text{ succeeds}|U = r) \leq 1$,

$$\sum_{r \in \mathcal{R}^m} |\mathcal{R}|^{-m} \Pr(A \text{ succeeds}|U = r)^2 \leq$$

$$\sum_{r \in \mathcal{R}^m} |\mathcal{R}|^{-m} \Pr(A \text{ succeeds}|U = r) = \Pr(A \text{ succeeds}) \ ,$$

(9)

because U is uniformly distributed over $\mathcal{R}$. Returning to the definition of $a_k(r)$, and using once again the uniformity of U, one sees that

$$|\mathcal{R}|^m a_k(r)^2 \leq |\mathcal{R}| \left(\prod_{1 \leq i < k} |\mathcal{R}| \Pr(G = r_i)^2 \right) |\Pr(U = r_k) - \Pr(G = r_k)|^2 \ .$$

(10)

Now, one needs to note that

$$\sum_{r_i \in \mathcal{R}} |\mathcal{R}| \Pr(G = r_i)^2 =$$

$$\sum_{r_i \in \mathcal{R}} |\mathcal{R}| \left(\left(\Pr(G = r_i) - \frac{1}{|\mathcal{R}|} \right)^2 - 1/|\mathcal{R}|^2 + (2/|\mathcal{R}|) \Pr(G = r_i) \right) =$$

$$|\mathcal{R}| \delta_2(G)^2 + 1 \ .$$

Thus, the inequality (10) becomes,

$$\sum_{r \in \mathcal{R}^m} |\mathcal{R}|^m a_k(r)^2 \leq$$

$$|\mathcal{R}| \left(1 + |\mathcal{R}| \delta_2(G)^2 \right)^{k-1} \sum_{r_k \in \mathcal{R}} |\Pr(U = r_k) - \Pr(G = r_k)|^2 =$$

$$|\mathcal{R}| \left(1 + |\mathcal{R}| \delta_2(G)^2 \right)^{k-1} \delta_2(G)^2 \ .$$

Returning to inequality (7), and using (8) and (9), we finally get:

$$|\Pr(A \text{ succeeds}) - \Pr(A_G \text{ succeeds})| \leq$$

$$\sum_{k=1}^{m} \left(|\mathcal{R}| \left(1 + |\mathcal{R}| \delta_2(G)^2 \right)^{k-1} \delta_2(G)^2 \right)^{1/2} (\Pr(A \text{ succeeds}))^{1/2} =$$

$$(|\mathcal{R}| \Pr(A \text{ succeeds}))^{1/2} \frac{(1 + |\mathcal{R}| \delta_2(G)^2)^{m/2} - 1}{(1 + |\mathcal{R}| \delta_2(G)^2)^{1/2} - 1} \delta_2(G) \ .$$

$\square$

Algorithms for Multi-exponentiation

Bodo Möller

Technische Universität Darmstadt, Fachbereich Informatik
moeller@cdc.informatik.tu-darmstadt.de

Abstract. This paper compares different approaches for computing power products $\prod_{1 \leq i \leq k} g_i^{e_i}$ in commutative groups. We look at the conventional simultaneous exponentiation approach and present an alternative strategy, interleaving exponentiation. Our comparison shows that in general groups, sometimes the conventional method and sometimes interleaving exponentiation is more efficient. In groups where inverting elements is easy (e.g. elliptic curves), interleaving exponentiation with signed exponent recoding usually wins over the conventional method.

1 Introduction

A common task in implementations of many public-key cryptosystems is multi-exponentiation in some commutative group G, i.e. evaluating a product

$$\prod_{1 \leq i \leq k} g_i^{e_i}$$

where $k \geq 2$ is a small integer, each g_i is an element of G, and each e_i is an integer (typically a few hundred up to a few thousand bits long). We require that the e_i be non-negative (otherwise, invert g_i). Example groups include $(\mathbb{Z}/n\mathbb{Z})^*$ for some integer n, e.g. for verification of ElGamal [11] or DSA [17] signatures; groups of rational points on elliptic curves over finite fields, e.g. for verification of ECDSA [1] signatures; and class groups of imaginary-quadratic orders, e.g. for verification of RDSA [2][7] signatures. We have $k = 2$ for DSA and ECDSA verification and $k = 3$ for ElGamal and RDSA verification. Larger values of k appear in protocols of Brands [4]. In the present paper, we allow $k = 1$ as well for algorithms; efficiency considerations may ignore this case. It is well known that in general it is unnecessarily inefficient to compute the powers $g_i^{e_i}$ separately and then multiply them. Instead, specific algorithms for multi-exponentiation are usually applied.

We assume that the e_i consist of independent random bits up to a respective maximum bit-length b_i; i.e., e_i is a uniformly distributed random integer in the interval $[0, 2^{b_i} - 1]$. (In practice the actual distribution may differ, but for typical cases this simplified assumption is reasonably close.) In this setting, we consider general algorithms for arbitrary exponents; we do not examine algorithms based on tailor-made addition chains in $\mathbb{Z}^k$ for given $e_1, \ldots, e_k$ (cf. [3]). (Note that even if an exponent is fixed in a cryptographic protocol, it is sometimes desirable to

S. Vaudenay and A. Youssef (Eds.): SAC 2001, LNCS 2259, pp. 165–180, 2001.
© Springer-Verlag Berlin Heidelberg 2001

perform computations using varying exponents in order to thwart side-channel attacks that try to use timings [12] or power consumption measurements [13] or other extra data to gain knowledge on secret exponents. To avoid constant exponents, g^e can be rewritten as $g^{n \cdot \text{ord}(g)+e}$ and $\prod g_i^{e_i}$ can be rewritten as $(\prod g_i)^n \prod g_i^{e_i-n}$ for arbitrary integers n.)

Like window-based algorithms for single exponentiations, the algorithms that we analyse work in two stages: First, in the *precomputation stage,* an auxiliary table of group elements is computed from the elements g_i; then, in the *evaluation stage,* the final result is computed using these auxiliary values.

The usual approach for multi-exponentiation combines all input group elements g_i with each other in the precomputation stage ([11], [20], [21]); then the evaluation stage looks at all exponents simultaneously. In the present paper, we discuss an alternative approach where the g_i are treated separately in the precomputation stage. In this approach, the evaluation stage uses an interleaving of the generators and exponents for the various i rather than handling multiple i simultaneously.

We collectively refer to the multi-exponentiation methods described in [11], [20] and [21] as *"simultaneous exponentiation"*. Section 2 describes these methods. Section 3 presents two variants of our alternative approach, which we dub *"interleaving exponentiation":* a basic method and an alternative method that can be used in groups where inverting elements is easy. In section 4, we compare the efficiency of simultaneous exponentiation methods and interleaving exponentiation methods. Section 5 discusses variants that can be advantageous when all bases g_i are fixed.

In specific groups, additional useful efficiently computable endomorphisms are available besides squaring and possibly inversion (see e.g. [19]); this may lead to better multi-exponentiation algorithms for these groups. Such special groups are out of the scope of the present paper.

1.1   Notation

We write $e[j]$ for bit j of a non-negative integer e. For negative j, we define that $e[j] = 0$. We write $e[j \ldots j']$ for the integer consisting of the concatenation of bits j down to j' of e; e.g., if $e = 10111_2 = 23$, then $e[3 \ldots 1] = 011_2 = 3$ and $e[1 \ldots -2] = 1100_2 = 12$.

2   Simultaneous Exponentiation Methods

We look at two multi-exponentiation methods using simultaneous exponentiation (as opposed to interleaving exponentiation, which is introduced in section 3): Straus's 2^w-ary method (section 2.1) and the sliding window method of Yen, Laih, and Lenstra (section 2.2). (The method known as "Shamir's trick" appears as a special case of both of these.)

As noted in the introduction, all algorithms that we consider are related and work in two stages: First, the *precomputation stage* prepares an auxiliary table of group elements; then, the *evaluation stage* computes the final result using this table. For comparing different methods, we examine the two stages separately.

When examining simultaneous exponentiation algorithms, we assume that b_i is the same for all i. Let at least one e_i be non-zero, and let b be the bit-length of the longest of the e_i. Parameter w is always a positive integer, the "window size"; larger window sizes make the precomputation stage less efficient, but speed up the evaluation stage. It is not possible to give a general rule for selecting an optimal w (cf. section 4).

Relevant features of the precomputation stage are the number of group operations required for computing the auxiliary table, and the number of table entries. For group operations, we differentiate between squarings and general multiplications, since the former often can be computed more efficiently. The precomputed tables will always contain the values g_1, ..., g_k, all of which are trivially available and hence can be neglected. It will be visible that computing each additional table entry requires one multiplication or, for some of the table entries in the simultaneous 2^w-ary method, one squaring. In addition to this, k squarings are needed by the simultaneous sliding window method if $w > 1$.

The evaluation stage requires both squarings and multiplications. For each multi-exponentiation method, we look at the number of squarings and the expected number of general multiplications for given k, b, and w. w is assumed to be small in comparison to b (otherwise the precomputation stage would become unreasonably expensive).

It should be noted that a slight optimisation for the precomputation stage is possible in all methods by first looking which table entries are actually needed (either during the evaluation stage, or because other precomputed table entries that are needed in the evaluation stage depend on them) and limiting precomputation to these. As this optimisation will usually only have a small effect in practice, we neglect it in our comparisons.

For the number of squarings in the evaluation stage, we assume that the following optimisation is used: As initially variable A is 1_G (the neutral element of G) in all algorithms, squarings can easily be avoided until a different value has been assigned to A.

Formulas for the expected number of multiplications during the evaluation stage given in the following are actually asymptotics for large b/w rather than precise values (we do not take into account the special probability distributions encountered at both ends of the exponents). As in practice w will be much smaller than b, the error is negligible for our purposes.

Just as squarings can be eliminated in the evaluation stage while A is 1_G, the first multiplication of A by a table entry can be replaced by an assignment. This minor optimisation is not used in our figures below; note that it applies similarly to all algorithms discussed in this paper (and does not affect asymptotics), so comparisons between different methods remain just as valid.

2.1   Simultaneous 2^w-Ary Method

The simultaneous 2^w-ary exponentiation method [20] (see also [15]) looks at w bits of each of the exponents for each evaluation stage group multiplication, i.e. kw bits in total. The special case where $w = 1$ is also known as "Shamir's trick" since it was described in [11] with a reference to Shamir.

Precomputation Stage. Precompute $\prod_{1 \leq i \leq k} g_i^{E_i}$ for all non-zero k-tuples $(E_1, \ldots, E_k) \in \{0, \ldots, 2^w - 1\}^k$.

Number of non-trivial table entries: $2^{kw} - 1 - k$. Of these, $2^{k(w-1)} - 1$ can be computed by squaring other table entries (all the E_i are even). The remaining $2^{kw} - 2^{k(w-1)} - k$ entries require one general multiplication each.

No additional squarings are required.

Evaluation Stage.

$$A \leftarrow 1_G$$
$$\text{for } j = \lfloor (b-1)/w \rfloor w \text{ down to } 0 \text{ step } w \text{ do}$$
$$\quad \text{for } n = 1 \text{ to } w \text{ do}$$
$$\quad\quad A \leftarrow A^2$$
$$\quad \text{if } \left(e_1[j+w-1 \ldots j], \ldots, e_k[j+w-1 \ldots j] \right) \neq (0, \ldots, 0) \text{ then}$$
$$\quad\quad A \leftarrow A \cdot \prod_i g_i^{e_i[j+w-1 \ldots j]} \quad \{\text{multiply } A \text{ by table entry}\}$$
$$\text{return } A$$

Number of squarings: $\left\lfloor \dfrac{b-1}{w} \right\rfloor w$.

Expected number of multiplications: $b \cdot \dfrac{1 - \frac{1}{2^{kw}}}{w}$.

2.2   Simultaneous Sliding Window Method

The simultaneous sliding window exponentiation method of Yen, Laih, and A. Lenstra [21] is an improvement of the 2^w-ary method described in section 2.1. Due to the use of a sliding window, table entries are required only for those tuples $(E_1, \ldots, E_k)$ where at least one of the E_i is odd. (Note that while values g_i^2 no longer appear in the precomputed table, the precomputation stage now needs them as intermediate values unless $w = 1$.) Also the expected number of multiplications required in the evaluation stage is reduced. Like the 2^w-ary method, this method looks at w bits of each of the exponents for each evaluation stage group multiplication (kw bits in total). For $w = 1$, this again is "Shamir's trick". For $k = 1$, this is the usual sliding window method for a single exponentiation (see e.g. [15]).

Precomputation Stage. Precompute $\prod_{1 \leq i \leq k} g_i^{E_i}$ for all k-tuples $(E_1, \ldots, E_k)$ $\in \{0, \ldots, 2^w - 1\}^k$ where at least one of the E_i is odd.

Number of non-trivial table entries (multiplications): $2^{kw} - 2^{k(w-1)} - k$.

Number of squarings: k if $w > 1$; none otherwise.

Evaluation Stage.

$A \leftarrow 1_G$
$j \leftarrow b - 1$
while $j \geq 0$ **do**
    if $\forall i \in \{1, \ldots, k\}\colon e_i[j] = 0$ **then**
        $A \leftarrow A^2$; $j \leftarrow j - 1$
    else
        $j_{\text{new}} \leftarrow \max(j - w, -1)$
        $J \leftarrow j_{\text{new}} + 1$
        while $\forall i \in \{1, \ldots, k\}\colon e_i[J] = 0$ **do**
            $J \leftarrow J + 1$
        $\{\text{now } j \geq J > j_{\text{new}}\}$
        for $i = 1$ **to** k **do**
            $E_i \leftarrow e_i[j \ldots J]$
        while $j \geq J$ **do**
            $A \leftarrow A^2$; $j \leftarrow j - 1$
        $A \leftarrow A \cdot \prod_i g_i^{E_i}$ {multiply A by table entry}
        while $j > j_{\text{new}}$ **do**
            $A \leftarrow A^2$; $j \leftarrow j - 1$
return A

Number of squarings: $b - w$ *up to* $b - 1$.
Expected number of multiplications: $b \cdot \dfrac{1}{w + \sum_{n \geq 1} \frac{1}{2^{kn}}} = b \cdot \dfrac{1}{w + \frac{1}{2^k - 1}}$.

3 Interleaving Exponentiation Methods

Here, we look at two interleaving exponentiation algorithms: A basic algorithm suitable for arbitrary groups (section 3.1) and a special variant using signed exponent recoding that can be applied if inverting elements is easy (section 3.2).

The comments in the introduction to section 2 apply similarly, with the exception that we no longer assume all the b_i to be identical. Instead of a single window size w, in this section we have k possibly different window sizes w_i ($1 \leq i \leq k$) used for the respective parts of the multi-exponentiation; each w_i is a small positive integer. Again we assume that initial squarings are eliminated while A is 1_G.

Note that for the algorithms described in this section, the precomputed table has disjoint parts for different bases g_i. If multiple multi-exponentiations have to be performed and some of the bases g_i appear again, then the corresponding parts of earlier precomputed tables can be reused.

3.1 Basic Interleaving Exponentiation Method

The basic interleaving exponentiation method is a generalization of the sliding window method for a single exponentiation (see e.g. [15]), to which it corresponds in case $k = 1$.

Precomputation Stage. For $i = 1, \ldots, k$, precompute g_i^E for all odd E such that $1 \leq E \leq 2^{w_i} - 1$.

Number of non-trivial table entries (multiplications): $\left(\sum_{1 \leq i \leq k} 2^{w_i - 1} \right) - k$.

Number of squarings: $\#\{i \in \{1, \ldots, k\} \mid w_i > 1\}$.

Evaluation Stage.

$$
\begin{aligned}
&A \leftarrow 1_G \\
&\textbf{for } i = 1 \text{ to } k \textbf{ do} \\
&\quad window\_handle_i \leftarrow \text{nil} \\
&\textbf{for } j = b - 1 \text{ down to } 0 \textbf{ do} \\
&\quad A \leftarrow A^2 \\
&\quad \textbf{for } i = 1 \text{ to } k \textbf{ do} \\
&\quad\quad \textbf{if } window\_handle_i = \text{nil and } e_i[j] = 1 \textbf{ then} \\
&\quad\quad\quad J \leftarrow j - w_i + 1 \\
&\quad\quad\quad \textbf{while } e_i[J] = 0 \textbf{ do} \\
&\quad\quad\quad\quad J \leftarrow J + 1 \\
&\quad\quad\quad \{\text{now } j \geq J > j - w_i \text{ and } J \geq 0\} \\
&\quad\quad\quad window\_handle_i \leftarrow J \\
&\quad\quad\quad E_i \leftarrow e_i[j \ldots J] \\
&\quad\quad \textbf{if } window\_handle_i = j \textbf{ then} \\
&\quad\quad\quad A \leftarrow A \cdot g_i^{E_i} \ \{\text{multiply } A \text{ by table entry}\} \\
&\quad\quad\quad window\_handle_i \leftarrow \text{nil} \\
&\textbf{return } A
\end{aligned}
$$

Number of squarings: $b - \max_i w_i$ up to $b - 1$.
Expected number of multiplications:

$$
\sum_{1 \leq i \leq k} b_i \cdot \frac{1}{w_i + \sum_{n \geq 1} \frac{1}{2^n}} = \sum_{1 \leq i \leq k} b_i \cdot \frac{1}{w_i + 1}.
$$

3.2   wNAF-Based Interleaving Exponentiation Method

In some groups, elements can be inverted very efficiently so that division is not significantly more expensive than multiplication. (Inversion is cheap in case of elliptic curves or class groups of imaginary quadratic number fields, but not in $(\mathbb{Z}/n\mathbb{Z})^*$.) This can be exploited for making exponentiation algorithms more efficient by recoding the exponents into a signed representation. We use a technique introduced for single exponentiations independently in [18] and in [16] and apply it to the task of multi-exponentiation.

Given an exponent e_i and a window size w_i, we need a *width-$(w_i + 1)$ non-adjacent form* (*width-(w_i+1) NAF* or *wNAF*) of e_i, which is an array $N_i[b_i], \ldots, N_i[0]$ of integers such that

- each $N_i[j]$ is either 0 or odd with an absolute value less than 2^{w_i};
- $e_i = \sum_{0 \leq j \leq b_i} N_i[j] \cdot 2^j$;
- at most one of any $w_i + 1$ consecutive components of the array is non-zero.

A width-$(w_i + 1)$ NAFs always exists and is uniquely determined; it can be computed by the following algorithm [19]:

```
c ← e_i
j ← 0
while c > 0 do
    if c[0] = 1 then
        u ← c[w_i ... 0]
        if u[w_i] = 1 then
            u ← u − 2^(w_i+1)
        c ← c − u
    else
        u ← 0
    N_i[j] ← u; j ← j + 1
    c ← c/2
while j ≤ b_i do
    N_i[j] ← 0; j ← j + 1
return N_i[b_i], ..., N_i[0]
```

The maximum possible index for a non-zero component of the wNAF of a B-bit integer is B; i.e., the length of the wNAF without leading zeros may exceed the length of the binary expansion by one. The average density (proportion of non-zero components) in width-$(w_i + 1)$ NAFs is $1/(w_i + 2)$ for $B \to \infty$ [19].

Precomputation Stage. For $i = 1, \ldots, k$, precompute g_i^E for all odd E such that $1 \leq E \leq 2^{w_i} - 1$. (As inversion in G is assumed to be easy, this makes g_i^{-E} available as well.)

Number of non-trivial table entries (multiplications): $\left(\sum_{1 \leq i \leq k} 2^{w_i - 1} \right) - k$.

Number of squarings: $\#\{i \in \{1, \ldots, k\} \mid w_i > 1\}$.

Evaluation Stage.

```
A ← 1_G
for i = 1 to k do
    N_i[b], ..., N_i[b_i + 1] ← 0, ..., 0
    N_i[b_i], ..., N_i[0] ← width-(w_i + 1) NAF of e_i
for j = b down to 0 do
    A ← A^2
    for i = 1 to k do
        if N_i[j] ≠ 0 then
            A ← A · g_i^{N_i[j]}   {multiply A by [inverse of] table entry}
return A
```

Number of squarings: $b - \max_i w_i$ up to b.

Expected number of multiplications: $\sum_{1 \leq i \leq k} b_i \cdot \frac{1}{w_i + 2}$.

We can compare wNAFs with the sliding window technique of the basic interleaving exponentiation algorithm. Windows can be represented by components of an array as in the wNAF approach: In the algorithm description of section 3.1, E_i provides component values; array indexes are given by $window\_handle_i$. With the array filled in accordingly, we can use the same evaluation stage algorithm as in the wNAF-based method. The average density is $1/(w_i + 1)$ (each window covers w_i bits, and the number of additional zero bits between neighbouring windows is 1 on average). With wNAFs, the average density goes down to $1/(w_i + 2)$ for exactly the same precomputation. Thus using wNAFs effectively increases the window size by one.

4   Comparison of Simultaneous and Interleaving Exponentiation Methods

There is no general rule for selecting window sizes for the multi-exponentiation algorithms that we have looked at. Various factors have to be considered: First of all, absolute memory constraints can impose limits on possible window sizes. Second, even if a particular window size appears to minimise the total amount of computation for a multi-exponentiation, sometimes slightly smaller windows may improve the actual performance; this is because larger window sizes mean larger precomputed tables, i.e. possibly additional memory allocation overhead and less effective memory caching. Last but not least, implementations can use different representations for group elements during different stages of the multi-exponentiation: For instance, extra effort may be spent during the precomputation stage in order to obtain representations of precomputed elements that speed up multiplication with them in the evaluation stage (e.g. affine rather than projective representations of points on elliptic curves [8]).

These effects, however, do not mean that we cannot compare algorithms without looking at concrete cases: We can compare different aspects separately (table size, precomputation stage efficiency, evaluation stage efficiency) and look if an algorithm wins on all counts.

For the following comparisons, we assume that all maximum exponent lengths b_i are the same (an assumption that we made in section 2 on simultaneous exponentiation methods, but not in section 3 on interleaving exponentiation methods). As before, let b be the length of the largest of the exponents e_i.

In section 4.1, we compare the simultaneous 2^w-ary method with the basic interleaving method and show that the latter is usually more efficient for $k = 2$ if squarings are about as costly as multiplications. In section 4.2, we compare the simultaneous sliding window method with the wNAF-based interleaving method and show that the latter is more efficient for $k = 2$ and $k = 3$, assuming that computing and storing the wNAFs is not too costly. Section 4.3 briefly discusses the alternative multi-exponentiation method from [10] and shows that is obviated by our interleaving exponentiation methods. Finally, in section 4.4, we look at some concrete figures for the number of multiplications required by different methods for example values of k and b.

4.1 Comparison between the Simultaneous 2^w-Ary Method and the Basic Interleaving Method

While the simultaneous sliding window method is more efficient than the simultaneous 2^w-ary method, this section focuses on the latter. The reasons is that the 2^w-ary method is often used in practice (e.g. [6]), possibly because it is perceived to be simpler to implement. The basic interleaving exponentiation method is not too complicated (in particular, indexes into the precomputed table are easy handle), and as we will see, it is often more efficient than the simultaneous 2^w-ary exponentiation method. So when the intention is to avoid the simultaneous sliding window method, the basic interleaving method appears preferable for many applications.

Assume that, given k and b, a certain w turns out to provide optimal efficiency for the simultaneous 2^w-ary exponentiation method (section 2.1) when performed in a specific environment. Then the precomputation table requires $2^{kw} - 1 - k$ non-trivial entries, $2^{k(w-1)} - 1$ of which can be computed with one squaring each (while each of the remaining entries requires one general multiplication).

For the basic interleaving exponentiation method (section 3.1), we can use uniform window sizes $w_1 = \ldots = w_k = kw$. Then the precomputation table has $k2^{kw-1} - k$ non-trivial entries, each of which requires one general multiplication; also k additional squarings are needed (unless $k = w = 1$).

Thus in case $k = 2$, the table grows from $2^{2w} - 3$ to $2^{2w} - 2$ non-trivial entries, and instead of $2^{2w} - 3$ group operations of which $2^{2(w-1)} - 1$ are squarings, we need 2^{2w} group operations of which only 2 are squarings. If squarings are about as expensive as general multiplications, then for $k = 2$ the overall cost of precomputation is comparable for these two multi-exponentiation methods.

The number of squarings in the evaluation stage is always nearly b for both methods. The expected number of general multiplications in the evaluation stage is smaller for the interleaving method (except if $k = w = 1$, in which case both algorithms do exactly the same): Dividing the value for the basic interleaving exponentiation method by the value for the simultaneous 2^w-ary exponentiation method yields

$$\frac{k}{kw+1} \cdot \frac{w}{1 - \frac{1}{2^{kw}}} = \frac{kw}{kw+1} \cdot \frac{2^{kw}}{2^{kw} - 1},$$

and this is less than 1 for $kw > 1$ (the minimum is $64/75$ at $kw = 4$).

Note that using $w_1 = \ldots = w_k = kw$ is not necessarily an optimal choice of window sizes for the basic interleaving exponentiation method; using smaller or larger windows might lead to better performance. (Indeed, if we look just at the number of operations and ignore memory usage, then there is no reason why window sizes should depend on k.) While the above proof only covers the case $k = 2$, there are actually many other cases where the basic interleaving method is more efficient than the simultaneous 2^w-ary method, even if general multiplications are much more expensive than squaring; see table 1 in section 4.4. Also note that the precomputation effort grows exponentially in k in simultaneous methods, but not in interleaving methods.

4.2   Comparison between the Simultaneous Sliding Window Method and the wNAF-Based Interleaving Method

Similarly to section 4.1, assume that a certain w provides optimal efficiency for the simultaneous sliding window exponentiation method (section 2.2) for given k and b. In the following analysis, we require $k > 1$. The precomputation table has $2^{kw} - 2^{k(w-1)} - k$ non-trivial entries, each of which requires one general multiplication to compute. In addition to this, k squarings are required for precomputation unless $w = 1$.

For the wNAF-based interleaving exponentiation method (section 3.2), we can use window sizes $w_1 = \ldots = w_k = kw - 1$. This leads to a precomputation table with $k2^{kw-2} - k$ non-trivial entries, requiring one general multiplication each. In addition to this, we need k squarings unless $kw = 2$.

The difference between the number of non-trivial tables entries (and general multiplications) for these two methods is

$$\left(2^{kw} - 2^{k(w-1)} - k\right) - \left(k2^{kw-2} - k\right) = 2^{kw}\left(1 - 2^{-k} - \frac{k}{4}\right).$$

This is positive for $k \le 3$ and negative for $k \ge 4$. Thus, with the w_i chosen like this, the precomputation stage of the wNAF-based interleaving exponentiation method is more efficient if $k = 2$ or $k = 3$ (except for the case $k = 3$, $w = 1$, where the wNAF-based interleaving exponentiation method saves one general multiplication, but requires three additional squarings).

The evaluation stage requires close to b squarings for both methods. The expected number of general multiplications is smaller for the wNAF-based interleaving method: $b/(w + 1/k)$ instead of $b/(w + \frac{1}{2^k - 1})$.

The wNAF-based interleaving method with this choice of window sizes will often provide better performance than the simultaneous sliding window method for $k \ge 4$ as well: If additional memory allocation is not a problem, then the efficiency gain of the evaluation stage usually compensates for the growth of the precomputed table.

Similar to the situation in the preceding section, $w_1 = \ldots = w_k = kw - 1$ is not necessarily an optimal choice, and smaller or larger window sizes might be better (see section 4.4).

4.3   Comparison between the Dimitrov-Jullien-Miller Multi-Exponentiation Method and Interleaving Exponentiation

A multi-exponentiation method for the case $k = 2$ requiring two precomputed values (in addition to g_1 and g_2) if inverting is easy, or six precomputed values if inversions have to be done during the precomputation stage, was described by Dimitrov, Jullien, and Miller in [10]. This algorithm is related to the simultaneous sliding window exponentiation method of Yen, Laih, and Lenstra [21] (section 2.2 in the present paper), but uses signed recoding of exponents in order to reduce the size of the precomputed table. While the Yen-Laih-Lenstra method with a window size of 1 requires an expected number of $b \cdot 0.75$ general multiplications during the evaluation stage, the new method requires only about $b \cdot 0.534$ multiplications according to [10] (the number of squarings stays about the same). Yen-Laih-Lenstra with a window size of 2 needs only $b \cdot 3/7 \approx b \cdot 0.429$ multiplications (table 3 of [10] erroneously assumes a value of $b \cdot 0.625$), but has the disadvantage of requiring more precomputed elements, which may be a problem in some constrained environments.

We do not examine the algorithm of [10] in detail; note that it is outperformed by the wNAF-based interleaving method of section 3.2 with $w_1 = w_2 = 2$ if inversion is cheap (two precomputed values, $b \cdot 0.5$ multiplications) and by the basic interleaving method of section 3.1 with $w_1 = w_2 = 3$ otherwise (six precomputed values, $b \cdot 0.5$ multiplications).

4.4   Examples

As noted before, endless variations are possible for defining optimisation goals. In this section, we ignore memory usage and squarings and the issue of different element representations; we make comparisons based just on the expected number of general multiplications required by the various methods, precomputation and evaluation stage combined. (Window sizes are chosen such that this cost measure is minimised.) Note that the number of squarings is approximately the same for the simultaneous sliding window method, the basic interleaving method, and the wNAF-based interleaving method: No more than k squarings are needed in the precomputation stage, and close to b squarings are needed in the evaluation stage. The simultaneous 2^w-ary method requires $2^{k(w-1)} - 1$ squarings for precomputation and again close to b evaluation stage squarings; so ignoring the cost of squaring tends to favour this method.

Table 1 compares the number of general multiplications needed by these four methods for various k and b values. The entries for the most efficient methods in a particular configuration are printed in bold: For groups where inversion is easy so that the wNAF-based method can be used, it wins in all of these examples; for general groups, sometimes the simultaneous sliding window method and sometimes the basic interleaving method requires the least number of multiplications. (Remember that for $w = 1$ there is no difference between the simultaneous 2^w-ary method and the simultaneous sliding window method; for $w > 1$, the former is always less efficient.)

5   Multi-exponentiation with Fixed Bases

When many multi-exponentiations use the same bases $g_1, \ldots, g_k$, it is sufficient to execute the precomputation stage just once, and we can try to make the evaluation stage more efficient by investing more work in precomputation. We cannot easily reduce the number of general multiplications in the evaluation stage, but we can reduce the number of squarings by using exponent splitting (cf. [5] and [9]) or the Lim-Lee "comb" method [14]. (Which approach is the

Table 1. Expected number of general multiplications for a multi-exponentiation $\prod_{1\le i\le k} g_i^{e_i}$ with exponents up to b bits (c_1: simultaneous 2^w-ary method, c_2: simultaneous sliding window method, c_3: basic interleaving method, c_4: wNAF-based interleaving method)

k		$b=160$	$b=256$	$b=512$	$b=1024$	$b=2048$
1	c_1	44.5 $(w=4)$	64.6 $(w=5)$	114.2 $(w=5)$	199.0 $(w=6)$	353.3 $(w=7)$
	c_2	**39.0** $(w=4)$	**57.7** $(w=5)$	**100.3** $(w=5)$	**177.3** $(w=6)$	**319.0** $(w=7)$
	c_3	**39.0** $(w_i=4)$	**57.7** $(w_i=5)$	**100.3** $(w_i=5)$	**177.3** $(w_i=6)$	**319.0** $(w_i=7)$
	c_4	**33.7** $(w_i=4)$	**49.7** $(w_i=4)$	**88.1** $(w_i=5)$	**159.0** $(w_i=6)$	**287.0** $(w_i=6)$
2	c_1	85.0 $(w=2)$	130.0 $(w=2)$	214.0 $(w=3)$	382.0 $(w=3)$	700.0 $(w=4)$
	c_2	78.6 $(w=2)$	119.7 $(w=2)$	**199.6** $(w=3)$	**353.2** $(w=3)$	660.4 $(w=3)$
	c_3	78.0 $(w_i=4)$	115.3 $(w_i=5)$	200.7 $(w_i=5)$	354.6 $(w_i=6)$	**638.0** $(w_i=7)$
	c_4	**67.3** $(w_i=4)$	**99.3** $(w_i=4)$	**176.3** $(w_i=5)$	**318.0** $(w_i=6)$	**574.0** $(w_i=6)$
3	c_1	131.8 $(w=2)$	179.0 $(w=2)$	305.0 $(w=2)$	557.0 $(w=2)$	1061.0 $(w=2)$
	c_2	127.7 $(w=2)$	172.5 $(w=2)$	291.9 $(w=2)$	530.9 $(w=2)$	1008.7 $(w=2)$
	c_3	117.0 $(w_i=4)$	173.0 $(w_i=5)$	301.0 $(w_i=5)$	531.9 $(w_i=6)$	957.0 $(w_i=7)$
	c_4	**101.0** $(w_i=4)$	**149.0** $(w_i=4)$	**264.4** $(w_i=5)$	**477.0** $(w_i=6)$	**861.0** $(w_i=6)$
4	c_1	161.0 $(w=1)$	251.0 $(w=1)$	491.0 $(w=1)$	746.0 $(w=2)$	1256.0 $(w=2)$
	c_2	161.0 $(w=1)$	251.0 $(w=1)$	483.7 $(w=2)$	731.5 $(w=2)$	**1227.0** $(w=2)$
	c_3	**156.0** $(w_i=4)$	230.7 $(w_i=5)$	401.3 $(w_i=5)$	**709.1** $(w_i=6)$	1276.0 $(w_i=7)$
	c_4	**134.7** $(w_i=4)$	**198.7** $(w_i=4)$	**352.6** $(w_i=5)$	**636.0** $(w_i=6)$	**1148.0** $(w_i=6)$
5	c_1	**181.0** $(w=1)$	274.0 $(w=1)$	522.0 $(w=1)$	1018.0 $(w=1)$	2010.0 $(w=1)$
	c_2	**181.0** $(w=1)$	274.0 $(w=1)$	522.0 $(w=1)$	1018.0 $(w=1)$	1994.7 $(w=2)$
	c_3	195.0 $(w_i=4)$	288.3 $(w_i=5)$	501.7 $(w_i=5)$	**886.4** $(w_i=6)$	1595.0 $(w_i=7)$
	c_4	**168.3** $(w_i=4)$	**248.3** $(w_i=4)$	**440.7** $(w_i=5)$	**795.0** $(w_i=6)$	**1435.0** $(w_i=6)$
6	c_1	**214.5** $(w=1)$	**309.0** $(w=1)$	**561.0** $(w=1)$	1065.0 $(w=1)$	2073.0 $(w=1)$
	c_2	**214.5** $(w=1)$	**309.0** $(w=1)$	**561.0** $(w=1)$	1065.0 $(w=1)$	2073.0 $(w=1)$
	c_3	234.0 $(w_i=4)$	346.0 $(w_i=5)$	602.0 $(w_i=5)$	**1063.7** $(w_i=6)$	1914.0 $(w_i=7)$
	c_4	**202.0** $(w_i=4)$	**298.0** $(w_i=4)$	**528.9** $(w_i=5)$	**954.0** $(w_i=6)$	**1722.0** $(w_i=6)$
7	c_1	278.8 $(w=1)$	**374.0** $(w=1)$	**628.0** $(w=1)$	**1136.0** $(w=1)$	2152.0 $(w=1)$
	c_2	278.8 $(w=1)$	**374.0** $(w=1)$	**628.0** $(w=1)$	**1136.0** $(w=1)$	2152.0 $(w=1)$
	c_3	273.0 $(w_i=4)$	403.7 $(w_i=5)$	702.3 $(w_i=5)$	1241.0 $(w_i=6)$	2233.0 $(w_i=7)$
	c_4	**235.7** $(w_i=4)$	**347.7** $(w_i=5)$	**617.0** $(w_i=5)$	**1113.0** $(w_i=6)$	**2009.0** $(w_i=6)$
8	c_1	406.4 $(w=1)$	502.0 $(w=1)$	757.0 $(w=1)$	**1267.0** $(w=1)$	2287.0 $(w=1)$
	c_2	406.4 $(w=1)$	502.0 $(w=1)$	757.0 $(w=1)$	**1267.0** $(w=1)$	2287.0 $(w=1)$
	c_3	312.0 $(w_i=4)$	461.3 $(w_i=5)$	802.7 $(w_i=5)$	1418.3 $(w_i=6)$	2552.0 $(w_i=7)$
	c_4	**269.3** $(w_i=4)$	**397.3** $(w_i=4)$	**705.1** $(w_i=5)$	1272.0 $(w_i=6)$	**2296.0** $(w_i=6)$
9	c_1	661.7 $(w=1)$	757.5 $(w=1)$	1013.0 $(w=1)$	**1524.0** $(w=1)$	**2546.0** $(w=1)$
	c_2	661.7 $(w=1)$	757.5 $(w=1)$	1013.0 $(w=1)$	**1524.0** $(w=1)$	**2546.0** $(w=1)$
	c_3	351.0 $(w_i=4)$	**519.0** $(w_i=5)$	903.0 $(w_i=5)$	1595.6 $(w_i=6)$	2871.0 $(w_i=7)$
	c_4	**303.0** $(w_i=4)$	**447.0** $(w_i=4)$	**793.3** $(w_i=5)$	1431.0 $(w_i=6)$	2583.0 $(w_i=6)$
10	c_1	1172.8 $(w=1)$	1268.8 $(w=1)$	1524.5 $(w=1)$	**2036.0** $(w=1)$	**3059.0** $(w=1)$
	c_2	1172.8 $(w=1)$	1268.8 $(w=1)$	1524.5 $(w=1)$	**2036.0** $(w=1)$	**3059.0** $(w=1)$
	c_3	390.0 $(w_i=4)$	576.7 $(w_i=5)$	1003.3 $(w_i=5)$	1772.9 $(w_i=6)$	3190.0 $(w_i=7)$
	c_4	**336.7** $(w_i=4)$	**496.7** $(w_i=4)$	**881.4** $(w_i=5)$	**1590.0** $(w_i=6)$	**2870.0** $(w_i=6)$

most efficient depends on details of the situation such as exponent lengths, the permissible size of the precomputed table, the relative cost of squarings versus general multiplications, and whether the wNAF-based interleaving exponentiation method is applicable.)

Let m be an arbitrary positive integer. Assuming that fixed exponent length bounds b_i are known, we show how to evaluate power products $\prod_{1 \leq i \leq k} g_i^{e_i}$ in at most $m - 1$ evaluation stage squarings, using a precomputed table independent of the specific exponents e_i.

5.1   Exponent Splitting

Exponent splitting constructs a new power product representation by rewriting each factor as follows:

$$g_i^{e_i} = \prod_{0 \leq j < \lceil b_i/m \rceil} (g_i^{2^{jm}})^{e_i[jm+m-1\,\ldots\,jm]}$$

This leads to power products consisting of $\sum_{1 \leq i \leq k} \lceil b_i/m \rceil$ factors. Any multi-exponentiation method can be used for evaluating these power products.

It is evident that for the multi-exponentiation methods described in this paper, exponent splitting does not help if k is already large and there are many large exponents. (In this case, instead of using precomputation table entries for additional bases, window sizes should be increased; then the evaluation stage will require more squarings, but fewer general multiplications than with exponent splitting.)

5.2   Lim-Lee Precomputation

To apply the Lim-Lee "comb" method, for every i we choose w_i such that $b_i \leq w_i m$ and precompute

$$G_i(S) := g_i^{\sum_{j \in S} 2^j}$$

for all subsets $S \subseteq \{0, m, 2m, \ldots, (w_i - 1)m\}$. Note that then every exponent up to b_i bits of length can be written as

$$e_i = \sum_{0 \leq j < m} N_i[j] \cdot 2^j$$

where each $N_i[j]$ is an integer of the form $\sum_{j \in S} 2^j$ with S as above. Thus we can use interleaving exponentiation with an evaluation stage algorithm similar to section 3.2, but with a reduced number of iterations. The $N_i[j]$ values for each iteration need not be stored in advance, they can be extracted from the e_i by tapping their bits in comb-shaped patterns; hence the nickname of this method.

A refinement of this (also from [14]) is based on the observation that the precomputed table can be reduced in size in exchange for additional evaluation stage multiplications: Partition $\{0, m, 2m, \ldots, (w_i - 1)m\}$ into v_i subsets

$T_{i,1}, \ldots, T_{i,v_i}$; now each of the above sets S can be written as $\bigcup_{1 \le n \le v_i} S_n$ with $S_n = S \cap T_{i,n}$, and then we have $G_i(S) = \prod_{1 \le n \le v_i} G_i(S_n)$. Thus it suffices to precompute $G_i(S_n)$ for all non-zero subsets $\overline{S}_n \subseteq T_{i,n}$ for all n; from this precomputed data, $G_i(S)$ can be computed in at most $v_i - 1$ multiplications.

While Lim-Lee precomputation reduces the number of squarings, the expected number of general multiplications is larger than for the basic interleaving exponentiation method with a similarly sized precomputed table. (In the basic interleaving method, $2^{w_i-1} - 1$ non-trivial precomputed values suffice to make sure that each evaluation stage multiplication covers w_i exponent bits, and we can skip many additional zero bits thanks to the sliding window. With Lim-Lee precomputation, we need at least $2^W - 2$ non-trivial precomputed values to be able to cover W exponent bits with each evaluation stage multiplication, and we lose the advantage of a sliding window.) Thus if k is large, using Lim-Lee precomputation is a disadvantage.

Note that it is possible to use Lim-Lee precomputation for some of the bases and standard precomputation (as in section 3) for others. This does not help for multi-exponentiation in these mixed cases, but precomputed data can then profitably be reused for pure Lim-Lee cases.

Without going into details, we remark that the Lim-Lee method can be considered an application of exponent splitting using specific multi-exponentiation algorithms suited for small exponents. For example, if $k = 1$, the simple Lim-Lee method uses "Shamir's trick", i.e. simultaneous exponentiation with a window size of 1. Further algorithmic variations are possible.

6   Conclusion

In many cases, the basic interleaving exponentiation method compares favourably to the simultaneous 2^w-ary method, in particular if $k = 2$ and squarings are about as costly as general multiplications. In groups where inverting elements is easy, the wNAF-based interleaving exponentiation method is available; its efficiency is superior even to the sliding window variant of simultaneous exponentiation both in the precomputation stage and the evaluation stage if $k = 2$ or $k = 3$, and it is usually more efficient for larger k as well. In all cases, interleaving exponentiation provides the following advantages over simultaneous exponentiation:

- Improved efficiency if the bit-lengths of the exponents e_i differ significantly.
- More flexibility in choice of the size of the auxiliary table (and, hence, the time spent on precomputation), particularly if k is large.
- Better handling of situations where one or more of the g_i are fixed while others are variable between multiple multi-exponentiation: A corresponding part of the precomputation has to be done only once. (This is the case in DSA, ECDSA, and RDSA signature verification if multiple signatures are verified that are based on the same underlying parameters.)

Thus, depending on circumstances, either the simultaneous sliding window method or one of the interleaving exponentiation methods may be advantageous.

It is easy to implement interleaving exponentiation for variable k. As the the special case $k = 1$ of the basic and wNAF-based interleaving exponentiation methods yields the usual sliding windows exponentiation method and wNAF-based exponentiation method, respectively, this makes it unnecessary to implement these separately.

References

1. AMERICAN NATIONAL STANDARDS INSTITUTE (ANSI). Public key cryptography for the financial services industry: The elliptic curve digital signature algorithm (ECDSA). ANSI X9.62, 1998.
2. BIEHL, I., BUCHMANN, J., HAMDY, S., AND MEYER, A. A signature scheme based on the intractability of extracting roots. *Designs, Codes and Cryptography*. To appear.
3. BOS, J., AND COSTER, M. Addition chain heuristics. In *Advances in Cryptology – CRYPTO '89* (1989), G. Brassard, Ed., vol. 435 of *Lecture Notes in Computer Science*, pp. 400–407.
4. BRANDS, S. *Rethinking Public Key Infrastructures and Digital Certificates – Building in Privacy*. MIT Press, 2000.
5. BRICKELL, GORDON, MCCURLEY, AND WILSON. Fast exponentiation with precomputation. In *Advances in Cryptology – EUROCRYPT '92* (1993), R. A. Rueppel, Ed., vol. 658 of *Lecture Notes in Computer Science*, pp. 200–207.
6. BROWN, M., HANKERSON, D., LÓPEZ, J., AND MENEZES, A. Software implementation of the NIST elliptic curves over prime fields. In *Progress in Cryptology – CT-RSA 2001* (2001), D. Naccache, Ed., vol. 2020 of *Lecture Notes in Computer Science*, pp. 250–265.
7. BUCHMANN, J., AND HAMDY, S. A survey on IQ cryptography. In *Proceedings of Public Key Cryptography and Computational Number Theory, 2000*. To appear. Preprint available at http://www.informatik.tu-darmstadt.de/TI/Veroeffentlichung/TR/.
8. COHEN, H., ONO, T., AND MIYAJI, A. Efficient elliptic curve exponentiation using mixed coordinates. In *Advances in Cryptology – ASIACRYPT '98* (1998), K. Ohta and D. Pei, Eds., vol. 1514 of *Lecture Notes in Computer Science*, pp. 51–65.
9. DE ROOIJ, P. Efficient exponentiation using precomputation and vector addition chains. In *Advances in Cryptology – EUROCRYPT '94* (1995), T. Helleseth, Ed., vol. 950 of *Lecture Notes in Computer Science*, pp. 389–399.
10. DIMITROV, V. S., JULLIEN, G. A., AND MILLER, W. C. Complexity and fast algorithms for multiexponentiation. *IEEE Transactions on Computers 49* (2000), 141–147.
11. ELGAMAL, T. A public-key cryptosystem and a signature scheme based on discrete logarithms. *IEEE Transactions on Information Theory 31* (1985), 469–472.
12. KOCHER, P. C. Timing attacks on implementations of Diffie-Hellman, RSA, DSS, and other systems. In *Advances in Cryptology – CRYPTO '96* (1996), N. Koblitz, Ed., vol. 1109 of *Lecture Notes in Computer Science*, pp. 104–113.
13. KOCHER, P. C., JAFFE, J., AND JUN, B. Differential power analysis. In *Advances in Cryptology – CRYPTO '99* (1999), M. Wiener, Ed., vol. 1666 of *Lecture Notes in Computer Science*, pp. 388–397.
14. LIM, C. H., AND LEE, P. J. More flexible exponentiation with precomputation. In *Advances in Cryptology – CRYPTO '94* (1994), Y. G. Desmedt, Ed., vol. 839 of *Lecture Notes in Computer Science*, pp. 95–107.

15. MENEZES, A. J., VAN OORSCHOT, P. C., AND VANSTONE, S. A. *Handbook of Applied Cryptography*. CRC Press, 1997.

16. MIYAJI, A., ONO, T., AND COHEN, H. Efficient elliptic curve exponentiation. In *International Conference on Information and Communications Security – ICICS '97* (1997), Y. Han, T. Okamoto, and S. Qing, Eds., vol. 1334 of *Lecture Notes in Computer Science*, pp. 282–290.

17. NATIONAL INSTITUTE OF STANDARDS AND TECHNOLOGY (NIST). Digital Signature Standard (DSS). FIPS PUB 186-2, 2000.

18. SOLINAS, J. A. An improved algorithm for arithmetic on a family of elliptic curves. In *Advances in Cryptology – CRYPTO '97* (1997), B. S. Kaliski, Jr., Ed., vol. 1294 of *Lecture Notes in Computer Science*, pp. 357–371.

19. SOLINAS, J. A. Efficient arithmetic on Koblitz curves. *Designs, Codes and Cryptography 19* (2000), 195–249.

20. STRAUS, E. G. Problems and solutions: Addition chains of vectors. *American Mathematical Monthly 71* (1964), 806–808.

21. YEN, S.-M., LAIH, C.-S., AND LENSTRA, A. K. Multi-exponentiation. *IEE Proceedings – Computers and Digital Techiques 141* (1994), 325–326.

Two Topics in Hyperelliptic Cryptography

Florian Hess[1], Gadiel Seroussi[2], and Nigel P. Smart[1]

[1] Computer Science Dept., Woodland Road,
Bristol University, Bristol, BS8 1UB, UK.
{florian, nigel}@cs.bris.ac.uk
[2] Hewlett-Packard Labs, 1501 Page Mill Road,
Palo Alto, CA. 650-857-1501, USA.
seroussi@hpl.hp.com

Abstract. In this paper we address two important topics in hyperelliptic cryptography. The first is how to construct in a verifiably random manner hyperelliptic curves for use in cryptography in generas two and three. The second topic is how to perform divisor compression in the hyperelliptic case. Hence, in both cases we generalise concepts used in the more familiar elliptic curve case to the hyperelliptic context.

1 Introduction

Elliptic curve cryptography was co-invented in 1985 by V. Miller [13] and N. Koblitz [11]. Cryptography based on elliptic curves is especially attractive due to the supposed difficulty of the discrete logarithm problem in the group of rational points on an elliptic curve. In 1989 Koblitz generalised this concept to hyperelliptic curves [12]. In hyperelliptic cryptography the hard problem on which the security is based is the discrete logarithm problem in the divisor class group of the curve.

Whilst elliptic curve cryptography is starting to become commercially deployed, hyperelliptic cryptography is still at the stage of academic interest. This is mainly due to the greater complexity of the underlying arithmetic and the fact that the protocols have been less standardised. One main problem in the hyperelliptic case, as argued in [16], is that it is currently very hard to generate hyperelliptic curves for use in cryptography which do not have any added extra structure. [1] Another problem is that the supporting algorithms which exist in

---

[1] There is a new general point counting algorithm by Kedlaya [10] for hyperelliptic curves in small odd characteristic. However, it is believed that this algorithm can be extended to the even characteristic case. At present the authors know of no implementation of this algorithm and so cannot we comment on its practical efficiency.

Just before submitting the final version of this paper to the conference proceedings, Pierrick Gaudry informed us that the AGM method presented at the rump session of EUROCRYPT 2001 can now be used to compute the group order of a Jacobian of a hyperelliptic curve in genus two over a field of characteristic two. Indeed the AGM method is practical for cryptographically sized Jacobians. Hence, the AGM method for genus two should therefore be preferred to ours since it allows a truly random curve to be used rather than one from a special family.

S. Vaudenay and A. Youssef (Eds.): SAC 2001, LNCS 2259, pp. 181–189, 2001.

the elliptic curve case have not been fully developed in the hyperelliptic case. In this paper we generalise two such techniques from the setting of elliptic curve cryptography to the setting of hyperelliptic curves.

In the first we give a method to produce hyperelliptic curves in genus two and three which are generated in a verifiably random manner. In the second we give a method to perform divisor compression.

The first contribution is needed to produce suitable curves in a trusted manner. In elliptic curve cryptography, one way to choose a curve is to generate curves at random until one satisfies the correct security requirements. However, someone else then using the system needs to trust that you did not construct a special curve which has some weakness that only you know about. To overcome this problem various standards bodies, e.g. [1], have proposed that the curve is generated in the following manner:

1. Generate in any manner a 160 bit string, S.
2. Using SHA-1 on this string generate some elliptic curve E in a known deterministic manner.
3. Compute the group order N using either the Schoof-Elkies-Atkin algorithm or one of the extensions to Satoh's algorithm, see [3], [14], [15] and [18].
4. If the curve passes the known security checks then publish the triple

$$(S, E, N),$$

otherwise return to the first step.

Under the assumption that SHA-1 is a one-way function the above method of curve generation prevents the choice of special elliptic curves with secret weaknesses. An elliptic curve chosen in the above way is said to have been chosen "verifiably at random" since any third party given the triple (S, E, N) can check very quickly that not only is the group order N correct but that the curve could not have been created with a known weakness since it would have been computationally impossible to reverse engineer the value of S which gave E using the above algorithm.

We show how the above algorithm can be used to generate verifiably random hyperelliptic curves in characteristic two for use in cryptography. Our method does not produce random hyperelliptic curves taken from the totality of all hyperelliptic curves but produces hyperelliptic curves which have verifiably been constructed in a random manner from a certain well defined subset of all hyperelliptic curves. In other words it is computationally infeasible for us to have created a special curve with some hidden weakness. However, we stress that since our method produces random hyperelliptic curves from a special family it is possible that the curves constructed by our method have a weakness which we are not aware of. For further details of how special the families we construct actually are the reader should consult the paper [6].

Previous attempts at generating cryptographically strong hyperelliptic curves have been based on analogues from the elliptic case, namely generalisations of the SEA algorithm or the CM method. In [8] a first attempt at an analogue of the

SEA algorithm for hyperelliptic curves of genus two is reported on. The authors manage to compute the order of a random hyperelliptic curve of genus two of group order roughly 2^{126}. However, this takes them many days of computing time. In practice one would need to repeat their method a large number of times before a suitable curve for use in cryptography was determined. Whilst the method in [8] is to be preferred over ours, it can only be used when (and if) the algorithms become sufficiently fast. Our method on the other hand, as we have already stated, is practical using today's knowledge and technology.

A number of authors have looked at using an analogue of the CM method to generate hyperelliptic curves for use in cryptography, [17], [19] and [5]. However, this has a number of draw backs compared to our method above. Firstly, the existing literature on applying the CM method to hyperelliptic curves only applies to large odd characteristic and not characteristic two as our method does. Secondly, the set of curves produced by the CM method in practice, if one could implement it in characteristic two, would be from a far more restricted set than the set of curves generated by our method.

Our second contribution is to give a method in all characteristics to perform divisor compression. In the elliptic curve case it is common practice to use a technique called point compression to reduce the sizes of the public keys being transported by fifty percent. This is done by noticing that an elliptic curve point (x, y) can be represented by x and a bit to decide which value of y to use. This is particularly important when deploying ECC in an environment where bandwidth is constrained. We will show that the elliptic curve point compression techniques can be naturally generalised to the hyperelliptic setting.

The first author would like to thank J. Cannon for his support while this work was in preparation.

2   Producing Hyperelliptic Curves

Our technique of producing hyperelliptic curves verifiably at random is based on the method of Weil restriction of scalars as outlined in [9]. In this technique one takes an elliptic curve E over the field $K = \mathbb{F}_{q^n}$, where q is a power of two and then one constructs a hyperelliptic curve H over the subfield $k = \mathbb{F}_q$. Since the groups $E(K)$ and $\mathrm{Jac}_k(H)$ are related by a group homomorphism one can easily compute, in certain cases, the group order of $\mathrm{Jac}_k(H)$.

To fix notation we are trying to generate a hyperelliptic curve H over the field $\mathbb{F}_q$, of genus g and of group order $N = 2^l p$, where p is a prime. Before giving our technique for the generation of hyperelliptic curves we need to summarise the main security requirements for our curve.

- $p > 2^{160}$. This is to protect against Pohlig-Hellman, Pollard-rho and Baby-Step/Giant-Step attacks.
- $g < 4$. This is to protect against the method of Gaudry [7].
- $q = 2^r$, where r is prime. This is to protect against using Weil descent on $\mathrm{Jac}_{\mathbb{F}_q}(H)$.

- The smallest $s \geq 1$ such that $q^s \equiv 1 \pmod{p}$ should be greater than $20g$. This is to protect against the Tate-pairing attack [4].

Note, there are no other conditions which give curves with a known weakness and all the above conditions can be easily checked given the curve and its group order.

In [9] a method is given for finding a group homomorphism from an elliptic curve defined over $\mathbb{F}_{q^n}$ to a hyperelliptic curve H defined over $\mathbb{F}_q$. The technique given is completely deterministic, although the resulting model for H is not in the standard form, an issue which we shall return to below. The method of [9] uses a set of Artin-Schreier extensions, the number of distinct extensions being given by an integer m, which satisfies $1 \leq m \leq n$. For the exact definition of m see [9], all that we shall require is that $m = n$ and that the genus of the resulting hyperelliptic curve is either 2^{m-1} or $2^{m-1} - 1$. In our applications we are able to control precisely when we obtain genus 2^{m-1} or genus $2^{m-1} - 1$.

Since we wish to produce hyperelliptic curves with Jacobians of the same group order as $E(K)$ we need to choose elliptic curves so that

$$n = 2^{m-1} \text{ or } n = 2^{m-1} - 1.$$

Since one of our security requirements on g is that it should be less than four, these conditions are easy to satisfy.

For cryptographic purposes it is advantageous to produce a model for the hyperelliptic curve of the form

$$H : Y^2 + H(X)Y = F(X)$$

where $\deg H(X) \leq g$ and $\deg F(X) = 2g + 1$. Such a model will be called "reduced" and we shall now describe a deterministic method to turn the hyperelliptic model, produced by the method of [9], into a reduced model. This is important, and was not addressed in [9]. If we wish to generate hyperelliptic curves verifiably at random we require a deterministic mapping from the elliptic curve to a reduced model of a hyperelliptic curve.

Assume that a fixed representation has been chosen for the finite fields of size q^n and q. Using this fixed representations we can define (lexicographical) orders in the finite fields, hence orders on polynomials, matrices etc. Utilising normalisation of polynomials, polynomial division, Hermite normal forms and other such reduction techniques we are then able to always consider the smallest (or the same) object having a desired property.

Taking the model for H produced by the method in [9] we then move the smallest rational point to infinity. A reduced hyperelliptic equation is then obtained by computing the minimal polynomial over the rational subfield of a function of smallest odd pole order at infinity and with no other poles.

Since the algorithm, outlined above, to proceed from an elliptic curve to a reduced model for a hyperelliptic curve is completely deterministic, all we need do to produce a verifiably "random" hyperelliptic curve is to find an elliptic curve verifiably at "random" with the required properties.

2.1   Genus Two

Take a finite field of the form $K = \mathbb{F}_{q^2}$ where q is 2 raised to a prime exponent. We construct, using the technique from [1] a verifiably random elliptic curve of the form

$$Y^2 + XY = X^3 + aX^2 + b$$

where $a, b \in K$, with group order equal to $2p$ where p is a prime number. Note that since p is a prime number and q is 'large', in the Weil descent we almost always obtain $m = 2$ and so the resulting hyperelliptic curve will have genus two. Then using the technique of Weil descent we can construct a hyperelliptic curve over the field $k = \mathbb{F}_q$ which has group order divisible by p. Since the Weil restriction of E and $\mathrm{Jac}_k(H)$ have the same dimension, they are therefore isogenous. But they then have the same number of points over k and so $\mathrm{Jac}_k(H)$ will have group order exactly $2p$.

2.2   Genus Three

For genus three we need to proceed in a slightly different way. First we choose a finite field of the form $K = \mathbb{F}_{q^3}$ where again q is 2 raised to a prime exponent. Then we take an random 160-bit string and pass it through SHA-1 to obtain a field element $v \in \mathbb{F}_{q^3}$ using the methods of [1]. Setting $b = v + v^q$ we see that

$$\mathrm{Tr}_{K/k}(b) = 0.$$

We then compute the elliptic curve

$$Y^2 + XY = X^3 + X^2 + b$$

and its group order. This is repeated until we find a group order equal to $2p$ where p is a prime. Then using the arguments of [9] we will obtain a hyperelliptic curve of genus three. Although we are not choosing elliptic curves completely at random from all elliptic curves defined over K, we are choosing them uniformly at random from a subset of size q^2. Just as before, we will have that $\mathrm{Jac}_k(H)$ has group order exactly $2p$.

Our technique for constructing hyperelliptic curves for use in cryptography is dominated by the time needed to apply the Schoof-Elkies-Atkin (SEA) algorithm or the algorithm of Satoh to a set of elliptic curves, until one with the correct cryptographic properties is determined. The step of transforming the elliptic curve into a hyperelliptic curve only takes a few seconds. Hence, to compute a single hyperelliptic curve of genus two with the correct cryptographic properties takes, for a Jacobian of size roughly 2^{190}, on the order of a couple of minutes. The main computational task is to repeatedly apply the SEA/Satoh algorithm until a suitable elliptic curve is found. Of course, exact times depend strongly on the details of the SEA/Satoh implementation

Finally to end this section we give a typical example:

n=166
Elliptic Curve : K is defined by $w^{166} + w^{37} + 1 = 0$

$$S = \text{E4D1C989A8999ED0EF8AC7D691E5D8ADDAD481F5},$$
$$a = \text{3951AD54028E7E3CF2D437A4186CCB53BF5DD39196},$$
$$b = \text{140463F3747C98BAE9D9D31EAF3FCE65ADF80AEA26},$$
$$N = \text{3FFFFFFFFFFFFFFFFFFFFF730032E01F3184452AA1A}.$$

Hyperelliptic Curve : k is defined by $t^{83} + t^7 + t^4 + t^2 + 1 = 0$.

$$H(X) = \text{6C935CFDD963AD086B738}X^2 + \text{103FEA81D67CBF0210A96}X$$
$$+ \text{47242588808C36BFBE701},$$
$$F(X) = \text{660212F23F5C16AE899A9}X^5 + \text{6CAEC90C545CF269FE5B1}X^4$$
$$+ \text{5A55B3786562759A427E0}X^3 + \text{32C4479705A4CEBF1FEA3}X^2$$
$$+ \text{7F018AAEC622917758194}X + \text{2BDCB9CD696E5142054C8}.$$

3 Divisor Compression

As noted previously point compression in the elliptic curve case is an important tool used to save around fifty percent of the bandwidth in transferring/storing public keys and in Diffie-Hellman key exchange. Before describing our analogous method in the hyperelliptic setting we shall describe the exact data format normally used for divisors on hyperelliptic curves. For more details on what follows the reader should consult the papers by Cantor [2] and Koblitz [12]. In this section we shall work with arbitrary characteristic fields.

A hyperelliptic curve of genus g, over a field k of characteristic p, we will assume is given by an equation of the form

$$Y^2 + H(X)Y = F(X),$$

where $H(X), F(X) \in k[X]$, $\deg H(X) \leq g$ and $\deg F(X) = 2g + 1$. For applications it is common to assume that either p is very large or equal to two. If p is large we usually assume that $H(X) = 0$. Notice that in characteristic two the ramified places lying above $p(X) \in k[X]$ are exactly those for which $p(X)$ divides $H(X)$.

The group elements, upon which our cryptographic protocols operate, are effective reduced divisors of degree less than or equal to g. Such a divisor can be represented by the pair

$$D = (a(X), b(X)),$$

where $a(X), b(X) \in k[X]$, $\deg b(X) < \deg a(X) \leq g$, $a(X)$ is monic and

$$b(X)^2 + H(X)b(X) - F(X) \equiv 0 \pmod{a(X)}.$$

The zero in the group is represented by the pair $(1, 0)$. That the divisor is reduced means that no ramified place occurs in the support of D with multiplicity greater than one, and that if a place $\mathfrak{p}$ occurs in the support of D then the image of p under the hyperelliptic involution does not. In many protocols one needs to transmit divisors, naively this requires at most g elements of k to represent $a(X)$ and at most g elements of k to represent $b(X)$.

However, given $a(X)$ there are only a small number of possible values for $b(X)$ which could correspond to $a(X)$. We shall show how one can recover the correct $b(X)$ from only $a(X)$, and at most an additional g bits of information.

Our first task is to decide a canonical order on the irreducible polynomials of degree less than or equal to g, which are defined over k. This is done by fixing a field representation and using the lexicographic order used for a similar purpose in Section 2.

When we are either compressing or decompressing we first factorize $a(X)$ into its irreducible factors and order them. Since factorisation of polynomials can be performed in random polynomial time, and in applications the degree of $a(X)$ will be quite small (usually less than four) this factorisation stage is no barrier to our method.

For example when $g = 2$ we need to factorize a degree two polynomial. This factors either when a certain trace is zero, for the even characteristic case, or when the discriminant is a square, for the odd characteristic case. In either characteristic we can easily deduce the factorisation when the polynomial is reducible using standard techniques for solving quadratic equations over finite fields. Similar considerations apply when $g = 3$.

Each irreducible factor $p(X)$ of $a(X)$ will correspond to at most two prime divisors on H:

$$D_p = (p(X), q(X)) \text{ and } D_p' = (p(X), -q(X) - H(X) \pmod{p(X)}),$$

where $q(X)$ is the polynomial of least degree such that

$$q(X)^2 + H(X)q(X) - F(X)$$

is divisible by p. Since the divisor we are compressing or decompressing is reduced we know that only one of these two possibilities is in the support of D. Hence, for each prime divisor of $a(X)$ we need only specify one bit of information to determine whether D_p or D_p' is in the support of D. The only questions remaining are how to produce this bit and how to recover the correct value of $b(X)$, given $a(X)$ and the resulting bits.

3.1   Compression

The basic idea is to execute the following steps for every distinct irreducible factor $p(X)$ of $a(X)$, this gives the bits β_p.

1. If $p(X)$ is ramified in $k(H)$ set $\beta_p = 0$.
2. If the characteristic of k is odd, and so $H(X) = 0$, then let β_p denote the parity of the smallest non-zero coefficient of $b(X) \pmod{p(X)}$.
3. If the characteristic of k is even then we set

$$\bar{t}(X) = b(X)/H(X) \pmod{p(X)},$$

notice that the inversion of $H(X)$ modulo $p(X)$ can be accomplished since $p(X)$ is unramified and so $\gcd(p(X), H(X)) = 1$, We then let β_p denote the least significant bit of the constant term of $\bar{t}(X)$.

Hence, the compressed form of the divisor D is $\{a(X), s\}$ where s is the bit string containing the β_p for each irreducible factor of $a(X)$. The bit string is ordered with respect to the ordering on the distinct irreducible factors of $a(X)$.

3.2 Decompression

Suppose $p(X)^k$ exactly divides $a(X)$, then if we can recover $b(X)$ modulo $p(X)^k$ for all irreducible factors $p(X)$ of $a(X)$ we can then recover $b(X)$ either via the Chinese Remainder Theorem or by adding together the local components for each prime $p(X)$.

Since $(a(X), b(X))$ is a reduced divisor, we know that if $p(X)$ is ramified then the value of k above is one, and recovering $b(X)$ modulo $p(X)$ is trivial, since it will be equal to zero modulo $p(X)$.

We now turn to the case where $p(X)$ is not ramified. Then recovering $b(X)$ modulo $p(X)^k$, is trivially done once we know $b(X) \pmod{p(X)}$. This recovery of $b(X)$ modulo $p(X)^k$ from $b(X) \pmod{p(X)}$ can be accomplished in one of two ways:

1. Using Hensel's Lemma.
2. By multiplying the divisor $(p(X), b(X) \pmod{p(X)})$ by k.

So we have reduced the decompression problem to determining the value of

$$b(X) \pmod{p(X)}$$

given $p(X)$ and the bit β_p.

Since $p(X)$ is irreducible, the algebra $k[X]/p(X)$ is a field and we can apply well known techniques to solve quadratic equations in a field to determine a candidate value $\bar{b}(X)$ for $b(X) \pmod{p(X)}$. To check whether $\bar{b}(X)$ is the correct value we compute the value of the bit β_p, as in the compression algorithm, assuming that $\bar{b}(X)$ is correct. If this value agrees with the supplied value then we know that $\bar{b}(X) = b(X) \pmod{p(X)}$, otherwise we set $b(X) = -\bar{b}(X) - H(X) \pmod{a(X)}$.

Finally, note that the above algorithms for divisor compression and decompression are only slightly more complicated than those used in the elliptic curve case.

References

1. X9.62 Public Key Cryptography For The Financial Services Industry: The Elliptic Curve Digital Signature Algorithm (ECDSA). *American National Standards Institute*, 1999.
2. D.G. Cantor. Computing in the Jacobian of a hyperelliptic curve. *Math. Comp.*, **48**, 95–101, 1987.
3. M. Fouquet, P. Gaudry and R. Harley. An extension of Satoh's algorithm and its implementation. *J. Ramanujan Math. Soc.*, **15**, 281–318, 2000.
4. G. Frey and H.-G. Rück. A remark concerning m-divisibility and the discrete logarithm problem in the divisor class group of curves. *Math. Comp.*, **62**, 865–874, 1994.
5. G. Frey, Applications of arithmetical geometry to cryptographic constructions, Preprint, 2000.
6. S.D. Galbraith. Limitations of constructive Weil descent. To appear *Proceedings of a Conference on Cryptography and Computational Number Theory, Warsaw, Sept 2000*.
7. P. Gaudry. An algorithm for solving the discrete log problem on hyperelliptic curves. *Advances in Cryptology, EUROCRYPT 2000*, Springer-Verlag LNCS 1807, 19–34, 2000.
8. P. Gaudry and R. Harley. Counting points on hyperelliptic curves over finite fields. *ANTS-IV*, Springer-Verlag LNCS 1838, 313–332, 2000.
9. P. Gaudry, F. Hess and N.P. Smart. Constructive and destructive facets of Weil descent on elliptic curves. To appear *J. Cryptology*.
10. K. Kedlaya. Counting points on hyperelliptic curves using Monsky-Washnitzer cohomology. Preprint 2001.
11. N. Koblitz. Elliptic curve cryptosystems. *Math. Comp.*, **48**, 203–209, 1987.
12. N. Koblitz. Hyperelliptic cryptosystems. *J. Cryptology*, **1**, 139–150, 1989.
13. V. Miller. Use of elliptic curves in cryptography. *Advances in Cryptology, CRYPTO - '85*, Springer-Verlag LNCS 218, 47–426, 1986.
14. T. Satoh. The canonical lift of an ordinary elliptic curve over a finite field and its point counting. *J. Ramanujan Math. Soc.*, **15**, 247–270, 2000.
15. B. Skjernaa. Satoh's algorithm in characteristic two. Preprint 2000.
16. N.P. Smart. On the performance of hyperelliptic cryptosystems. *Advances in Cryptology, EUROCRYPT '99*, Springer-Verlag, LNCS 1592, 165–175, 1999.
17. A.-M. Spallek, *Kurven vom Geschlecht 2 und ihre Anwendung in Public-Key-Kryptosystemen*, PhD Thesis, IEM Essen, 1994.
18. F. Vercauteren, B. Preneel and J. Vandewalle. A memory efficient version of Satoh's algorithm. *Advances in Cryptology, EUROCRYPT 2001*, Springer-Verlag, LNCS 2045, 1–13, 2001.
19. A. Weng, Constructing hyperelliptic curves of genus 2 suitable for cryptography, Preprint, 2000.

A Differential Attack
on Reduced-Round SC2000*

Håvard Raddum and Lars R. Knudsen

Department of Informatics, University of Bergen, N-5020 Bergen, Norway
haavardr@ii.uib.no, lars@ramkilde.com

Abstract. SC2000 is a 128-bit block cipher with key length of 128, 192 or 256 bits, developed by Fujitsu Laboratories LTD. For 128-bit keys, SC2000 consists of 6.5 rounds, and for 192- and 256-bit keys it consists of 7.5 rounds. In this paper we demonstrate two different 3.5-round differential characteristics that hold with probabilities 2^{-106} and 2^{-107}. These characteristics can be used to extract up to 32 bits of the first and last round keys in a 4.5-round variant of SC2000.

1 Introduction

SC2000 [1,5] is a 128-bit block cipher designed by Fujitsu Laboratories LTD, and accepts keys of 128, 192 and 256 bits. The cipher has been submitted as a candidate for the Nessie project [2], and was presented at the Nessie workshop in Leuven in November 2000, and at FSE2001 in Yokohama in April 2001. In the submission the designers analysed SC2000 against differential cryptanalysis [3], and gave lower bounds on the complexities of a differential attack based on characteristics. However, this search for differential characteristics does not necessarily reveal those with the highest probabilities. We found two different characteristics over 3.5 rounds, which can be used to extract 32 of the bit s in both the first and the last round key in a 4.5-round variant (the definition of a half round will become clear below). These characteristics have probabilities 2^{-106} and 2^{-107}.

The paper is organised as follows. In Section 2 we give a brief description of the SC2000 algorithm. In Section 3 we give the best characteristic we found for the most complicated part of the Feistel round function, used in the cipher round function. In Section 4 we create the different characteristics based on the findings of Section 3. In Section 5 we extract bits from the first and last round keys by using these characteristics, and we conclude in Section 6 with some remarks on the design of SC2000.

---

* This work was supported by the European Union fund IST-1999-12324 - Nessie. The information in this document is provided as is, and no warranty is given or implied that the information is fit for any particular purpose. The user thereof uses the information at its sole risk and liability.

S. Vaudenay and A. Youssef (Eds.): SAC 2001, LNCS 2259, pp. 190–198, 2001.

2   Description of SC2000

The plaintext block in SC2000 is broken into four 32-bit words. The plaintext words are first XORed with a round key, and then passed through a layer of 32 parallel 4-bit S-boxes. We will call one of these 4-bit S-boxes S4 in this paper. The input to S4 are the bits that are in the same position in each of the words, see Fig. 1. After this the block is XORed with another round key. The XOR with a round key, the 32 executions of S4, and the XOR with a different round key is what we call one half round. The block is now broken into halves, and passed

Fig. 1. How the 4–bit S–box works on the bits in position j.

through a two-round Feistel network. The round function in the Feistel network is depicted in Fig. 2. Each of the two words that are sent into the round function are first passed through a layer of two 6-bit and four 5-bit S-boxes, called S6 and S5 respectively. To create diffusion, each word is then regarded as a vector of length 32 over $GF(2)$, and pre-multiplied with M, a 32x32 matrix over $GF(2)$. Let the two words of output from the multiplication of M be a_1 and a_2. These words are now mixed in a linear function to create the two words of output from the Feistel round function, b_1 and b_2, as follows.

$$b_1 = (a_1 \wedge m) \oplus a_2$$

$$b_2 = (a_2 \wedge \bar{m}) \oplus a_1$$

m is a 32-bit constant, $\bar{m}$ its bitwise complement, and the $\wedge$ denotes the logical AND operation. b_1 and b_2 are now XORed onto the two words in the other half, and the halves are swapped. The other half is then passed through the Feistel round function and XORed onto the first half, but there is no swap after the second Feistel round. This concludes one round of SC2000. For 128-bit keys the cipher consists of six full rounds, plus the first half of the seventh round. For 192- and 256-bit keys the cipher consists of seven full rounds, plus the first half of the eighth round. The constants m used in each round are 55555555_x in the odd numbered rounds and 33333333_x in the even numbered rounds. One round of the cipher is shown in Fig. 3.

We omit the details of the key scheduling. The key schedule in SC2000 is quite complex, and our attack does not depend on how the key schedule works. We note however that the key schedule appears to be very strong, the knowledge of one round key does not seem to leak any information about any other round key, or about the key selected by the user.

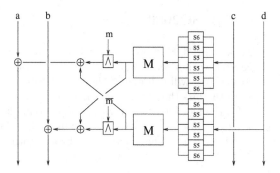

Fig. 2. The Feistel round function.

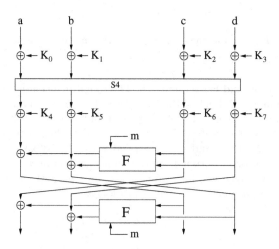

Fig. 3. The round function of SC2000.

3   Searching for Differential Characteristics

In [1] and [5] the designers have performed some differential cryptanalysis of
SC2000. However, as shown in the sequel, the designers' search for characteristics
was not sufficient, and several differentials exist with probabilities exceeding the
bounds of the designers.

In order to explain how we found the differential used in our attack, we first
define the *support* of an n-bit string $w = (w_1, w_2, \ldots, w_n)$, written $\chi(w)$ to be
the set of coordinates where w has a non-zero value.

$$\chi(w) = \{i | w_i \neq 0\}$$

We concentrated on the two first components of the F-function (see Fig. 2), namely the layer of S5's and S6's, and the multiplication with M. The F-function takes two 32-bit words as input, but they are not mixed with each other until after these two steps are executed, so we focused on only one of the input words. The idea was to find a differential δ with low Hamming weight that mapped to a differential ϵ after the S-boxes and the M-multiplication, such that $\chi(\epsilon) \subseteq \chi(\delta)$. To help us with this we first computed the two differential distribution tables for S5 and S6.

We searched through all 32-bit words of Hamming weight six or less, and for each word we did the following. The word (or differential) ϵ was assumed to be the output of the M-multiplication. Since this is a linear component, we multiplied ϵ with M^{-1} to find the α that would map to ϵ through M. By looking up in the two distribution tables, we then checked whether ϵ could be mapped to α through the layer of the S5 and S6 S-boxes.

We found 11 differentials of Hamming weight five, which only had four of the six S-boxes active and were mapped to themselves with some non-zero probability. None of the ϵ's of weight four or less could not be mapped to themselves, but for each of them we checked how many of the S-boxes that failed to do the required mapping from ϵ to α. We found one ϵ of weight two, namely $\epsilon = 40200000_x$ that mapped to the corresponding $\alpha = f7d30017_x$ in four of the six S-boxes. By adding a 1-bit in the differences going into the two remaining S-boxes, we were able to produce $\delta = \delta_0 = 40220001_x$ of weight four that is mapped to α with probability 2^{-18} (The probabilities are 2^{-5} for the two S6's, and 2^{-4} for the two active S5's). In fact, there are eight different δ's of weight four that can map to ϵ. In one of the two S-boxes that require a non-zero difference we can add the 1-bit in two different ways, and in the other S-box we can add the 1-bit in four different ways. The seven other δ's are

$$\delta_1 = 40220004_x, \delta_2 = 40220010_x, \delta_3 = 40220020_x, \delta_4 = 40300001_x,$$

$$\delta_5 = 40300004_x, \delta_6 = 40300010_x, \delta_7 = 40300020_x$$

Now the idea was to send the differentials 0 and δ into the F-function, and have δ map to ϵ. In the third and last part of the F-function, the AND operation with the fixed masks does not effect the 0 difference. The AND operation applied to ϵ will turn ϵ into $\epsilon_2 = 00200000_x$ when the mask 33333333_x or $aaaaaaaa_x$ is used, and turn ϵ into $\epsilon_4 = 40000000$ when the mask 55555555_x or $cccccccc_x$ is used. Finally, ϵ will be XORed onto the 0 difference and 0 will be XORed onto the difference that is either ϵ_2 or ϵ_4. In total we have the differential characteristics $(\delta, 0) \xrightarrow{F} (\epsilon_4, \epsilon)$ and $(0, \delta) \xrightarrow{F} (\epsilon, \epsilon_2)$ in a round where the mask 55555555_x is used, and the differential characteristics $(\delta, 0) \xrightarrow{F} (\epsilon_2, \epsilon)$ and $(0, \delta) \xrightarrow{F} (\epsilon, \epsilon_4)$ when the mask 33333333_x is used. Each of these characteristics have probability 2^{-18}.

4   Building the One-Round Characteristics

Let us now see how we can use δ and ϵ to create a differential characteristic through several rounds of SC2000.

4.1   Two One-Round Differential Characteristics

The first one-round characteristic, explained below, is shown in Figure 4, where we have omitted the additions of round keys since they do not affect the analysis. s Let $(\delta, 0, 0, 0)$ be the difference in the blocks before the two Feistel rounds.

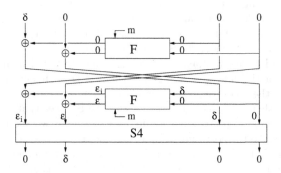

Fig. 4. A one–round differential characteristic with probability 2^{-30}.

First the two rightmost words are sent through the F-function. They have difference $(0, 0)$, so the output will have difference $(0, 0)$ with probability 1. This $(0, 0)$-difference is XORed onto the left half, and the halves are swapped so the difference before the second Feistel round is $(0, 0, \delta, 0)$. The right halves with difference $(\delta, 0)$ are then sent into the F-function, and with probability 2^{-18} the difference after multiplication with M will be $(\epsilon, 0)$. After this ϵ will meet one of the masks 55555555_x or 33333333_x, so the output of F will be (ϵ_2, ϵ) or (ϵ_4, ϵ). These outputs are XORed onto the left halves, and since there is no swap, the difference of the blocks becomes $(\epsilon_2, \epsilon, \delta, 0)$ or $(\epsilon_4, \epsilon, \delta, 0)$ before the S4 layer.

Since δ has weight four and $\chi(\epsilon_i) \subseteq \chi(\epsilon) \subseteq \chi(\delta)$, there will only be four active S-boxes in the layer of the 32 S4's. Two of them will have input difference 2_x, one will have input difference 6_x, and one will have input difference e_x. All the differences 2_x, 6_x and e_x can go to the difference 4_x through S4, each with probability 2^{-3}. So with probability $(2^{-3})^4 = 2^{-12}$ we get the characteristic $(\epsilon_i, \epsilon, \delta, 0) \xrightarrow{S4} (0, \delta, 0, 0)$ through S4. All together, we get the following characteristic with probability $2^{-18} \cdot 2^{-12} = 2^{-30}$.

$$(\delta, 0, 0, 0) \xrightarrow{F-F-S4} (0, \delta, 0, 0)$$

The other useful one-round characteristic is the one that starts with the difference $(0, \delta, 0, 0)$. After the first Feistel round with the swap the difference becomes $(0, 0, 0, \delta)$. With probability 2^{-18} the right half difference $(0, \delta)$ becomes $(0, \epsilon)$ after multiplication with M. The output difference of F will then be (ϵ, ϵ_2) or (ϵ, ϵ_4) and after the XOR with the left halves the difference will be $(\epsilon, \epsilon_i, 0, \delta)$. Two of the input differences to S4 will now be 1_x, one of them will be 9_x and one will be d_x. The 1_x and d_x differences can lead to the difference 8_x with probability 2^{-3}, and the 9_x difference goes to 8_x with probability 2^{-2}. This gives us the following one-round characteristic with probability 2^{-29}.

$$(0, \delta, 0, 0) \xrightarrow{F-F-S4} (\delta, 0, 0, 0)$$

4.2   Concatenating the One-Round Characteristics

The differential characteristics above start and end just before the Feistel rounds. The cipher itself begins with the application of the S4 layer, but the characteristics we build by concatenating the one-round characteristics will start after the first half of the first round. The next section explains how to use these characteristics.

The characteristic $(\delta, 0, 0, 0) \xrightarrow{F-F-S4} (0, \delta, 0, 0)$ can be concatenated with $(0, \delta, 0, 0) \xrightarrow{F-F-S4} (\delta, 0, 0, 0)$. By doing this, we get the following differential characteristic through three and a half rounds with probability 2^{-107}.

$$(\delta,0,0,0) \xrightarrow{F-F-S4} (0,\delta,0,0) \xrightarrow{F-F-S4} (\delta,0,0,0) \xrightarrow{F-F-S4} (0,\delta,0,0) \xrightarrow{F-F} (\epsilon,\epsilon_4,\delta,0)$$

The other characteristic is the one starting with input $(0, \delta, 0, 0)$.

$$(0,\delta,0,0) \xrightarrow{F-F-S4} (\delta,0,0,0) \xrightarrow{F-F-S4} (0,\delta,0,0) \xrightarrow{F-F-S4} (\delta,0,0,0) \xrightarrow{F-F} (\epsilon_2,\epsilon,\delta,0)$$

This characteristic has probability 2^{-106}.

5   Extracting Bits from the First and Last Round Key

In this section we will explain how to extract up to 32 bits from both the first and last round key in a 4.5-round variant of SC2000.

5.1   How to Find 16 Key Bits of the First and Last Round Key

The characteristics in the previous section do not start with the plaintext difference, but with the difference after the first S4 layer. To use these characteristics, we create structures Σ of 2^{16} plaintexts as follows. Fix the bits going into the 28 S4's that are not affected by δ, and let the 16 bits going into the S4's determined by δ take on all 2^{16} values. Let $\Delta_8 = (\delta, 0, 0, 0)$, $\Delta_4 = (0, \delta, 0, 0)$, $\Omega_1 = (\epsilon, \epsilon_4, 0, \delta)$ and $\Omega_2 = (\epsilon_2, \epsilon, \delta, 0)$. For each plaintext $P \in \Sigma$, the plaintext $P \oplus \Delta_i$ is also in Σ, for $i = 4, 8$. In other words, of the $\binom{2^{16}}{2} \approx 2^{31}$ pairs in Σ there are 2^{15} pairs with difference Δ_4 and 2^{15} pairs with difference Δ_8. Encrypting the plaintexts in Σ through the first S4 layer does not change the structure in any essential way. The 112 fixed bits remain fixed, and the other 16 bits range over all 2^{16} values. So there will be 2^{15} pairs in Σ that have difference Δ_4, and 2^{15} pairs that have difference Δ_8 after encryption through the first S4 layer. A randomly chosen input difference to one S4 will have the possibility to go to the output difference 4_x with probability $1/2$, so the probability that a randomly chosen pair of texts from Σ will have the possibility of having difference Δ_4 after S4 is 2^{-4}. By the same argument, the probability that a pair of texts from Σ can have the difference Δ_8 after S4 is approximately 2^{-4}. In total, the probability that a randomly chosen pair of texts from Σ has difference Δ_4 or Δ_8 after S4 is 2^{-3}.

We call a pair of plaintexts that follows either of the two characteristics from Section 4 a *right* pair, and a pair of plaintexts that does not follow any of these characteristics a *wrong* pair.

The probability that a structure contains a right pair is $2^{15} \cdot (2^{-106} + 2^{-107}) = 3 \cdot 2^{-92}$. After encrypting 2^{93} structures, we expect to have $2^{93} \cdot 3 \cdot 2^{-92} = 6$ right pairs among the $2^{31} \cdot 2^{93} = 2^{124}$ pairs we get from the structures. We filter out most of the wrong pairs as follows.

Find potential good pairs by inserting the 2^{16} ciphertexts from one structure in a hash-table according to 20 bits in the first word (see [4]). The ciphertexts in a right pair will be inserted in the same position in the table. If a pair is a right pair, all of the 112 bits corresponding to the inactive S4's must be equal. This gives a filtering factor of 2^{-112}. If a pair of ciphertexts are equal in these 112 bits, check the differences in the four S4's corresponding to δ. If the pair is a right pair, it must be possible that the output difference from the last S4-layer has had input differences Ω_1 or Ω_2. As explained above, a random pair passes this test with probability 2^{-3}. If a ciphertext pair is a right pair, and had difference Ω_1 before the last S4, then the pair of plaintexts must have had the possibility to get the difference Δ_8 after the initial S4. A random pair passes this test with probability 2^{-4}. Likewise, a right pair that has difference Ω_2 before the last S4 must have had difference Δ_4 after the first S4, and the probability that this holds for a random pair is 2^{-4}.

With these steps we have a filtering factor of 2^{-119}. After using this filtering procedure on the 2^{124} different pairs we expect to be left with $2^{124} \cdot 2^{-119} = 32$ pairs, among which we expect six right pairs.

The main part of the work to generate 16 potentially right pairs comes from the 2^{109} encryptions required. The memory requirements to get the 16 potentially good pairs is small. In addition to the potential right pairs, we only need to hold 2^{16} plaintexts and the corresponding 2^{16} ciphertexts in memory at the same time.

The rest of the attack follows along the lines of a standard differential attack. For each of the ciphertexts in the 16 potentially right pairs, guess on the 16 key bits from the last round key corresponding to the active S-boxes. For each guess, decrypt the ciphertext bits in the active S4's, if the decrypted values have one of the input differences Ω_1 or Ω_2, suggest these bits as part of the last round key.

The correct value will be suggested for each right pair. For each Ω_i, there will be 2 - 4 4-bit values suggested for each S-box, and the values suggested by each S-box can be combined in 2^4 - 4^4 different ways. Since we accept both Ω_1 and Ω_2 as input difference, we will get 32 - 512 suggestions for the 16 key bits. The right value will be suggested for every right pair, i.e. six times. The suggestions of the wrong values are expected to be distributed more or less uniformly over the 2^{16} different values, so it is highly unlikely that any wrong value will be suggested six times. Take the most suggested value as the correct bits in the last round key.

We find 16 bits of the first round key in the same manner. Guess on the 16 bits corresponding to the active S4's in the first round, and for each guess, encrypt each pair of plaintexts through the active S4's. If a pair gets the difference Δ_4 or Δ_8, suggest the value as part of the first round key. Again there will be 32 - 512 suggestions for every pair. We expect the correct value to be suggested six times, and the incorrect values to be more or less uniformly distributed over the 2^{16} values. Again we take the most suggested value as the correct one.

5.2   How to Find Another 16 Bits

We can repeat the attack described above for a different δ, say for $\delta_5 = 40300004_x$, to find 16 bits of the first and last round keys. Among the key bits we will find, eight of them will be the same as we found using δ_0, because the two active S-boxes defined by ϵ will overlap in both attacks. So repeating the attack with δ_5 will only yield eight new bits in the two round keys. After this we have found 24 bits of each key. The last eight bits we can get are the ones that correspond to the S4's defined by 00000010_x and 00000020_x. They can be found by repeating the attack with δ's using these S-boxes, like $\delta_2 = 40220010_x$ and $\delta_3 = 40220020_x$. Repeating the attack four times gives an overall complexity of 2^{111}.

6   Conclusions

For a 4.5 round variant of SC2000, we have shown how to find 32 bits of both the first and the last round key, using 2^{111} chosen plaintexts. The strong key schedule in SC2000 prevents us from actually breaking 4.5 rounds by searching exhaustively for the remaining 96 bits in the first or last round key, since we can not easily deduce the other round keys from them.

This paper may teach us a different lesson, though. Several places in [1], the designers hint that SC2000 can be thought of as an advanced Feistel cipher. The layer of 4-bit S-boxes between every other Feistel round can be regarded as a cryptographically stronger component than the swap of halves found in ordinary Feistel ciphers. This S-box layer certainly gives better confusion than a simple swap, but it introduces another weakness not found in regular Feistel ciphers.

It was shown in [5] that in SC2000 it is possible to have a differential characteristic that feeds every other Feistel round with a 0-difference. This is not possible in a regular Feistel cipher. In this paper we have extended the search done in [5] to two-round iterative characteristics. This resulted in characteristics with higher probabilities than what was found in [5].

Having an S-box layer instead of a swap between some rounds might be a good idea, but one should be careful to make sure that any cryptographically good property of the swap is not lost when replacing it. The designers state that one of the design criteria for S4 is that except for the all-zero difference, an input difference $(\alpha_0, \alpha_1, 0, 0)$ can not lead to an output difference $(\beta_0, \beta_1, 0, 0)$, and an input difference $(0, 0, \alpha_2, \alpha_3)$ can not lead to an output difference $(0, 0, \beta_2, \beta_3)$. This is to make sure that there is some form of "swap" involved when going through S4. However, one should also demand that if α_L and α_R are two non-zero 2-bit values, then the input difference (α_L, α_R) will always lead to an output difference (β_L, β_R) where both β_L and β_R are non-zero.

References

1. http://www.cosic.esat.kuleuven.ac.be/nessie/workshop/
2. http://www.cryptonessie.org/
3. E. Biham and A. Shamir. *Differential Cryptanalysis of the Data Encryption Standard.* Springer Verlag, 1993.
4. L.R. Knudsen and T. Berson. Truncated differentials of SAFER. In Gollmann D., editor, *Fast Software Encryption, Third International Workshop, Cambridge, UK, February 1996, LNCS 1039*, pages 15–26. Springer Verlag, 1995.
5. Shimoyama et al. The Block Cipher SC2000. *Fast Software Encryption, Eighth International Workshop, Yokohama, Japan, April 2001*, preproceedings, pages 326–340.

On the Complexity of Matsui's Attack

Pascal Junod

Security and Cryptography Laboratory,
Swiss Federal Institute of Technology, CH-1015 Lausanne, Switzerland
pascal.junod@epfl.ch

Abstract. Linear cryptanalysis remains the most powerful attack against DES at this time. Given 2^{43} known plaintext-ciphertext pairs, Matsui expected a complexity of less than 2^{43} DES evaluations in 85 % of the cases for recovering the key. In this paper, we present a theoretical and experimental complexity analysis of this attack, which has been simulated 21 times using the idle time of several computers. The experimental results suggest a complexity upper-bounded by 2^{41} DES evaluations in 85 % of the case, while more than the half of the experiments needed less than 2^{39} DES evaluations. In addition, we give a detailed theoretical analysis of the attack complexity.

Keywords: linear cryptanalysis, DES

1   Introduction

Linear cryptanalysis against DES [10] has been introduced by Matsui [6,7] and remains at this time the most powerful attack against this cipher. A single experimental implementation [7] has been carried out. During this attempt, Matsui managed to break a DES key in about 50 days on 12 powerful computers, the plaintext-ciphertext pairs generation lasting 40 days and the exhaustive search for the remaining unknown bits taking the last 10 days. It was noticed that the second phase performed faster than one could expect theoretically.

Although several authors have studied, generalized and applied the linear cryptanalysis concept in several ways, little work concerning its success probability and its complexity has been done, and while it is widely accepted that linear cryptanalysis of DES, given 2^{43} known plaintext-ciphertext pairs, has a success probability of 85 % within a complexity of 2^{43} DES evaluations, it was conjectured that this value is pessimistic [9,3].

Motivated by this fact, by the parallel implementation concept of Biham [1] and the actual 64-bit processor performances, we propose in this paper a theoretical and experimental complexity analysis. By using a fast DES routine implemented for the Intel MMX architecture, the production part of the attack has been run several time, virtually breaking a total of 21 keys.

This paper is organized as follows: in §2, we recall some theoretical background on the attack. In §3, we describe briefly the design of the fast DES routine and the attack implementation. In §4, we discuss and complete the success probability and complexity model. In §5, we discuss some issues on the linear expression

S. Vaudenay and A. Youssef (Eds.): SAC 2001, LNCS 2259, pp. 199–211, 2001.
© Springer-Verlag Berlin Heidelberg 2001

biases, the piling-up approximation and the wrong-key randomization hypothesis, comparing the known theoretical results to our experimental ones and finally we give in §6 our experimental results.

2 Matsui's Attack

In this paper, we deal with the improved attack [7] proposed by Matsui against DES. The attack's core is unbalanced linear expressions, i.e. equations involving a modulo two sum of plaintext and ciphertext bits on the left and a modulo two sum of key bits on the right. Such an expression is unbalanced if it is satisfied with probability [1] $p = \frac{1}{2} + \kappa\epsilon$ with $0 < \epsilon \leq \frac{1}{2}$ and $\kappa \in \{-1, 1\}$ when the plaintexts and the key are independent and chosen uniformly at random and where κ depends on the key value.

Given some plaintext bits $P_{i_1}, \ldots, P_{i_r}$, ciphertext bits $C_{j_1}, \ldots, C_{j_s}$ and key bits $K_{k_1}, \ldots, K_{k_t}$, and using the notation $X_{l_1} \oplus X_{l_2} \oplus \ldots \oplus X_{l_u} = X_{[l_1, \ldots, l_u]}$, we can write a linear expression $\mathcal{L}$ as

$$\mathcal{L} : P_{[i_1, \ldots, i_r]} \oplus C_{[j_1, \ldots, j_s]} = K_{[k_1, \ldots, k_t]} \tag{1}$$

Matsui's improved attack operates on 14 rounds using two biased linear expressions which collect statistical information on 26 bits out of the first and last round subkeys. The remaining 30 unknown key bits have to be searched exhaustively. The linear expression (1) involves thus two terms of F-function and can be rewritten as

$$\mathcal{L} : P_{[i_1, \ldots, i_r]} \oplus C_{[j_1, \ldots, j_s]} \oplus F^{(1)}_{[l_1, \ldots, l_u]}\left(P, K^{(1)}\right) \oplus$$
$$F^{(16)}_{[m_1, \ldots, m_v]}\left(C, K^{(16)}\right) = K_{[k_1, \ldots, k_t]} \tag{2}$$

where $F^{(1)}_{[l_1, \ldots, l_u]}\left(P, K^{(1)}\right)$ is the modulo two sum of some bits resulting from the F-function output in the first round and $K^{(1)}$ is the subkey of round 1. A similar notation is used for the last F-function.

The attack main idea is related to the following assumption:

Assumption 1 (Wrong-key randomization hypothesis [3]). *For any linear expression $\mathcal{L}$ operating on n rounds for which*

$$\left| \Pr\left[\mathcal{L} = 0 \mid K^{(1)} = k^{(1)}, \ldots, K^{(n)} = k^{(n)}\right] - \frac{1}{2} \right|$$

is large for virtually all values $k^{(1)}, \ldots, k^{(n)}$ of the round keys, the following is true: for virtually all possible full keys $(k^{(1)}, \ldots, k^{(n)})$ and for all estimates k of

---

[1] In the literature, this non-linearity measure is often called *linear probability*, and expressed as $\mathrm{LP}^f(a, b) = (2\Pr[a \cdot x = b \cdot f(x)] - 1)^2$, where a and b are the masks selecting the plaintext and ciphertext bits, respectively. In this paper, we will refer to the *bias* ϵ for simplicity reasons.

the last round key,

$$\frac{\left| \Pr\left[\mathcal{L} = 0 \mid \mathsf{K} = k_r \right] - \frac{1}{2} \right|}{\left| \Pr\left[\mathcal{L} = 0 \mid \mathsf{K} = \hat{k} \right] - \frac{1}{2} \right|} \gg 1 \quad \forall \hat{k} \neq k_r \tag{3}$$

where k_r is the right key.

Intuitively, the decryption of the first and the last round with wrong subkey candidates can be considered as two rounds more of encryption. Thus, the plaintext and the ciphertext will be less dependent, and the linear expressions less biased. The first linear cryptanalysis phase (see Fig. 1) consists in evaluating the bias of both linear expressions for all possible subkey candidates and for all known plaintext-ciphertext pairs. In a second phase (Fig. 2), the two lists of subkey candidates corresponding each to a linear expression are sorted in a maximum-likelihood manner, combined, and the missing bits are finally searched exhaustively for each pair of subkey candidate until the right key is found.

The complexity $\mathcal{C}$ of the attack is then related to the number of needed DES encryptions in the exhaustive search part while its success probability $\mathcal{P}_{\mathcal{C}}$ within a given complexity $\mathcal{C}$ is *also* related to the success while guessing the right part of both linear expressions.

1: N = number of known plaintext-ciphertext pairs at disposal.
2: **for** linear expressions $\mathcal{L}_1$ and $\mathcal{L}_2$ **do**
3:   **for all** subkey candidates $\hat{k}_i, 1 \leq i \leq 2^{12}$ **do**
4:     $C_{\hat{k}_i}$ = number of times out of N where left part of (2) is equal to 0 when
       $\mathsf{K} = \hat{k}_i$.
5:   **end for**
6: **end for**

Fig. 1. Matsui's algorithm 2 [7] (phase 1)

3 Implementation of the Attack

The linear cryptanalysis attack against DES, except the exhaustive search part, has been implemented as described in [7]. After having determined the rank of the right subkey candidate in the final list, it is not difficult to compute [2] the expected complexity (in DES function evaluations) of the exhaustive search part:

$$\mathrm{E}[\hat{\mathcal{C}}] = (r - 1) \cdot 2^{30} + 2^{29}$$

---

[2] The strategy used to combine the two lists of 13-bit subkey candidates is Matsui's proposed one [7]: sort the pairs by increasing $r = i \cdot j$ (see lines 12-13 of Fig. 2), where i and j are the respective ranks in the 13-bit subkey lists.

1: **for** linear expressions $\mathcal{L}_1$ and $\mathcal{L}_2$ **do**

2:     Sort the $C_{\hat{k}_i}$'s by decreasing $\left| \frac{N}{2} - C_{\hat{k}_i} \right|$ and rename them $C_j^*, 1 \leq j \leq 2^{12}$.

3:     **for** $1 \leq j \leq 2^{12}$ **do**

4:         /* κ is defined in Sect. 2 (expected bias of $\mathcal{L}$) */

5:         **if** $\left(C_j^* - \frac{N}{2} \right) \kappa > 0$ **then**

6:             Guess $\mathsf{K}_{[k_1,\ldots,k_t]} = 0$

7:         **else**

8:             Guess $\mathsf{K}_{[k_1,\ldots,k_t]} = 1$

9:         **end if**

10:     **end for**

11: **end for**

12: Form 2^{24} $(C_i^*, C_j^*)_r$ pairs where $r := i \cdot j$.

13: Sort them by increasing r and rename them $D_k, 1 \leq k \leq 2^{24}$.

14: **for** $1 \leq k \leq 2^{24}$ **do**

15:     Fix the key bits given by D_k and search exhaustively the remaining 30 bits of K until the right key is found.

16: **end for**

Fig. 2. Matsui's algorithm 2 [7] (phase 2)

where r is the rank in the list D of subkey candidates. The complexity's estimation error has thus a maximal value of 2^{29} DES evaluations, which is negligible almost all the time.

The computational most intensive part of the attack being data encryption, the involved DES routine speed is a key parameter regarding the time needed to process 2^{43} plaintexts. We have thus implemented a very fast DES routine using the bitslicing concept [1] and some attack-related optimizations. Our routine has been designed for the Intel MMX architecture which has eight 64-bit registers at disposal. Although this platform has several drawbacks regarding a bitsliced implementation [8], it has the advantage of being very common.

Kwan's gate representation of the S-boxes [5] builds the core of the implementation, the other parts of the cipher (key schedule, permutations, ...) being hard-coded. By eliminating parts of the cipher unrelated to the attack and by using advanced optimization techniques like instruction pairing, prefetching of the data and code unrolling, we managed to get an encryption speed of 183 Mbps on an Intel Pentium III clocked at 666 MHz. This represents 232.7 clock cycles for encrypting one block of data. One can hardly compare this number with existing good implementations [3], because of the optimizations related to the attack; however, using classical available implementations for our purposes would have resulted in poorer performances.

---

[3] A DES routine was implemented for similar purposes in [12] on other platforms; they report 62 Mbps on a Ultra SPARC 200 MHz and 336 Mbps on a Alpha 21164A 500 MHz. The significant speed difference on the latter platform is due to the large number of available 64-bit registers (and thus to a lesser number of slow memory accesses).

The attack has run 21 times, using the idle time of 8 to 16 computers; this represents between 3 and 6 days for a single run.

4   Success Probability

In this section, we address a general way to characterize the probability distribution of the rank of the right 13-bit subkey in the list of candidates given by a linear approximation $\mathcal{L}$.

4.1   Rank Probability

As the complexity $\mathcal{C}$ of the attack is closely related to the rank of the right subkey in the candidates list, we address first the problem of estimating the rank distribution.

Let $W_1, \ldots, W_n$ be n independent and identically distributed continuous random variables having $f_W(x)$ and $F_W(x)$ as common density function and distribution function, respectively. Let R be a continuous random variable independent of the W_i's and having $f_R(x)$ and $F_R(x)$ as density and distribution function. Sort these $n+1$ random variables in non-increasing order and rename them $Z_{(1)} > Z_{(2)} > \ldots > Z_{(n+1)}$. Finally, let Ψ be a discrete random variable taking values on $\{1, \ldots, n+1\}$ which models the rank of R in the sorted list: $\Psi = \psi \Leftrightarrow Z_{(\psi)} = R$. The distribution of Ψ and its expected value are given by the following theorem, whose proof is given in Appendix A.

Theorem 1. *Under previous assumptions and for $1 \leq \psi \leq n \in \mathbb{N}$, the distribution function of Ψ is equal to*

$$\Pr[\Psi \leq \psi] = \int_{-\infty}^{+\infty} B_{n+1-\psi,\psi}(F_W(x)) f_R(x) dx$$

and

$$\mathrm{E}[\Psi] = 1 + n \left(1 - \int_{-\infty}^{+\infty} f_R(x) F_W(x) dx\right)$$

where

$$B_{a,b}(x) = \frac{\Gamma(a+b)}{\Gamma(a)\Gamma(b)} \int_0^x t^{a-1}(1-t)^{b-1} dt$$

is the incomplete beta function of order (a, b).

In order to be able to compute the densities of the estimated biases[4], we first have to make the following assumptions [13]; the two first ones are heuristic in nature, while the last one is motivated by the law of large numbers. C_{k_r} (C_{k_w}) will denote a random variable modeling the counter value (as defined at line 4 of Fig. 1) in the case of a right (wrong) subkey candidate and N is the number of known plaintext-ciphertext pairs.

---

[4] The mean and standard deviation of the counters and the respective biases of the linear expression being linearly related, we will use in the following the bias terminology.

Assumption 2. *The bias*

$$B_w = \left| \frac{1}{2} - \frac{C_{k_w}}{N} \right|$$

of a linear expression evaluated with wrong subkey candidates has a distribution independent of the key value.

Assumption 3. *The bias*

$$B_r = \left| \frac{1}{2} - \frac{C_{k_r}}{N} \right|$$

of a linear expression evaluated with the right subkey candidates has a distribution independent of the distribution defined in Assumption 2 and independent of the key value.

Assumption 4. *The distributions of $\frac{C_{k_r}}{N}$ and $\frac{C_{k_w}}{N}$ are well approximated by a normal law.*

We denote in the following the normal law density with mean μ and variance σ^2 by $\phi_{(\mu,\sigma^2)}$ and the corresponding cumulative distribution function by $\Phi_{(\mu,\sigma^2)}$. Because the cryptanalyst ignores the linear expression's right part, she is more interested in the absolute value of the biases. Noting that if X is a normal law $\phi_{(\mu,\sigma^2)}$, the density of $Y = |X - a|$, $a \leq \mu$ is given by $f_Y^{(\mu,\sigma^2)}(y, a) = \phi_{(\mu,\sigma^2)}(y + a) + \phi_{(\mu,\sigma^2)}(a - y)$ for $0 \leq y \leq +\infty$, the bias densities in case of wrong and right subkey candidates are respectively given by

$$f_W(x) = f^{(\mu_w, \sigma_w^2)}(x, \tfrac{1}{2}) \tag{4}$$
$$f_R(x) = f^{(\mu_r, \sigma_r^2)}(x, \tfrac{1}{2}) \tag{5}$$

with

$$\mu_r = \mathrm{E}\left[\frac{C_{k_r}}{N}\right] = \frac{1}{2} + \kappa \epsilon_r \qquad \mu_w = \mathrm{E}\left[\frac{C_{k_w}}{N}\right] = \frac{1}{2} + \kappa \epsilon_w$$
$$\sigma_r^2 = \mathrm{Var}\left[\frac{C_{k_r}}{N}\right] \approx \frac{1}{4N} \qquad \sigma_w^2 = \mathrm{Var}\left[\frac{C_{k_w}}{N}\right] \approx \frac{1}{4N}$$

where $\kappa \in \{-1, +1\}$ depends of the unknown key bits and C_{k_r} (C_{k_w}) is the random variable modeling the value of the counter corresponding to the (a) right (wrong) subkey. Fig. 4 gives some numerical evaluations of Theorem 1 for these densities while the following table gives the expected rank for various amounts of known plaintext-ciphertext pairs at disposal. Here, we assume that $\epsilon_r = 1.19 \cdot 2^{-21}$ is equal to the piling-up lemma approximation and that $\epsilon_w = 0$. We note that Theorem 1 gives exactly the same values as Matsui's experimental computations [7] regarding the cumulative rank probability of the right subkey candidate.

N	2^{43}	$2^{42.5}$	2^{42}	2^{41}	2^{40}
$E[\Psi]$	71.3	182.5	361.9	847.3	1311.6

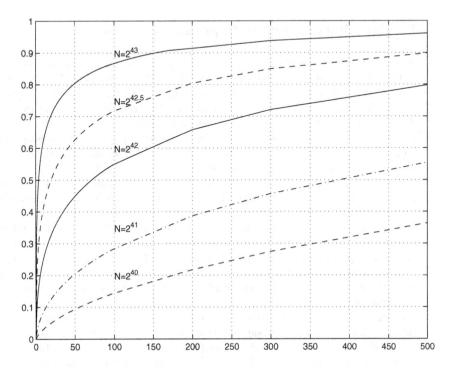

Fig. 3. Rank distribution $\Pr[\Psi \leq \psi]$ for various amounts N of plaintext-ciphertext pairs.

4.2   Success Probability

The attack's success probability $\mathcal{P}_{\mathcal{C}}$ within a given complexity $\mathcal{C}$ is also dependent on the error probability while guessing the bit of information about $K_{[k_1,...,k_t]}$. Using the same assumptions as during the previous computations, it is easy to compute this error probability (in the case where $\kappa = +1$ and $K_{[k_1,...,k_t]} = 0$, the other ones being symmetric).

$$p_{wg} = \Pr\left[\text{``}K_{[k_1,...,k_t]} \text{ wrongly guessed''}\right] = \Phi_{(\mu_r, \sigma_r^2)}\left(\frac{N}{2}\right) \qquad (6)$$

The following table gives some numerical approximations for various N:

N	2^{43}	$2^{42.5}$	2^{42}	2^{41}	2^{40}
p_{wg}	0.0004	0.0023	0.0086	0.0462	0.1170

5   Experimental Linear Expressions Biases

A key parameter regarding the linear cryptanalysis success is of course the bias of the involved linear expression(s). As it is infeasible to compute the exact bias of a linear expression, one uses implicit assumptions, such as the wrong-key randomization one and the independence of data between two successive rounds. The incidence of these assumptions has been well discussed in the literature [9,2,3,4]. Although several situations where these assumptions can fail have been suggested and discussed, it is accepted that the linear expression real bias should be well approximated in case the of DES.

The experimental results go in this direction. We have computed the sample means of the experimental biases $\hat{B}_r$ and $\hat{B}_w$, which can be compared to the expected values of densities (4) and (5).

In case of right key, the sample mean is equal to $5.5 \cdot 10^{-7}$ with a standard deviation of $0.2 \cdot 10^{-7}$. This value has to be compared with the one given by the piling-up approximation and (5), $E[B_r] = 5.674 \cdot 10^{-7}$. As a first observation, one can note that the *linear hull effect* [9] is not visible for DES, the mean experimental bias being not perceptibly greater than the piling-up lemma approximation.

Our experiments provide furthermore a good opportunity to confirm the validity of Assumption 1. The sample mean in case of wrong subkey candidates, averaged over all the wrong subkeys and all experiments, is equal to $1.38 \cdot 10^{-7}$ with a standard deviation of $0.03 \cdot 10^{-7}$. This value has to be compared with $E[B_w] = 1.345 \cdot 10^{-7}$ given by $\epsilon_w = 0$ and (4). Obviously, as one could expect, the mean seems to be slightly greater than for a perfect cipher and thus the plaintext and ciphertext are still correlated. However, the bias values for the wrong candidates are not on the same scale as those for the right candidates, confirming the validity of Assumption 1 for DES.

6   Experimental Results

It is widely accepted that linear cryptanalysis of DES, given 2^{43} known plaintext-ciphertext pairs, has a success probability of $\mathcal{P}_{\mathcal{C}_A} = 85\%$ within a complexity of $\mathcal{C}_A = 2^{43}$ DES encryptions, which are values given in [7]. Our experimental results suggest a lower complexity.

6.1   Rank and Guessing Error Probabilities

Each of the 21 experiments provides two statistical samples. Following table summarizes our results about the ranks of the right subkey candidates for various

N	2^{43}	$2^{42.5}$	2^{42}	2^{41}	2^{40}
$\psi \leq 5$	20 (22)	13 (13.5)	7 (7.6)	0 (2.3)	0 (0.8)
$\psi \leq 10$	27 (25.8)	16 (17.1)	9 (10.5)	2 (3.6)	0 (1.3)
$\psi \leq 50$	33 (33.6)	26 (26.2)	18 (18.8)	5 (8.6)	2 (3.9)
$\psi \leq 150$	38 (37.7)	34 (32.3)	24 (25.7)	10 (14.3)	5 (7.7)
$\psi \leq 300$	42 (39.4)	39 (35.7)	31 (30.3)	17 (19.2)	14 (11.6)
$\psi \leq 600$	42 (40.8)	40 (38.5)	35 (34.6)	25 (24.7)	22 (16.8)
$E[\psi]$	38 (71)	129 (182)	302 (362)	654 (847)	1121 (1312)

amounts N of known plaintext-ciphertext pairs and compare them to the theoretical expectations (values in smaller characters) given by Theorem 1. We observe that Theorem 1 seems to give a pessimistic rank expected value. It is difficult to explain this fact because of the small statistical sample size. Furthermore, we have noticed that Theorem 1 is very sensitive numerically. For instance, the expected rank $E[\Psi]$ is equal to 113 and to 39 when we assume that $\epsilon_r = 1.1 \cdot 2^{-21}$ and $\epsilon_r = 1.3 \cdot 2^{-21}$, respectively.

The experimental results regarding the remaining bit guessing error probability are summarized in the following table. The number n_{wg} of cases where the guessing phase was unsuccessful is reported, together with the theoretical expected values given by (6) which are given in smaller characters. One can see that (6) is

N	2^{43}	$2^{42.5}$	2^{42}	2^{41}	2^{40}
n_{wg}	0 (0.02)	0 (0.10)	0 (0.36)	0 (1.94)	1 (4.91)

a bit pessimistic, which can be explained a new time by the arguments developed below. We note furthermore that the success probability $\mathcal{P}_C$ of the linear cryptanalysis of DES within a given complexity C seems not to be so dependent on the guessed bit of information about the key and that the key factor regarding $\mathcal{P}_C$ is the given upper bound C.

6.2   Complexity of the Attack

An exhaustive table of our experimental results regarding the complexity is given in Appendix B. Key facts (mean, median, maximal and minimal $\hat{C}$) are summarized in the following table where a value of x means 2^x DES evaluations: Our experimental results lead to the following observations:

- Given 2^{43} known plaintext-ciphertext pairs, our experiments have a complexity of less than 2^{41} DES evaluation with a success probability of 86 % where more than the half of the cases have a complexity less than 2^{39}. Furthermore, if an attacker is ready to decrease her success probability, the

N	2^{43}	$2^{42.5}$	2^{42}	2^{41}	2^{40}
$\mu_{\hat{C}}$	41.4144	47.1516	48.9504	50.2121	51.4154
$\hat{C}_{med}$	38.1267	41.8023	44.2949	48.5492	51.0533
$\hat{C}_{min}$	32.1699	29.0000	36.5157	43.8552	41.9750
$\hat{C}_{max}$	45.4059	51.2973	52.3671	52.1953	53.1000

complexity drops dramatically (less than 2^{34} DES evaluations with a success probability of 10 %).

- Given $2^{42.5}$ known plaintext-ciphertext pairs (i.e. with 30 % less pairs), half of the experiments have a complexity less than 2^{42} DES evaluations.
- With only 2^{40} pairs at disposal, the complexity is far lower than an exhaustive search.

Even if we have to take these experimental results carefully because of the relative small number of statistical samples, they suggest strongly a lower complexity than expected by Matsui in [7] and we risk the following conjecture:

Proposition 1. *Given 2^{43} known plaintext-ciphertext pairs, it is possible to recover a DES key using Matsui's linear cryptanalysis within a complexity of 2^{41} DES evaluations with a success probability of 85 %.*

7   Conclusion

The first goal of this research was to perform an experimental linear cryptanalysis of DES as many times as possible in order to get a better insight into the real complexity and success probability of this attack. Using a very fast DES function developed for the Intel MMX architecture, we have simulated Matsui's attack 21 times.

Our experimental results suggest a lower complexity than estimated by Matsui. Given 2^{43} known plaintext-ciphertext pairs, the complexity was upper-bounded by 2^{41} DES evaluations with a success probability of 85 %. This has to be compared with the estimated 2^{43}.

We give furthermore a detailed theoretical analysis of the rank probability of the right subkey in the list of candidates, confirming Matsui's experimental results, and we discuss the validity of our theoretical model towards the experimental results, together with several issues regarding past research.

Acknowledgments

The author would like to thank Serge Vaudenay for having suggested this research topic and the SAC'01 anonymous referees for many helpful comments.

References

1. E. Biham, *A fast new DES implementation in software*, FSE '97, LNCS, vol. 1267, Springer-Verlag, 1997, pp. 260–272.
2. U. Blöcher and M. Dichtl, *Problems with the linear cryptanalysis of DES using more than one active S-box per round*, FSE '94, LNCS, vol. 1008, Springer-Verlag, 1995, pp. 265–274.
3. C. Harpes, G. Kramer, and J.L. Massey, *A generalization of linear cryptanalysis and the applicability of Matsui's piling-up lemma*, Advances in Cryptology - EuroCrypt '95, LNCS, vol. 921, Springer-Verlag, 1995, pp. 24–38.
4. Z. Kukorelly, *The piling-up lemma and dependent random variables*, Cryptography and coding: 7th IMA conference, LNCS, vol. 1746, Springer-Verlag, 1999.
5. M. Kwan, *Reducing the gate count of bitslice DES*, http://eprint.iacr.org/2000/051.ps, 2000.
6. M. Matsui, *Linear cryptanalysis method for DES cipher*, Advances in Cryptology - EuroCrypt '93, LNCS, vol. 765, Springer-Verlag, 1993, pp. 386–397.
7. _____, *The first experimental cryptanalysis of the Data Encryption Standard*, Advances in Cryptology - Crypto '94, LNCS, vol. 839, Springer-Verlag, 1994, pp. 1–11.
8. L. May, L. Penna, and A. Clark, *An implementation of bitsliced DES on the pentium MMX ^{TM} processor*, Information Security and Privacy: 5th Australasian Conference, ACISP 2000, LNCS, vol. 1841, Springer-Verlag, 2000.
9. K. Nyberg, *Linear approximation of block ciphers*, Advances in Cryptology - EuroCrypt '94, LNCS, vol. 950, Springer-Verlag, 1995, pp. 439–444.
10. National Bureau of Standards, *Data encryption standard*, U. S. Department of Commerce, 1977.
11. A. Rényi, *Probability theory*, Elsevier, 1970.
12. T. Shimoyama and T. Kaneko, *Quadratic relation of s-box and its application to the linear attack of full round DES*, Advances in Cryptology - Crypto '98, LNCS, vol. 1462, Springer-Verlag, 1998, pp. 200–211.
13. S. Vaudenay, *An experiment on DES statistical cryptanalysis*, 3rd ACM Conference on Computer and Communications Security, ACM Press, 1996, pp. 139–147.

A   Proof of Theorem 1

As a first step, let's consider the following situation: let $W_1, W_2, \ldots, W_n$ be n independent and identically distributed continuous random variables having f_W as density function and F_W as distribution function. We arrange the values of $W_1, W_2, \ldots, W_n$ in strictly[5] increasing order and denote them by $W_{(1)} < W_{(2)} < \ldots < W_{(n)}$. The distribution function $F_{W_{(i)}}$ of $W_{(i)}$ is given by the following Lemma whose proof can be found in [11].

Lemma 1. *Under previous assumptions, the distribution function of the i-th smallest random variable is*

$$F_{W_{(i)}}(x) = B_{i,n-i+1}\left(F\left(x\right)\right)$$

---

[5] The probability that equal values occur is 0.

where

$$B_{a,b}(x) = \frac{\Gamma(a+b)}{\Gamma(a)\Gamma(b)} \int_0^x t^{a-1}(1-t)^{b-1}\, dt$$

is the incomplete beta function of order (a, b).

By using the previous Lemma and the independence between the involved random variables, we can compute $F_\Psi(x)$ as follows:

$$\Pr\left[\Psi \le \psi\right] = \Pr\left[W_{(\psi)} < R\right]$$

$$= \int_{-\infty}^{+\infty} \int_{-\infty}^{y} f_{W_{(\psi)}}(x) f_R(y)\, dx\, dy$$

$$= \int_{-\infty}^{+\infty} B_{n+1-\psi,\psi}\left(F_W(y)\right) f_R(y)\, dy$$

By definition, we have

$$\mathrm{E}\left[\Psi\right] = \sum_{\psi=1}^{n+1} \psi \cdot \Pr\left[\Psi = \psi\right]$$

$$= \Pr\left[\Psi = 1\right] + \sum_{\psi=2}^{n+1} \psi \left(\Pr\left[\Psi \le \psi\right] - \Pr\left[\Psi \le \psi-1\right]\right)$$

$$= n+1 - \sum_{\psi=1}^{n} \Pr\left[\Psi \le \psi\right]$$

where

$$\sum_{\psi=1}^{n} \Pr\left[\Psi \le \psi\right] = \sum_{\psi=1}^{n} \int_{-\infty}^{+\infty} B_{n+1-\psi,\psi}\left(F_W(y)\right) f_R(y)\, dy$$

$$= \int_{-\infty}^{+\infty} f_R(y) \sum_{\psi=1}^{n} B_{n+1-\psi,\psi}\left(F_W(y)\right)\, dy$$

It is easy to see that

$$\sum_{\psi=1}^{n} B_{n+1-\psi,\psi}\left(F_W(y)\right) = n \int_0^{F_W(y)} \sum_{i=0}^{n-1} \binom{n-1}{i} t^i (1-t)^{n-1-i}\, dt$$

$$= n \int_0^{F_W(y)} dt$$

$$= n F_W(y)$$

and we can thus conclude with

$$E\left[\Psi\right] = n + 1 - \int_{-\infty}^{+\infty} f_R(y) \sum_{\psi=1}^{n} B_{n+1-\psi,\psi}\left(F_W(y)\right) dy$$

$$= n + 1 - n \int_{-\infty}^{+\infty} f_R(y) F_W(y)\, dy$$

$$= 1 + n \left(1 - \int_{-\infty}^{+\infty} f_R(y) F_W(y)\, dy\right)$$

B   Complete Experimental Results

This table gives the experimental results regarding the complexity $\hat{C}$ of each run of the attack for various amounts of plaintext-ciphertext pairs.

Exp	$N = 2^{43}$	$N = 2^{42.5}$	$N = 2^{42}$	$N = 2^{41}$	$N = 2^{40}$
1	39.1836	38.4818	45.0307	51.3802	51.0533
2	33.2479	41.6346	43.6383	48.0928	43.1913
3	38.6055	41.8023	43.9622	48.5492	51.6012
4	38.1267	34.6147	41.3351	48.7240	51.2041
5	37.4878	29.0000	36.5157	46.1991	52.3685
6	34.0444	44.2753	46.6834	48.5221	50.1937
7	36.4676	45.5732	44.2949	47.3010	51.2913
8	36.1189	44.7722	41.4091	51.6338	52.1143
9	40.3515	47.0565	48.6184	52.1953	53.1000
10	41.6540	41.8682	45.7429	47.9120	41.9750
11	45.4059	51.2973	51.9932	51.8155	52.1972
12	36.1189	43.6633	46.7256	50.3949	49.2317
13	36.4009	36.1189	43.2183	47.0756	46.7680
14	39.0042	42.6736	44.3057	44.7116	47.3256
15	37.6330	39.8572	47.6536	49.5244	52.6439
16	38.9204	36.6653	41.5447	49.1082	49.9939
17	33.5236	38.8502	43.3128	46.1030	48.6798
18	39.8478	47.4938	52.3671	50.6770	50.3675
19	32.1699	31.8074	40.5093	43.8552	48.4968
20	40.7503	38.3729	40.3734	45.2436	52.3101
21	41.8721	44.9063	45.4147	52.0730	52.8571

Random Walks Revisited: Extensions of Pollard's Rho Algorithm for Computing Multiple Discrete Logarithms

Fabian Kuhn[1] and René Struik[2]

[1] Departement Informatik, ETH Zentrum,
CH-8092 Zürich, Switzerland,
fkuhn@iiic.ethz.ch
[2] Certicom Research, 5520 Explorer Drive, 4th Floor,
Mississauga, Ontario, Canada L4W 5L1
rstruik@certicom.com

Abstract. This paper extends the analysis of Pollard's rho algorithm for solving a single instance of the discrete logarithm problem in a finite cyclic group G to the case of solving more than one instance of the discrete logarithm problem in the same group G. We analyze Pollard's rho algorithm when used to iteratively solve *all* the instances. We also analyze the situation when the goal is to solve *any one* of the multiple instances using any DLP algorithm.

1 Introduction

The security of many public-key cryptographic systems is based on the discrete logarithm problem (DLP). Examples are the Diffie-Hellman key agreement protocol and the ElGamal encryption and signature schemes.

The DLP can be defined as follows: Let g be a generator of a finite cyclic group $G = \langle g \rangle$ of order N. For the general DLP, we have to find an integer x ($0 \leq x < N$) such that $g^x = h$, where h is chosen uniformly at random from G (written $h \in_R G$). The integer x is called the discrete logarithm of h to the base g, denoted $\log_g h$. If N is composite, one can compute $x \bmod p^k$ in the subgroup of order p^k for each prime power p^k dividing N. Then, one can compute x by application of the Chinese Remainder Theorem. Further, calculating the discrete logarithm in the subgroup of order p^k can be reduced to finding the discrete logarithm in the group of prime order p (see [7]). For these reasons, we only consider the DLP in groups of prime order N.

Shoup [10] gave a lower bound for the running time for computing discrete logarithms by generic algorithms (probabilistic or deterministic) in groups of prime order. The time needed to solve the DLP with a non-negligible probability is $c\sqrt{N}$ group operations for some constant c. The best algorithm known for solving the general DLP is Pollard's rho algorithm [8]. It does not only match Shoup's lower bound, but also needs very little memory and is parallelizable with a linear speed-up (see [6]). For many groups of cryptographic interest, such

S. Vaudenay and A. Youssef (Eds.): SAC 2001, LNCS 2259, pp. 212–229, 2001.

as the multiplicative group of a finite field (see [1]), and the Jacobians of hyperelliptic curves of high genus (see [2]), there are subexponential-time algorithms known for the DLP that are more efficient than Pollard's rho algorithm. However, Pollard's rho algorithm is the best algorithm known for solving the DLP in some groups such as the group of points on an elliptic curve, and the Jacobian of genus 2 and 3 hyperelliptic curves. Thus, the results in this paper are particularly relevant to the DLP in elliptic curve groups and in genus 2 and 3 hyperelliptic curves.

This paper extends the analysis of Pollard's rho algorithm for solving a single instance of the discrete logarithm problem in a finite cyclic group G to the case of solving more than one instance of the discrete logarithm problem in the same group G. Pollard's rho algorithm is reviewed in §2. In §3, we provide a runtime analysis in an idealized model and do an exact analysis of possible time-memory trade-offs for the parallelized version. When using Pollard's rho algorithm to iteratively solve *all* n instances of the DLP in the same group, the data that is gathered during the calculation of a single discrete logarithms can be used to compute subsequent discrete logarithms. Thus, the additional time needed for every new DLP may be smaller than the time needed to solve the one before. A careful analysis for this case is provided in §4. In §5 we consider the case where the goal is to solve *any one* of a set of n DLPs in the same group using any DLP algorithm.

2   Pollard's Rho Algorithm

2.1   Basic Idea

Pollard's rho algorithm is based on the birthday paradox. If we randomly choose elements (with replacement) from a set of N numbered elements, we only need to choose about $\sqrt{N}$ elements until we get one element twice (called a *collision*). This can be applied to find discrete logarithms as follows. By choosing $a, b \in_R [0, N-1]$, one obtains a random group element $g^a h^b$. Such group elements are randomly selected until we get a group element twice. If $g^{a_i} h^{b_i}$ and $g^{a_j} h^{b_j}$ represent the same group element then $a_i + b_i x \equiv a_j + b_j x \pmod{N}$, whence

$$x = (a_j - a_i)(b_i - b_j)^{-1} \bmod N \quad \text{for } b_i \not\equiv b_j \pmod{N}. \tag{1}$$

Let T be the random variable describing the number of group elements chosen until the first collision occurs. We denote the probability that $T > k$ by p_k. We have

$$p_k = 1 \left(1 - \frac{1}{N}\right)\left(1 - \frac{2}{N}\right)\cdots\left(1 - \frac{k-1}{N}\right) \approx \left(1 - \frac{k-1}{2N}\right)^k \approx e^{-\frac{k^2}{2N}}. \tag{2}$$

For $k \in O(\sqrt{N})$, the relative error of the above approximation is $O(N^{-1/2})$. As shown in Appendix B, the expected value of T is $E(T) \approx \sqrt{\pi N/2}$. The first collision can be found by simply storing all the randomly selected group elements until a repeat is detected. However, this simple-minded method has an expected storage requirement of $\sqrt{\pi N/2}$ group elements.

2.2   The Single Processor Case

The question now is how to detect a collision without having to store $\sqrt{\pi N/2}$ group elements. In Pollard's rho algorithm, this is done by means of a random function[1] $f : G \rightarrow G$. For actual implementations, f is chosen such that it approximates a random function as closely as possible. Further, it should be calculated with a single group multiplication and map an element $g^a h^b$ to an element $g^c h^d$ so that c and d can easily be computed from a and b. The originally suggested function by Pollard (for $\mathbb{Z}_p^*$) can be generalized towards arbitrary cyclic groups as

$$f(x) = \begin{cases} hx & \text{if } x \in \mathcal{S}_1; \\ x^2 & \text{if } x \in \mathcal{S}_2; \\ gx & \text{if } x \in \mathcal{S}_3. \end{cases}$$

Here, $\mathcal{S}_1, \mathcal{S}_2$ and $\mathcal{S}_3$ are three sets of roughly the same size which form a partition of G. In [12,13], Teske shows that this function is not random enough and gives a better function:

$$f(x) = x \cdot g^{m_s} h^{n_s}, \text{ if } x \in \mathcal{M}_s \quad \text{for } s \in \{1, \dots, r\} \text{ and } r \approx 20.$$

Here again, the $\mathcal{M}_s$ are of roughly the same size and form a partition of G. But this time, G is partitioned into more than three subsets. For both functions, it is of course necessary that determining the subset $\mathcal{M}_i$, resp. $\mathcal{S}_i$, to which a group element belongs is very efficient.

By starting at a random point $g^{a_0} h^{b_0}$ and iteratively applying a random function, random points $g^{a_i} h^{b_i}$ are generated. Because the group is finite, we eventually arrive at a point for the second time. The sequence of subsequent points then cycle forever. From §2.1 we know that the first repeat happens after an expected $E(T) \approx \sqrt{\pi N/2}$ function applications. With very little time and space overhead, it is possible to detect such a cycle with Floyd's cycle-finding algorithm (or with an improved variant by Brent [3]).

2.3   Parallelization of Pollard's Rho Algorithm

Unfortunately, iteratively applying a function is an inherently serial process and cannot efficiently be parallelized. If m processors run the Pollard-rho algorithm as described above, the speed-up when compared to the single processor case is only about $\sqrt{m}$. For, if the processors run the algorithm individually, the probability that none of them has found a collision after k steps is $p_k^m \approx e^{-k^2 m/(2N)}$. This leads to an expected time of $(\sqrt{\pi N/2})/\sqrt{m}$ for finding the first collision.

---

[1] A random function is a function that is chosen uniformly at random from the set of all functions $f : G \rightarrow G$.

If, however, the processors communicate with each other, we can do better. If we could detect and use any collision which occurs between two processors, the speed-up to the single processor case would be a factor m because m processors calculate m times as many points as a single processor does.

In [6], Wiener and van Oorschot presented a very elegant way of parallelizing Pollard's rho alorithm which is based on distinguished points. A distinguished point is a group element with an easy testable property. An often used distinguishing property is whether a point's binary representation has a certain number of leading zeros. Each processor starts the iteration at a different random element (but all have the same iteration function). As soon as the iteration hits a distinguished point, this point will be sent to a central server and the processor starts a new iteration. The server stores all collected points (a_i, b_i and $y_i = g^{a_i} h^{b_i}$) in a hash table. As soon as the server has received the same point twice, it has two representations $g^{a_i} h^{b_i}$ and $g^{a_j} h^{b_j}$ for a group element and can calculate the discrete logarithm x of h as given in (1).

As soon as a point occurs in two iterations, the remainder of those two iteration trails will be the same and thus lead to the same distinguished point. Therefore, by performing the iterations, all processors calculate random group elements of the form $g^a h^b$ and as soon as the same element has been calculated twice, we are going to get the same distinguished point twice, as well. If the two representations of the point, where the trails collided, are different, the representations of this distinguished point are different too, and we are therefore able to calculate x.

3   Analysis of the Parallelized Pollard's Rho Algorithm

For our analysis, we make the following assumptions (cf. §3.3):

1. The iterative function really behaves like a random mapping and thus generates uniformly distributed random group elements.
2. All collisions are useful, i.e. the collision reveals two representations $g^{a_i} h^{b_i}$ and $g^{a_j} h^{b_j}$ of a group element with $b_i \not\equiv b_j \pmod{N}$.
3. All trails lead to distinguished points (i.e., we neglect the existence of iteration paths which eventually run into a cycle that does not contain a distinguished point).

We denote the number of processors by m. The proportion of the points that constitute distinguished points is called θ (i.e., there are θN distinguished points). Additionally, for the analysis, we assume that all processors operate at the same speed.

3.1   Running Time

The runtime of the Pollard-rho algorithm can be divided into two statistically independent phases. First, all processors have to calculate points until a collision occurs. We already know that an expected $\sqrt{\pi N/2}$ points must be calculated for this part of the algorithm. Because all m processors calculate their points independently, the expected time for this part is $\sqrt{\pi N/2}/m$ function iterations.

After a collision, the iteration has to be continued until it arrives at a distinguished point. Because for each function application, the probability to come to a distinguished point is θ, the number of steps from a collision to its detection is a geometrically distributed random variable with expected value $1/\theta$.

The iteration function is such that the time for one function application is equal to the time of one group operation plus a negligible overhead. Thus, the overall expected value for the running time of the parallel Pollard-rho algorithm is $E(T) = (\sqrt{\pi N/2})/m + 1/\theta$ group operations.

3.2   Memory Requirements

Essentially, the only memory needed for the parallel version of Pollard's rho algorithm is that for storing the distinguished points on the server[2]. For every iteration, the server has to store one distinguished point. The length of a trail portion between distinguished points is geometrically distributed with parameter θ. Therefore, the expected length of such an iteration trail is $\frac{1}{\theta}$. This means that for the whole duration of the algorithm, all processors will send a distinguished point to the server every $\frac{1}{\theta}$ group operations on average. Therefore, because the time until a collision occurs and the average length of the trails are assumed to be statistically independent, the expected space needed on the server is $E(S) = m\theta E(T) = \theta\sqrt{\pi N/2} + m$ distinguished points. Note that for each distinguished point, we have to store the group element $g^a h^b$ and the integers a and b. Therefore, the actual space needed to store one distinguished point is $O(\log N)$ bits.

For the memory requirements to be as small as possible, we have to choose θ as small as possible. But of course, if θ gets smaller, the time overhead $\frac{1}{\theta}$ to detect a collision gets bigger. In order to keep the overall running time in $O(\sqrt{N}/m)$, we have to choose θ in $O(m/\sqrt{N})$. Therefore, we choose θ as $\theta = \alpha m/(\sqrt{\pi N/2})$. The expected values for time and space then become $E(T) = (1 + \frac{1}{\alpha})\sqrt{\pi N/2}/m$ and $E(S) = m(1 + \alpha)$. We see that there is a time-space trade-off. But even if we choose the constant α quite big, the space requirements are still small. Therefore, the limiting factor for solving discrete logarithms with the parallel rho algorithm is definitely time.

Remark 1. We have assumed that all distinguished points are collected by a single server. However, it is possible to parallelize the server side with no communication overhead. Assume that k servers collect the distinguished points. One could split up the distinguished point set $\mathcal{D}$ into k disjoint subsets $\mathcal{D}_i$ of roughly the same size. Server i would then only collect the points of $\mathcal{D}_i$. When a client gets a distinguished point, it would have to check to which subset $\mathcal{D}_i$ it belongs and send it to the appropriate server. Checking if a new distinguished

---

[2] All clients also need to store a description of the iteration function. This, however, requires only $O(\log N)$ bits per client.

point has already been computed previously can be done independently on each server.

3.3 Assumptions of the Analysis

At the beginning of §3, we made three assumptions on which we based our time and space analysis. We will now elaborate on how realistic these assumptions are in an actual implementation.

Randomness of the function: For our analysis, we assumed that the iteration function is perfectly random and therefore produces uniformly distributed group elements. In [12,13], Teske shows that the function suggested by her behaves practically like a truly random function if the group elements are partitioned into about 20 subsets.

All collisions are useful: A collision reveals two representations of the form $g^{a_i}h^{b_i}$ and $g^{a_j}h^{b_j}$ of the same group element. If $b_i \not\equiv b_j \pmod{N}$, the collision can be used to calculate x. Because the b_i are random elements of $\mathbb{Z}_N$, the probability for this is $1 - \frac{1}{N}$. Therefore, the probability that a collision is not useful is $\frac{1}{N}$ and thus negligible.

Each iteration reaches a distinguished point: In [9], Schulte-Geers shows that the distinguished point set must be at least of size $c\sqrt{N}$ while c should not be too small. This is intuitively clear, since the only way for an iteration not to arrive at a distinguished point is to end up in a cycle without distinguished points, the expected length of which is $\sqrt{\pi N/8}$. The condition is certainly met by our distinguished point set (c is $\alpha m\sqrt{2/\pi}$ in our case). Schulte-Geers also finds that if we choose θ as described in §3.2, the proportion of starting points with iterations that end up in distinguished points is $1/(1 + \frac{\pi \mathcal{N}(0,1)^2}{2\alpha^2 m^2})$, where $\mathcal{N}(0,1)$ is a standard normally distributed random variable.

Further, Schulte-Geers shows that if $\theta \gg 1/\sqrt{N}$, only a negligible number of starting points will miss the distinguished point set. We could meet this requirement by setting α to $O(\log N)$. The space requirements still remain very small.

Additionally, van Oorschot and Wiener [6] suggest to abandon all trail portions without a single distinguished point that are longer than k/θ, k times their expected lengths. The proportion of time wasted through abandoned trails can be estimated[3] as $k(1 - \theta)^{k/\theta} \approx ke^{-k}$ which is very small.

3.4 Statistical Analysis

Until now, we have only considered expected values for time and space. We will now have a look at the probability distributions of these.

---

[3] Here, we assume the length of the trail portions between subsequent distinguished points to be geometrically distributed. Note that this model is slightly inaccurate since it implies that all such trail portions eventually lead to a distinguished point. For reasonably chosen values of θ, the model will do, however, since the probability of ending up in cycles without distinguished points is, indeed, very small.

As already explained, the time for finding a discrete logarithm with parallel Pollard-rho can be divided in two phases, the time until a collision occurs and the time needed for its detection. We will first treat those phases individually. As seen in §2.1, the probability that more than l points are needed for a collision is $p_l \approx e^{-l^2/2N}$. Because in time k, mk points are calculated, the probability that the time T_1 for the first phase is longer than k is $\Pr\{T_1 > k\} = p_{mk} \approx e^{-\frac{(mk)^2}{2N}}$. Because the time T_2 for the second phase of detecting a collision is geometrically distributed, the probability that $T_2 > k$ is $\Pr\{T_2 > k\} = (1-\theta)^k$. Therefore, the probabilities that T_1, resp. T_2 are bigger than β times their expected values is:

$$\Pr\{T_1 > (\beta\sqrt{\pi N/2})/m\} \approx e^{-\beta^2\pi/4} \quad \text{and} \quad \Pr\{T_2 > \beta/\theta\} = (1-\theta)^{\beta/\theta} \approx e^{-\beta}. \tag{3}$$

For the probability for T_2, given in (3), note that θ is very small and that $\lim_{x\downarrow 0}(1-x)^{\beta/x} = e^{-\beta}$.

We want to avoid having to calculate exact probabilities for the overall time $T = T_1 + T_2$. Therefore, we assume that α in §3.2 is chosen sufficiently large to achieve a good running time. In this case, T_1 dominates the time T and we can approximate the probability that $T > \beta E(T)$ with $\Pr\{T_1 > \beta E(T_1)\}$. Taking Equation (3), we then get:

$$\Pr\{T > \beta E(T)\} \approx e^{-\beta^2\pi/4}. \tag{4}$$

Table 1 gives samples of the probabilities for various values of β.

Table 1. Probabilities for the running time of Pollard's rho algorithm.

β	1/100	1/10	1/3	1/2	1	3/2	2	3
$\Pr\{T > \beta E(T)\}$	1.000	0.992	0.916	0.822	0.456	0.171	0.043	0.001

For space, exactly the same analysis holds. In fact, the space needed is very close to $m\theta T$ where T is the actual running time. This is because the length of every iteration trail is geometrically distributed with parameter θ and the lengths of different trails are statistically independent. By application of the limit theorem, we get that the average length of the trails is very close to the expected length.

4   Solving Multiple Instances of the DLP

In this section, we consider the situation where one wants to solve multiple, say L, discrete logarithms in the same group (using the same generator). Hence, we have a set of L group elements $h_i = g^{x_i}$ (where $1 \leq i \leq L$) and we would like to find all exponents x_i. This can be done by solving each of the discrete logarithms individually, using the rho algorithm. A better approach, however, is to take advantage of the distinguished points gathered during the solution of the first k discrete logarithm problems using the rho algorithm, to speed up the solution of the $(k+1)^{st}$ discrete logarithm. As soon as we find a discrete logarithm $x_i = \log_g h_i$, we have a representation of the form g^c for all distinguished points $g^{a_j} h_i^{b_j}$ that were calculated in order to find x_i. The value of c is $c = (a_j + x_i b_j) \bmod N$. If we now find a collision between a distinguished point g^c and a new one of the form $g^a h_k^b$, we can calculate x_k as $x_k = (c - a)b^{-1} \bmod N$. This method was also suggested by Silverman and Stapleton [11], although a precise analysis has not been published. It seems obvious that the number of operations required for solving each new logarithm will become smaller, if one takes advantage of information gathered during previous computations. In this section, we will provide an exact analysis for this case.

The number of points we have to calculate with the rho algorithm to find L discrete logarithms is equal to the number of points we have to choose with replacement out of a set with N numbers until we have chosen L numbers at least twice (i.e. there are L collisions). We denote the expected value for the number of draws W to find L collisions by $E(W) = E_L$.

Theorem 1. We have $E_L \approx \sqrt{\pi N/2} \sum_{t=0}^{L-1} \frac{\binom{2t}{t}}{4^t}$ for $L \ll \sqrt[4]{N}$.

Proof: Suppose that an urn has N differently numbered balls. We consider an experiment where one uniformly draws n balls from this urn one at a time, with replacement, and lists the numbers. It is clear that if one obtains $k < n$ different numbers after n draws, then $n - k$ balls must have been drawn more than once (counting multiplicity), i.e., $n - k$ 'collisions' must have occurred. We will be mainly interested in the probability distribution of the number of collisions as a function of the number of draws.

Let $q_{n,k}$ denote the probability that one obtains exactly k differently numbered outcomes after n draws. For any fixed k-set, the number of ways to choose precisely k differently numbered balls in n draws equals $a(n, k)$, the number of surjections from an n-set to a k-set. Hence, the number of possibilities to choose exactly k different balls in n draws equals $\binom{N}{k} a(n, k)$ and, therefore,

$$q_{n,k} = \binom{N}{k} a(n,k)/N^n = \frac{a(n,k)}{k! N^{n-k}} \frac{N(N-1)\cdots(N-k+1)}{N \cdot N \cdots N} = \frac{S(n,k)}{N^{n-k}} p_k,$$

where $p_k = (1 - 1/N)(1 - 2/N) \cdots (1 - (k-1)/N)$ is the probability of drawing k differently numbered balls in k draws, and where $S(n, k) := a(n, k)/k!$ is a Stirling number of the second type (cf. Appendix A).

We now compute the expected number E_L of draws until one obtains precisely L collisions. Let $Q_{n,n-L}^+$ denote the probability that one requires more than n draws in order to obtain L collisions. Hence

$$Q_{n,n-L}^+ = \sum_{t=0}^{L-1} q_{n,n-t}. \tag{5}$$

Now, the probability that one needs exactly n draws in order to obtain L collisions is given by $Q^+_{n-1,n-1-L} - Q^+_{n,n-L}$. As a result, the expected number of draws that one needs in order to obtain L collisions is given by

$$E_L = \sum_{n=L}^{\infty} n(Q^+_{n-1,n-1-L} - Q^+_{n,n-L}) = (L-1) + \sum_{n=L-1}^{\infty} Q^+_{n,n-L}.$$

From equation (5) we infer that $Q^+_{n,n-(L+1)} = Q^+_{n,n-L} + q_{n,n-L}$, hence one obtains

$$E_{L+1} - E_L = 1 + \sum_{n=L}^{\infty} Q^+_{n,n-(L+1)} - \sum_{n=L-1}^{\infty} Q^+_{n,n-L} \tag{6}$$

$$= \sum_{n=L}^{\infty} q_{n,n-L} = \sum_{k=0}^{\infty} \frac{S(k+L,k)}{N^L} p_k. \tag{7}$$

Obviously, one has $E_0 = 0$, hence we can compute E_L via

$$E_L = \sum_{t=0}^{L-1} \sum_{k=0}^{\infty} \frac{S(k+t,k)}{N^t} p_k. \tag{8}$$

We will now approximate E_L based upon an approximation for $E_{t+1} - E_t$ (for $t < L$). It will turn out that the relative error of our approximation is negligible if $L < c_N \sqrt[4]{N}$ (here $0 < c_N < 1$ is a small constant). We will use the fact that for any fixed value of L, the Stirling number $S(k+L,k)$ is a polynomial in k of degree $2L$. More specifically, one has (cf. Lemma 1 of Appendix A) that

$$S(k+L,k) = \frac{1}{2^L L!} \sum_{j=0}^{2L} \varphi_j(L) k^{2L-j}, \quad \text{where } \varphi_j(L) \in \mathbb{Q}[x] \text{ has degree at most } 2j.$$

A substitution in Equation (6) now yields

$$E_{L+1} - E_L = \frac{1}{2^L L!} \sum_{j=0}^{2L} \frac{\varphi_j(L)}{\sqrt{N}^j} \sum_{k=0}^{\infty} \left(\frac{k}{\sqrt{N}} \right)^{2L-j} p_k. \tag{9}$$

We will now approximate this expression, using approximations for p_k and the function $\varphi_j(L)$. The inner summation can be approximated, using the approximation $p_k \approx e^{-k^2/2N}$ and a Riemann integral. We have

$$\sum_{k=0}^{\infty} \left(\frac{k}{\sqrt{N}} \right)^{2L-j} p_k \approx \sum_{k=0}^{\infty} \left(\frac{k}{\sqrt{N}} \right)^{2L-j} e^{-k^2/2N} \approx \sqrt{N} \int_{x=0}^{\infty} x^{2L-j} e^{-x^2/2} dx = \sqrt{N} I_{2L-j},$$

where I_t is the value of the integral determined in Lemma 2.[4] Substitution of this approximation in Equation (9) now yields

$$E_{L+1} - E_L \approx \sqrt{N} \frac{1}{2^L L!} \sum_{j=0}^{2L} \frac{\varphi_j(L)}{\sqrt{N}^j} I_{2L-j}$$

$$= \sqrt{N} \frac{1}{2^L L!} \left(\frac{(2L)!}{L! 2^L} \sqrt{\pi/2} + \sum_{j=1}^{2L} \frac{\varphi_j(L)}{\sqrt{N}^j} I_{2L-j} \right)$$

$$= \sqrt{\pi N/2} \frac{\binom{2L}{L}}{4^L} (1 + o(1)) \approx \sqrt{\pi N/2} \frac{\binom{2L}{L}}{4^L}.$$

The latter approximation follows from the fact that $\varphi_j(L) = 1$ and that for $j > 0$, $\varphi_j(L)$ is a polynomial in L of degree at most $2j$ without a constant term and, hence, $\varphi_j(L)/(\sqrt{N})^j \approx 0$ if $L \ll \sqrt[4]{N}$. Substituting this approximation in Equation (8), we now find that

$$E_L \approx \sum_{t=0}^{L-1} \sqrt{\pi N/2} \frac{\binom{2t}{t}}{4^t} = \sqrt{\pi N/2} \sum_{t=0}^{L-1} \frac{\binom{2t}{t}}{4^t}$$

$$= \sqrt{\pi N/2}(2L - 1) \frac{\binom{2L-2}{L-1}}{4^{L-1}} \approx (2/\sqrt{\pi})\sqrt{L}\sqrt{\pi N/2} = \sqrt{2LN}.$$

$\square$

Remark 2. The above result gives a highly accurate estimate of the expected time required to solve multiple instances of the discrete logarithm problem in the same underlying group. Unfortunately, this does not give direct insight in the probability distribution hereof. We should mention, however, that the same techniques used above to estimate expected values can also be used to estimate the vaiance of the probability distribution. It turns out that the variance, when compared to the expected time, is relatively low, especially if the number L of discrete logarithm problems one considers is not too small. Thus, the expected value of the running time of Theorem 1 is a good approximation of practically observed values (for L not too small). Full details will be provided in the full paper, space permitting.

We can conclude from Theorem 1 that computing discrete logarithms iteratively, rather than independently, is advantageous, since the workload involved in computing the $t + 1^{\text{st}}$ discrete logarithm, once the first t of these have been solved, now becomes only $4^{-t}\binom{2t}{t} \approx 1/\sqrt{\pi t}$ times as much as the workload $\sqrt{\pi N/2}$ required for computing a single discrete logarithm. Thus, we arrived at a total workload for computing L discrete logarithms iteratively of approximately $\sqrt{2NL}$ group operations, which is $(2/\sqrt{\pi}) \cdot \sqrt{L} \approx 1.128\sqrt{L}$ times as much

---

[4] It turns out that the relative error of this approximation is $O(() \log(N)/\sqrt{N})$. For details, cf. Lemma 4 and its subsequent remark.

as the workload for computing a single discrete logarithm. Thus, economies of scale apply: computing L discrete logarithms iteratively comes at an average cost per discrete logarithm of roughly $\sqrt{2N/L}$ group operations, rather than of approximately $\sqrt{\pi N/2}$ group operations (as is the case when computing discrete logarithms independently). Our results hold for $0 < L < c_N \sqrt[4]{N}$, where $0 < c_N < 1$ is some small constant.

Our extension of Pollard's rho algorithm is a generic algorithm for solving multiple instances of the discrete logarithm problem in finite cyclic groups. The low average workload [5] we obtained for computing multiple discrete logarithms seems to be counter-intuitive, since it seems to contradict Shoup's result [10], which gives a lower bound of $\Omega(\sqrt{N})$ group operations required by generic algorithms solving the discrete logarithm problem in groups of prime order N. The result is explained by observing that the low average workload is due to the fact that solving subsequent discrete logarithm problems requires relatively few operations, once the first few discrete logarithms have been computed. Thus, the bottleneck remains the computation of, e.g., the first discrete logarithm, which in our case requires roughly $\sqrt{\pi N/2} = \Omega(\sqrt{N})$ group operations. It should be noted, that Shoup's result does not apply directly, since he addresses the scenario of a single instance of the discrete logarithm problem, rather than that of multiple instances hereof, which we address. Thus, one cannot a priori rule out the existence of other generic algorithms that, given L instances of the discrete logarithm problem in a group of prime order N, solve an arbitrary one of these using only $O(\sqrt{N/L})$ group operations.

5   On the Complexity of DLP-like Problems

In the previous sections, we discussed the workload required for solving multiple instances of the discrete logarithm problem with respect to a fixed generator of a finite cyclic group of order N, using extensions of Pollard's rho algorithm. We found that computing discrete logarithms iteratively, rather than independently, is advantageous. In §5.1 we will consider the problem of solving 1 out of n instances of the discrete logarithm problem and several other relaxations of the classical discrete logarithm problem (DLP) and consider the computational complexity hereof. It turns out that these problems are all computationally as hard as DLP. In particular, it follows that generic algorithms for solving each of these relaxations of the discrete logarithm problem in a prime order group require $\Omega(\sqrt{N})$ group operations. In §5.2 we consider the generalization of the classical discrete logarithm problem of solving k instances hereof (coined kDLP). Again, we consider several relaxations of this so-called kDLP and discuss their computational complexity. It turns out that, similar to the case $k = 1$, these problems are all computationally as hard as solving kDLP. We end the section with a conjectured lower bound $\Omega(\sqrt{kN})$ on the complexity of generic algo-

---

[5] The average workload per discrete logarithm is $O(N^{3/8})$ group operations if one solves $L \approx c_N \sqrt[4]{N}$ discrete logarithm problems iteratively.

rithms for solving kDLP, which – if true – would generalize Shoup's result for DLP towards kDLP.

5.1   Complexity of Solving 1 of Multiple Instances of the DLP

We consider the following variations of the discrete logarithm problem:

1. (DLP-1) Solving a single instance of the discrete logarithm problem:

    System: Cyclic group G; generator g for G.
    Input: Group element $h \in_R G$.
    Output: Integer x such that $h = g^x$.

2. (DLP-2) Solving a single instance of the discrete logarithm problem (selected arbitrarily from a set of n instances of the discrete logarithm problem):

    System: Cyclic group G; generator g for G.
    Input: Group elements $h_1, \ldots, h_n \in_R G$.
    Output: Pair (j, x_j) such that $h_j = g^{x_j}$ and such that $1 \le j \le n$.

3. (DLP-3) Finding a discrete logarithm with respect to an arbitrary basis element (selected from a set of m basis elements):

    System: Cyclic group G; arbitrary generators $g_1, \ldots, g_m$ for G.
    Input: Group element $h \in_R G$.
    Output: Pair (i, x) such that $h = g_i^x$ and such that $1 \le i \le m$.

4. (DLP-4) Finding a linear equation in terms of the discrete logarithms of all group elements of a set of n instances of the discrete logarithm problem:

    System: Cyclic group G; generator g for G.
    Input: Group elements $h_1, \ldots, h_n \in_R G$.
    Output: A linear equation $\sum_{j=1}^{n} a_j \log_g h_j = b$ (with known values of $a_1, \ldots, a_n$ and b).

5. (DLP-5) Finding the differences of two discrete logarithms (selected arbitrarily from of a set of n instances of the discrete logarithm problem):

    System: Cyclic group G; generator g for G.
    Input: Group elements $h_1, \ldots, h_n \in_R G$.
    Output: Triple $(i, j, \log_g h_i - \log_g h_j)$, where $0 \le i \neq j \le n$ and where $h_0 := g$.

The following theorem relates the expected workloads required by optimal algorithms for solving the discrete logarithm problem (DLP-1) and for solving arbitrarily 1 out n instances of the discrete logarithm problem (DLP-2).

Theorem 2. *Let T_{DLP}, resp. $T_{DLP(1:n)}$, be the expected workload of an optimal algorithm for solving the discrete logarithm problem, resp. for arbitrary solving 1 out of n instances of the discrete logarithm problem. Then, one has $T_{DLP(1:n)} \leq T_{DLP} \leq T_{DLP(1:n)} + n$ (in group operations).*

Proof: The inequality $T_{DLP} \leq T_{DLP(1:n)} + n$ follows from a reduction of an instance of the DLP to an instance of the DLP(1:n). The other inequality follows from the observation that DLP=DLP(1:1). Let $h := g^x$ be a problem instance of DLP. We will reduce this to a problem instance $h_1, \ldots, h_n$ of DLP(1:n) as follows: for all $i, 1 \leq i \leq n$, select the numbers r_i uniformly at random from the set $\{0, \ldots, N-1\}$ and define $h_i := g^{r_i} h = g^{x+r_i}$. Note that all h_i are random, since all $x + r_i$ are random and independent. Now apply an oracle that solves DLP(1:n), to produce an output (j, x_j), with $x_j := \log_g h_j$ and with $1 \leq j \leq n$. Since $h_j = g^{r_j} h$ and since r_j is known, we get the required discrete logarithm x as $x \equiv x_j - r_j (\text{mod } N)$. $\square$

Corollary 1. *The problem of solving arbitrarily 1 out of n instances of the discrete logarithm is computationally as hard as solving the discrete logarithm problem, provided $n \ll T_{DLP}$. Moreover, any generic algorithm that solves this problem in a group of prime order N requires at least $\Omega(\sqrt{N})$ group operations.*

Proof: The bound $T_{DLP(1:n)} = \Omega(T_{DLP})$ follows from Theorem 2 and the inequality $n \ll T_{DLP}$. The lower bound on the required workload for a generic algorithm that solves[6] the relaxed discrete logarithm problem DLP(1:n) in groups of prime order N follows from the corresponding result for the discrete logarithm problem [10]. $\square$

Remark 3. In fact, one can show that Theorem 2 and Corollary 1 easily generalize to each of the problems DLP-1, ..., DLP-5 above. In particular, one has that each of the problems DLP-1, ..., DLP-5 is as hard as solving the discrete logarithm problem, provided $n, m \ll T_{DLP}$. Moreover, any generic algorithm that solves any of the the problems DLP-1, ..., DLP-5 in a group of prime order N requires at least $\Omega(\sqrt{N})$ group operations.

Remark 4. One can show that each of the problems DLP-1, ..., DLP-5 can be solved directly using Pollard's rho algorithm, with starting points of the randomized walks that are tailored to the specific problem at hand. In each case, the resulting workload is roughly $\sqrt{\pi N/2}$ group operations.

5.2 Complexity of Solving Multiple Instances of the DLP

In the previous section, we related the workload involved in solving various relaxations of the classical discrete logarithm problem. The main result was that the problem of solving arbitrarily 1 out of n instances of the discrete logarithm is computationally as hard as solving a single instance of the discrete logarithm

---

[6] with a probability bounded away from zero

problem. In this section, we consider the similar problem where we are faced with solving k given instances of the discrete logarithm problem.

We consider the following variations of the discrete logarithm problem kDLP:

1. (kDLP-1) Solving k instances of the discrete logarithm problem:

   System: Cyclic group G; generator g for G.
   Input: Group elements $h_1, \ldots, h_k \in_R G$.
   Output: k pairs (i, x_i) such that $h_i = g^{x_i}$.

2. (kDLP-2) Solving k instances of the discrete logarithm problem (selected arbitrarily from a set of n instances of the discrete logarithm problem):

   System: Cyclic group G; generator g for G.
   Input: Group elements $h_1, \ldots, h_n \in_R G$.
   Output: k pairs (j, x_j) such that $h_j = g^{x_j}$, where $j \in J$ and where J is a k-subset of $\{1, \ldots, n\}$.

3. (kDLP-3) Finding k discrete logarithms with respect to k arbitrary basis elements (selected from a set of m basis elements):

   System: Cyclic group G; arbitrary generators $g_1, \ldots, g_m$ for G.
   Input: Group element $h \in_R G$.
   Output: k pairs (i, x_i) such that $h = g_i^{x_i}$, where $i \in I$ and where I is a k-subset of $\{1, \ldots, m\}$.

4. (kDLP-4) Finding k linear equations in terms of the discrete logarithms of all group elements of a set of n instances of the discrete logarithm problem:

   System: Cyclic group G; generator g for G.
   Input: Group elements $h_1, \ldots, h_n \in_R G$.
   Output: A set of k linear equations $\sum_{j=1}^{n} a_{ij} \log_g h_j = b_i$ (with known values of a_{ij} and b_i).

5. (kDLP-5) Finding k differences of two discrete logarithms (selected arbitrarily from of a set of n instances of the discrete logarithm problem):

   System: Cyclic group G; generator g for G.
   Input: Group elements $h_1, \ldots, h_n \in_R G$.
   Output: A set of k triples $(i, j, \log_g h_i - \log_g h_j)$, where $0 \le i \neq j \le n$ and where $h_0 := g$.

One can show that the results of the previous subsection carry over to this section, as follows:

- Each of the problems kDLP-1, ..., kDLP-5 is as hard as solving k instances of the discrete logarithm problem, provided $kn, km, k^2 \ll T_{DLP}$.
- Any generic algorithm for solving k instances of the discrete logarithm problem in a group of prime order N require at least $\Omega(\sqrt{N})$ group operations.

- Each of the problems kDLP-1, ..., kDLP-5 can be solved directly using the extension of Pollard's rho algorithm presented in §4, with starting points of the randomized walks that are tailored to the specific problem at hand. In each case, the resulting workload is roughly $\sqrt{2Nk}$ group operations.

The proofs use constructions based on maximum distance separable codes (cf., e.g., [5]). Details will be included in the full paper.

The lower bound on the required workload for a generic algorithm that solves k instances of the discrete logarithm problem is not very impressive: it derives directly from Shoup's lower bound $\Omega(\sqrt{N})$ for solving a single discrete logarithm (i.e., $k = 1$) . It would be of interest to find a stronger lower bound in this case. Based on the workload involved in computing k discrete logarithm problems iteratively, we postulate that the 'true' lower bound is $\Omega(\sqrt{kN})$. We suggest this as an open problem.

Research Problem. Show that any generic algorithm that solves, with a probability bounded away from zero, k instances of the discrete logarithm problem in groups of prime order N requires at least $\Omega(\sqrt{kN})$ group operations.

References

1. L. Adleman and J. De Marrais, A Subexponential Algorithm for Discrete Logarithms over All Finite Fields, in *Advances of Cryptology –CRYPTO'93*, D.R. Stinson, Ed., pp. 147-158, Lecture Notes in Computer Science, Vol. 773, Berlin: Springer, 1993.
2. L. Adleman, J. DeMarrais and M. Huang, A Subexponential Algorithm for Discrete Logarithms over the Rational Subgroup of the Jacobians of Large genus Hyperelliptic Curves over Finite Fields, in *Algorithmic Number Theory*, pp. 28-40, Lecture Notes in Computer Science, Vol. 877, Berlin: Springer, 1994.
3. R.P. Brent, An Improved Monte Carlo Factorization Algorithm, *j-BIT*, Vol. 20, No. 2, pp. 176-184, 1980.
4. J.H. van Lint and R.M. Wilson, *A Course in Combinatorics*, Cambridge: Cambridge University Press, 1992.
5. F.J. MacWilliams and N.J.A. Sloane, *The Theory of Error-Correcting Codes*, Amsterdam: North-Holland, 1977.
6. P.C. van Oorschot and M.J. Wiener, Parallel Collision Search with Cryptanalytic Applications, *J. of Cryptology*, Vol. 12, pp. 1-28, 1999.
7. S. Pohlig and M. Hellman, An Improved Algorithm for Computing Logarithms in $GF(p)$ and its Cryptographic Significance, *IEEE Trans. Inform. Theory*, Vol. IT-24, pp. 106-111, January 1978.
8. J.M. Pollard, Monte Carlo Methods for Index Computation (mod p), *Mathematics of Computation*, Vol. 32, No. 143, pp. 918-924, July 1978.
9. E. Schulte-Geers, Collision Search in a Random Mapping: Some Asymptotic Results, presented at ECC 2000 – The Fourth Workshop on Elliptic Curve Cryptography, University of Essen, Germany, October 4-6, 2000.
10. V. Shoup, Lower Bounds for Discrete Logarithms and Related Problems, in *Advances in Cryptology – EUROCRYPT '97*, W. Fumy, Ed., Lecture Notes in Computer Science, Vol. 1233, pp. 256-266, Berlin: Springer, 1997.

11. R. Silverman and J. Stapleton, Contribution to ANSI X9F1 working group, December 1997.
12. E. Teske, Speeding up Pollard's Rho Method for Computing Discrete Logarithms, in *Proceedings of ANTS III – The 3rd International Symposium on Algorithmic Number Theory*, J.P. Buhler, Ed., Lecture Notes in Computer Science, Vol. 1423, pp. 351-357, Berlin: Springer, 1998.
13. E. Teske, On Random Walks for Pollard's Rho Method, *Mathematics of Computation*, Vol. 70, pp. 809-825, 2001.
14. E. Teske, Square-Root Algorithms for the Discrete Logarithm Problem (A Survey), Technical Report CORR 2001-07, Centre for Applied Cryptographic Research, University of Waterloo, 2001.
15. M.J. Wiener and R.J. Zuccherato, Faster Attacks on Elliptic Curve Cryptosystems, in *Proceedings of SAC'98 – Fifth Annual Workshop on Selected Areas in Cryptography*, E. Tavares, H. Meijer, Eds., Lecture Notes in Computer Science, Vol. 1556, pp. 190-200, Berlin: Springer, 1998.

A   Stirling Numbers

In this section, we introduce Stirling numbers. These numbers play an important role in several parts of combinatorics. We mention several properties of these numbers that will be used in the paper. For a detailed discussion, we refer to [4].

In the sequel, n and k denote non-negative integers. Let $a(n,k)$ denote the number of surjections from an n-set to a k-set. Obviously, one has $a(n,k) = 0$ if $k > n$ and $a(n,k) = n!$ if $n = k$. In general, one can use the principle of inclusion-exclusion to show that

$$a(n,k) = \sum_{i=0}^{k} (-1)^i \binom{k}{i} (k-i)^n.$$

Let $S(n,k)$ denote the number of ways to partition an n-set into k nonempty subsets. The numbers $S(n,k)$ are called Stirling numbers of the second kind. Obviously, one has that $S(n,k) = a(n,k)/k!$. Moreover, by a simple counting argument, one can show that

$$S(n,k) = \sum_{\substack{1a_1 + 2a_2 + \cdots + na_n = n \\ a_1 + a_2 + \cdots + a_n = k}} \frac{n!}{(1!)^{a_1}(2!)^{a_2}\ldots(n!)^{a_n}a_1!a_2!\ldots a_n!}. \tag{10}$$

Our main interest is in those Stirling numbers $S(n,k)$ for which $n-k$ is relatively small. The first few of these are

$$S(k,k) = 1;$$
$$S(k+1,k) = \tfrac{1}{2}(k+1)k = \tfrac{1}{2}(k^2 + k);$$
$$S(k+2,k) = \tfrac{1}{8}(k+2)(k+1)k(k+\tfrac{1}{3}) = \tfrac{1}{8}(k^4 + \tfrac{10}{3}k^3 + 3k^2 + \tfrac{2}{3}k);$$
$$S(k+3,k) = \tfrac{1}{48}(k+3)(k+2)(k+1)k(k^2 + k)$$
$$\qquad = \tfrac{1}{48}(k^6 + 7k^5 + 17k^4 + 17k^3 + 6k^2);$$
$$S(k+4,k) = \tfrac{1}{384}(k+4)(k+3)(k+2)(k+1)k(k^3 + 2k^2 + \tfrac{1}{3}k - \tfrac{2}{15})$$
$$\qquad = \tfrac{1}{384}(k^8 + 12k^7 + \tfrac{166}{3}k^6 + \tfrac{616}{5}k^5 + \tfrac{403}{3}k^4 + 60k^3 + \tfrac{4}{3}k^2 - \tfrac{16}{5}k).$$

Computing $S(k + L, k)$ for bigger values of L is quite cumbersome. Fortunately, however, one can express $S(k + L, k)$ as a polynomial in k with coefficients in L, as is demonstrated by the following lemma.

Lemma 1. *For all* $k, L \geq 0$, *one has* $S(k + L, k) = \frac{1}{2^L L!} \sum\limits_{j=0}^{2L} \varphi_j(L) k^{2L-j}$, *where* $\varphi_j(x) \in \mathbb{Q}[x]$ *has degree at most* $2j$ $(0 \leq j \leq 2L)$. *For* $j > 0$, *one has* $x|\varphi_j(x)$. *The first few coefficients are*

$$\varphi_0(x) = 1, \quad \varphi_1(x) = \frac{2}{3}x^2 + \frac{1}{3}x, \quad \varphi_2(x) = \frac{2}{9}x^4 + \frac{2}{3}x^3 - 2\frac{1}{18}x^2 - \frac{7}{16}x.$$

Proof: To be provided in the full version of this paper.   □

B   Approximations of Combinatorial Expressions

In this section, we provide approximations of some combinatorial expressions that will be used in the paper and indicate the accuracy of these approximations.

Lemma 2. *For all* $t \geq 0$, *define* $I_t := \int\limits_{x=0}^{\infty} x^t e^{-x^2/2} dx$. *Then*

$$I_{2t} = \frac{(2t)!}{t! 2^t} \sqrt{\pi/2} \text{ and } I_{2t-1} = (t-1)! 2^{t-1}.$$

Proof: The result follows using partial integration and an induction argument. For $t > 1$, one has $I_t = (t-1)I_{t-2}$, since

$$I_t = \int\limits_{x=0}^{\infty} x^t e^{-x^2/2} dx = -\int\limits_{x=0}^{\infty} x^{t-1} d(e^{-x^2/2})$$

$$= [-x^{t-1} e^{-x^2/2}]_{x=0}^{\infty} + (t-1) \int\limits_{x=0}^{\infty} x^{t-2} e^{-x^2/2} dx = (t-1)I_{t-2}.$$

Moreover, $I_0 = \sqrt{\pi/2}$ and $I_1 = 1$, since

$$I_0^2 = \left(\int\limits_{x=0}^{\infty} e^{-x^2/2} dx \right)^2 = \int\limits_{\varphi=0}^{\pi/2} \int\limits_{r=0}^{\infty} e^{-r^2/2} r \, dr \, d\varphi = \pi/2 \text{ and }$$

$$I_1 = -\int\limits_{x=0}^{\infty} d(e^{-x^2/2}) = 1.$$

The result now follows using induction.   □

Lemma 3. *Let $N > 0$, let $t \in \mathbb{N}$. Then*

$$\frac{1}{\sqrt{N}} \sum_{k=0}^{\infty} \left(\frac{k}{\sqrt{N}} \right)^t e^{-k^2/2N} \to I_t \quad (N \to \infty).$$

Proof: The result follows from the observation that the expression is a Riemann sum that converges to I_t. $\qquad \square$

Lemma 4. *Let $N > 0$, let $t \in \mathbb{N}$, and let $p_k := \prod_{i=0}^{k-1} (1 - i/N)$. Then*

$$\frac{1}{\sqrt{N}} \sum_{k=0}^{\infty} \left(\frac{k}{\sqrt{N}} \right)^t p_k \to I_t \quad (N \to \infty).$$

Proof: The result follows from Lemma 3, using the estimate $p_k \approx e^{-k^2/2N}$ while upper bounding the approximation error in the 'tail' of the summation. Details will be provided in the full paper. $\qquad \square$

Remark 5. One can show that convergence is as follows:

$$\frac{1}{\sqrt{N}} \sum_{k=0}^{\infty} \left(\frac{k}{\sqrt{N}} \right)^t p_k = (1 + \varepsilon) I_t, \quad \text{where } |\varepsilon| \in O(\log(N)/\sqrt{N}).$$

Hence, for big values of N (as in our applications) the approximation of the expression by I_t is almost precise.

Fast Normal Basis Multiplication
Using General Purpose Processors
(Extended Abstract)

Arash Reyhani-Masoleh[1] and M. Anwar Hasan[2]

[1] Centre for Applied Cryptographic Research,
Department of Combinatorics and Optimization,
University of Waterloo, Waterloo, Ontario, Canada N2L 3G1.
areyhani@cacr.math.uwaterloo.ca
[2] Department of Electrical and Computer Engineering,
University of Waterloo, Waterloo, Ontario, Canada N2L 3G1.
ahasan@ece.uwaterloo.ca

Abstract. For cryptographic applications, normal bases have received considerable attention, especially for hardware implementation. In this article, we consider fast software algorithms for normal basis multiplication over the extended binary field $GF(2^m)$. We present a vector-level algorithm which essentially eliminates the bit-wise inner products needed in the conventional approach to the normal basis multiplication. We then present another algorithm which significantly reduces the dynamic instruction counts. Both algorithms utilize the full width of the data-path of the general purpose processor on which the software is to be executed. We also consider composite fields and present an algorithm which can provide further speed-up and an added flexibility toward hardware-software co-design of processors for very large finite fields.

Keywords: Finite field multiplication, normal basis, software algorithms, ECDSA, composite fields.

1 Introduction

The extended binary finite field $GF(2^m)$ of degree m is used in important cryptographic operations, such as, key exchange, signing and verification. For today's security applications the minimum values of m are considered to be 160 in the elliptic curve cryptography and 1024 in the standard discrete log based cryptography. Elliptic curve crypto-systems use relatively smaller field sizes, but require considerable amount of field arithmetic for each group operation (i.e., addition of two points). In such crypto-systems, often the most complicated and expensive module is the finite field arithmetic unit. As a result, it is important to develop suitable finite field arithmetic algorithms and architectures that can meet the constraints of various implementation technologies, such as, hardware and software.

S. Vaudenay and A. Youssef (Eds.): SAC 2001, LNCS 2259, pp. 230–244, 2001.

For cryptographic applications, the most frequently used $GF(2^m)$ arithmetic operations are addition and multiplication. Compared to the former, the latter is much more complicated and time consuming operation. The complexity of $GF(2^m)$ multiplication depends very much on how the field elements are represented. For hardware implementation of a multiplier, the use of normal bases has received considerable attention and a number of hardware architectures and implementations have been reported (see for example [1], [2], [7], [20]). A majority of such efforts were motivated by the fact that certain normal bases, e.g., optimal bases, yield area efficient multipliers, and that the field squaring, which is heavily used in exponentiation and Frobenius mapping, is a simple cycle shift of the field element's coordinates and hence in hardware it is almost free of cost. However, the task of implementing a normal basis multiplier in hardware poses a number of challenges. For example, when one has to deal with very large fields, the interconnections among the various parts of the multiplier could be quite irregular which may slow down the clock speed. Also, normal basis multipliers are not easily scalable with m. Given a normal basis multiplier designed for $GF(2^{233})$, one cannot conveniently make it usable for $GF(2^{163})$ or $GF(2^{283})$.

Unlike hardware, so far software implementation of a $GF(2^m)$ multiplier using normal bases has not been very efficient. This is mainly due to a number of practical considerations. Most importantly, normal basis multiplication algorithms require inner products or matrix multiplications over the ground field $GF(2)$. Such computations are not directly supported by most of today's general purpose processors. These computations require bit-by-bit logical AND and XOR operations, which are not efficiently implemented using the instruction set supported by the processors. Also, when a high level programming language, such as, C is used, the cyclic shifts needed for field squaring operations, are not as efficient as they are in hardware.

In this article, we consider algorithms for fast software normal basis multiplication on general purpose processors. We discuss how the conventional bit-level algorithm for normal basis multiplication fails to utilize the full data-path of the processor and makes its software implementation inefficient. We then present a vector-level normal basis multiplication algorithm which eliminates the matrix multiplication over $GF(2)$ and significantly reduces the number of dynamic instructions. We then derive another scheme for normal basis multiplication to further improve the speed. We also consider normal basis multiplication over certain special classes of composite fields. We show that normal basis multipliers over such composite fields can provide an additional speed-up and a great deal of flexibility toward hardware-software co-design of very large finite field processors.

2 Preliminaries

2.1 Normal Basis Representation

It is well known that there exists a normal basis (NB) in the field $GF(2^m)$ over $GF(2)$ for all positive integers m. By finding an element $\beta \in GF(2^m)$ such that

$\{\beta, \beta^2, \cdots, \beta^{2^{m-1}}\}$ is a basis of $GF(2^m)$ over $GF(2)$, any element $A \in GF(2^m)$ can be represented as $A = \sum_{i=0}^{m-1} a_i \beta^{2^i} = a_0 \beta + a_1 \beta^2 + \cdots + a_{m-1} \beta^{2^{m-1}}$, where $a_i \in GF(2)$, $0 \le i \le m-1$, is the i-th coordinate of A. In this article, this normal basis representation of A will be written in short as $A = (a_0, a_1, \cdots, a_{m-1})$. In vector notation, element A will be written as $A = \underline{a} \cdot \underline{\beta}^T = \underline{\beta} \cdot \underline{a}^T$, where $\underline{a} = [a_0, a_1, \cdots, a_{m-1}]$, $\underline{\beta} = [\beta, \beta^2, \cdots, \beta^{2^{m-1}}]$, and T denotes vector transposition. Now, consider the following matrix

$$\mathbf{M} = \underline{\beta}^T \cdot \underline{\beta} = \left[\beta^{2^i + 2^j}\right]_{i,j=0}^{m-1}, \tag{1}$$

whose entries belong to $GF(2^m)$. Writing these entries with respect to the NB, one obtains the following.

$$\mathbf{M} = \mathbf{M}_0 \beta + \mathbf{M}_1 \beta^2 + \cdots + \mathbf{M}_{m-1} \beta^{2^{m-1}}, \tag{2}$$

where $\mathbf{M}_i$'s are $m \times m$ *multiplication matrices* whose entries belong to $GF(2)$. Let $H(\mathbf{M}_i)$, $0 \le i \le m - 1$, be the number of 1's (or Hamming weight) of $\mathbf{M}_i$. It is easy to verify that $H(\mathbf{M}_0) = H(\mathbf{M}_1) = \cdots = H(\mathbf{M}_{m-1})$. The number of logic gates needed for the implementation of a NB multiplier depends on $H(\mathbf{M}_i)$ which is referred to as the complexity of the normal basis. Let us denote this complexity as C_N. It was shown in [12] that $C_N \ge 2m - 1$. When $C_N = 2m - 1$, the NB is called an optimal normal basis (ONB).

Two types of ONBs were constructed by Mullin *et al.* [12]. Gao and Lenstra [5] showed that these two types are all the ONBs in $GF(2^m)$. As an extension of the work on ONBs, Ash et al. in [3] proposed low complexity normal bases of type t where t is a positive integer. These low complexity bases are referred to as *Gaussian Normal Basis* (GNB). When $t = 1$ and 2, the GNBs become the two types of ONBs of [3]. A type t GNB for $GF(2^m)$ exists if and only if $p = tm + 1$ is prime and $\gcd(\frac{tm}{k}, m) = 1$, where k is the multiplicative order of 2 modulo p [8]. More on this can be found in [3].

2.2   Conventional NB Multiplication Algorithm

Below we give the conventional normal basis multiplication algorithm as described by NIST in [13]. This algorithm is for t even only (the reader is referred to [8] for algorithm with t odd). The case of t even is of particular interest for implementing high speed crypto-systems based on Koblitz curves. Such curves with points over $GF(2^m)$ exist for $m = 163, 233, 283, 409, 571$, where normal bases have t even. Note that in the following algorithm, $p = tm + 1$, and $A \ll i$ (resp. $A \gg i$) denotes i-fold left (resp. right) cyclic shifts of the coordinates of A. The algorithm requires the input sequence $F(1), F(2), \cdots, F(p - 1)$ to be pre-computed using

$$F(2^i u^j \bmod p) = i, \ 0 \le i \le m - 1, \ 0 \le j < t, \tag{3}$$

where u is an integer of order $t \bmod p$.

Algorithm 1 (Bit-Level NB Multiplication)
  Input: A, $B \in GF(2^m)$, $F(n) \in [0,\, m-1]$ for $1 \le n \le p-1$
  Output: $C = AB$
  1.   Initialize $C = (c_0, c_1, \cdots, c_{m-1}) := 0$
  2.   For $i = 0$ to $m - 1$ {
  3.   For $n = 1$ to $p - 2$ {
  4. $c_i := c_i + a_{F(n+1)} b_{F(p-n)}$
  5.   }
  6. $A \ll 1,\ B \ll 1$
  7.   }

Software implementation of Algorithm 1 is not very efficient for the following reasons. First, in each execution of line 4, one coordinate of each of A and B are accessed. These accesses are such that their software implementation is rather unsystematic and typically requires more than one instruction. Secondly, in line 4 the mod 2 *multiplication* of the coordinates, which is implemented by bit level logical AND operation, is performed $m(p-2)$ times in total, and the mod 2 *addition*, which is implemented by bit level logical XOR operation, is performed $\frac{1}{4}m(p-2)$ times, on average, assuming that A and B are two random inputs. In the C programming language, these mod 2 multiplication and addition operations correspond to about $m(p-2)$ AND and $\frac{1}{4}m(p-2)$ XOR instructions[1], respectively.

3   Vector-Level NB Multiplication

In this section we discuss improvements to Algorithm 1 so that normal basis multiplication can be efficiently implemented in software. One crucial improvement is that most arithmetic operations are done on vectors instead of bits. This enables us to use the full data-path of the processor on which the software is executed. The assumption that t is even in Algorithm 1 is also used in the remaining discussion of this section.

Lemma 1. *For GNB of type t, where t is even, the sequence $F(n)$ of $p - 1$ integers as defined above is mirror symmetric around the center, i.e., $F(n) = F(p - n)$, $1 \le n \le p - 1$.*

Proof. In (3), t is the smallest nonzero integer such that $u^t \bmod p = 1$. Then $u^{\frac{t}{2}} \bmod p$ must be equal to -1 . For $0 \le i \le m - 1$ and $0 \le j \le t - 1$, let $n = 2^i u^j \bmod p$. Then $F(n) = F(2^i u^j \bmod p) = i$. Also, $F(2^i u^{\frac{t}{2}+j} \bmod p) = i$. Thus $F(n) = F(2^i u^{\frac{t}{2}+j} \bmod p) = F(-2^i u^j \bmod p) = F(p - n)$. $\square$

From (3) and Lemma 1, one has $F(1) = F(p - 1) = 0$. For $1 \le n \le p - 2$, let us define

$$\Delta F(n) = F(n + 1) - F(n) \ \bmod \ m. \qquad (4)$$

Now we have the following corollary.

---
[1] These are dynamic instructions which the underlying processor needs to execute.

Corollary 1. *For $\Delta F(n)$ as defined above and for t even, the following holds*

$$\Delta F(p-n) = m - \Delta F(n-1) \bmod m, \ 1 \le n \le p-2.$$

Proof. Using (4), one obtains $F(n+1) = \sum_{i=1}^{n} \Delta F(i)$. Applying Lemma 1 into (4), one can also write $\Delta F(p-n) = -\Delta F(n-1) \bmod m$, $2 \le n \le p-1$ which results in $\Delta F(p-n) = m - \Delta F(n-1)$, $2 \le n < \frac{p-1}{2}$, and $\Delta F(\frac{p-1}{2}) = 0$. $\square$

In Algorithm 1, the i-th coordinate of the product $C = AB$ is computed in its inner loop which can be written as follows

$$c_i = \sum_{n=1}^{p-2} a_{F(n+1)+i} b_{F(p-n)+i}, \ 0 \le i \le m-1. \tag{5}$$

Using Lemma 1 and equation (4), one can write

$$c_i = \sum_{n=1}^{p-2} a_{F(n+1)+i} b_{F(n)+i}, \ 0 \le i \le m-1, \tag{6}$$

$$= \sum_{n=1}^{p-2} a_{F(n)+\Delta F(n)+i} b_{F(n)+i}, \ 0 \le i \le m-1. \tag{7}$$

For a particular GNB, the values of $\Delta F(n)$, $1 \le n \le p-2$, are fixed and are to be determined only once, i.e., at the time of choosing the basis. Additionally, Corollary 1 implies that it is sufficient to store only half (i.e., $\frac{p-1}{2}$) of these $\Delta F(n)$'s. We now state the vector-level algorithm for t even as follows. A similar algorithm for odd values of t is given in [18].

Algorithm 2 (Vector-Level NB Multiplication)
    Input: $A, B \in GF(2^m)$, $\Delta F(n) \in [0, m-1]$, $1 \le n \le p-1$
    Output: $C = AB$
    1.   Initialize $S_A := A$, $S_B := B$, $C := 0$
    2.   For $n = 1$ to $p-2$ {
    3. $S_A \ll \Delta F(n)$
    4. $R := S_A \odot S_B$
    5. $C := C + R$
    6. $S_B \ll \Delta F(n)$
    7.   }

In line 4 of Algorithm 2, for $X, Y \in GF(2^m)$, $X \odot Y$ denotes the bit-wise AND operation between coordinates of X and Y, i.e., $X \odot Y = (x_0 y_0, x_1 y_1, \cdots, x_{m-1} y_{m-1})$. In order to obtain an overall computation time for a $GF(2^m)$ multiplication using Algorithm 2, the coordinates of the field elements can be divided into $\lceil \frac{m}{\omega} \rceil$ units where ω corresponds to the data-path width of the processor. We assume that the processor can perform bit-wise XOR and AND of two ω-bit operands using one single XOR instruction and one single AND instruction, respectively. Since the loop in Algorithm 2, has $p-2$ iterations, the total number of bit-wise AND and bit-wise XOR instructions are the same and is $(p-2)\left\lceil\frac{m}{\omega}\right\rceil =$ $(tm-1)\left\lceil\frac{m}{\omega}\right\rceil$. Also, this algorithm needs $2\,(p-2)\left\lceil\frac{m}{\omega}\right\rceil = 2\,(tm-1)\left\lceil\frac{m}{\omega}\right\rceil$ cyclic shifts. We assume that an i-fold, $1 \le i < \omega$, left/right shift can be emulated in the C programming language using a total of ρ instructions. The value of ρ is typically 4 when simple logical instructions, such as AND, SHIFT, and OR are used. We can now state the following theorem.

Theorem 1. *The dynamic instruction count for Algorithm 2 is given by*

$$\#Instructions \approx 2(1+\rho)\,(tm-1)\left\lceil\frac{m}{\omega}\right\rceil.$$

4   Efficient NB Multiplication over $GF(2^m)$

In this section, we develop another algorithm for normal basis multiplication. We also analyze the cost of this algorithm in terms of dynamic instruction counts and memory requirements and then compare them with those of similar other algorithms.

4.1   Algorithm

For the normal basis $\{\beta, \beta^{2^1}, \cdots, \beta^{2^{m-1}}\}$, let $\delta_j = \beta^{1+2^j}$, $j = 1, \cdots, v$, where $v = \left\lceil\frac{m-1}{2}\right\rceil$. Then one has the following result from [16].

Lemma 2. *Let A and B be two elements of $GF(2^m)$ and C be their product. Then*

$$C = \begin{cases} \sum_{i=0}^{m-1}\left[a_i b_i \beta^{2^{i+1}} + \left(\sum_{j=1}^{v} x_{i,j}\delta_j^{2^i}\right)\right], & for\ m\ odd \\ \sum_{i=0}^{m-1}\left[a_i b_i \beta^{2^{i+1}} + \left(\sum_{j=1}^{v-1} x_{i,j}\delta_j^{2^i}\right) + a_i b_{v+i}\delta_v^{2^i}\right], & for\ m\ even \end{cases}$$

where a_i's and b_i's are the NB coordinates of A and B, respectively. Also, indices and exponents are reduced $\bmod m$ and

$$x_{i,j} = a_i b_{i+j} + a_{i+j}b_i, \qquad 1 \le j \le v,\ 0 \le i \le m-1. \tag{8}$$

Let h_j, $1 \le j \le v$, be the number of 1's in the normal basis representation of δ_j. Let $w_{j,1}, w_{j,2}, \cdots, w_{j,h_j}$ denote the positions of 1's in the normal basis representation of δ_j, i.e.,

$$\delta_j = \sum_{k=1}^{h_j} \beta^{2^{w_{j,k}}}, \ 1 \le j \le v, \tag{9}$$

where $0 \le w_{j,1} < w_{j,2} < \cdots < w_{j,h_j} \le m-1$. Now, using (9) into Lemma 2, we have the following for m odd.

$$C = \sum_{i=0}^{m-1} a_i b_i \beta^{2^{i+1}} + \sum_{i=0}^{m-1}\sum_{j=1}^{v} x_{i,j}\left(\sum_{k=1}^{h_j} \beta^{2^{w_{j,k}}}\right)^{2^i}$$

$$
\begin{aligned}
&= \sum_{i=0}^{m-1} a_i b_i \beta^{2^{i+1}} + \sum_{i=0}^{m-1} \sum_{j=1}^{v} x_{i,j} \left(\sum_{k=1}^{h_j} \beta^{2^{i+w_{j,k}}} \right) \\
&= \sum_{i=0}^{m-1} a_i b_i \beta^{2^{i+1}} + \sum_{j=1}^{v} \sum_{k=1}^{h_j} \left(\sum_{i=0}^{m-1} \beta^{2^{i+w_{j,k}}} \right).
\end{aligned} \tag{10}
$$

Also, for even values of m, one has $v = \frac{m}{2}$ and $\delta_v = \delta_v^{2^{\frac{m}{2}}}$. This implies that in the normal basis representation of δ_v, its i-th coordinate is equal to its $(\frac{m}{2} + i \bmod m)$-th coordinate. Thus, h_v is even and one can write

$$
\delta_v = \sum_{k=1}^{\frac{h_v}{2}} (\beta^{2^{w_{v,k}}} + \beta^{2^{w_{v,k}+v}}), \quad v = \frac{m}{2}. \tag{11}
$$

Now, using (11) into Lemma 2 (for m even) and using (10), we have the following theorem, where all indices and exponents are reduced modulo m.

Theorem 2. Let A and B be two elements of $GF(2^m)$ and C be their product. Then

$$
C = \begin{cases}
\sum_{i=0}^{m-1} a_i b_i \beta^{2^{i+1}} + \sum_{j=1}^{v} \sum_{k=1}^{h_j} \left(\sum_{i=0}^{m-1} x_{i,j} \beta^{2^{i+w_{j,k}}} \right), & \text{for } m \text{ odd} \\
\sum_{i=0}^{m-1} a_i b_i \beta^{2^{i+1}} + \sum_{j=1}^{v-1} \sum_{k=1}^{h_j} \left(\sum_{i=0}^{m-1} x_{i,j} \beta^{2^{i+w_{j,k}}} \right) + F, & \text{for } m \text{ even}
\end{cases} \tag{12}
$$

where

$$
F = \sum_{k=1}^{\frac{h_v}{2}} \sum_{i=0}^{v-1} x_{i,v} (\beta^{2^{i+w_{v,k}}} + \beta^{2^{i+w_{v,k}+v}}), \quad \text{and } v = \frac{m}{2}.
$$

Note that for a normal basis, the representation of δ_j is fixed and so is $w_{j,k}$, $1 \le j \le v$, $1 \le k \le h_j$. Now, define

$$
\Delta w_{j,k} \triangleq w_{j,k} - w_{j,k-1}, \ 1 \le j \le v, \ 1 \le k \le h_j, \ w_{j,0} = 0, \tag{13}
$$

where $w_{j,k}$'s are as given in (9). For a particular normal basis, all $w_{j,k}$'s are fixed. Hence, all $\Delta w_{j,k}$'s need to be determined only at the time of choosing the basis. Using $\Delta w_{j,k}$'s, below we present an efficient NB (ENB) multiplication algorithm over $GF(2^m)$ for odd values of m. The corresponding algorithm for even values of m is shown in [18]. Also, an efficient scheme to compute $\Delta w_{j,k}$'s is presented in [18].

Algorithm 3 (ENB Multiplication for m Odd)
    Input: $A, B \in GF(2^m)$, $\Delta w_{j,k} \in [0, m-1]$, $1 \le j \le v$, $1 \le k \le h_j$, $v = \frac{m-1}{2}$
    Output: $C = AB$
    1.   Initialize $C := A \odot B$, $S_A := A$, $S_B := B$
    2. $C \gg 1$
    3.   For $j = 1$ to v {

$$4. \quad S_A \ll 1, S_B \ll 1$$
$$5. \quad T_A := A \odot S_B, T_B := B \odot S_A$$
$$6. \quad R := T_A + T_B$$
$$7. \quad \text{For } k = 1 \text{ to } h_j \ \{$$
$$8. \quad\quad R \gg \Delta w_{j,k}$$
$$9. \quad\quad C := C + R$$
$$10. \quad\quad \}$$
$$11. \quad \}$$

In the above algorithm, shifted values of A and B are stored in S_A and S_B, respectively. In line 6, $R \in GF(2^m)$ contains $(x_{0,j}, x_{1,j}, \cdots, x_{m-1,j})$, i.e., $\sum_{i=0}^{m-1} x_{i,j}\beta^{2^i}$. Also, right cyclic shift of R in lines 8, corresponds to $\sum_{i=0}^{m-1} x_{i,j}\beta^{2^{i+w_{j,k}}}$. After the final iteration, C is the normal basis representation of the required product AB. To illustrate the operation of the above algorithm, we present the following example.

Example 1. Consider the finite field $GF(2^5)$ generated by the irreducible polynomial $F(z) = z^5 + z^2 + 1$ and let α be its root, *i.e.*, $F(\alpha) = 0$. We choose $\beta = \alpha^5$, then $\{\beta, \beta^2, \beta^4, \beta^8, \beta^{16}\}$ is a type 2 GNB. Here $m = 5$, and $v = \frac{5-1}{2} = 2$. Using Table 2 in [12], one has

$$\delta_1 = \beta^3 = \beta + \beta^8, \quad h_1 = 2, [w_{1,k}]_{k=1}^{h_1} = [0, 3],$$
$$\delta_2 = \beta^5 = \beta^8 + \beta^{16}, \quad h_2 = 2, [w_{2,k}]_{k=1}^{h_2} = [3, 4].$$

Let $A = \beta^2 + \beta^4 + \beta^8 = (01110)$ and $B = \beta + \beta^4 + \beta^{16} = (10101)$ be two field elements. Table 1 shows contents of various variables of the algorithm as they are updated. The row with j being '-' is for the initialization step (i.e., line 1) of the algorithm.

Table 1. Contents of variables in Algorithm 3 for multiplication of $A = (01110)$ and $B = (10101)$.

j	S_A	S_B	T_A	T_B	k	$\Delta w_{j,k}$	R	C
-	01110	10101	-	-	-	-	-	00010
1	11100	01011	01010	10100			11110	
					1	0	11110	11100
					2	3	11011	00111
2	11001	10110	00110	10001			10111	
					1	3	11110	11001
					2	1	01111	10110

As it can be seen in Algorithm 3, all $\Delta w_{j,k}$'s have to be pre-computed. In the above example, they are determined by calculating δ_j's, which is essentially a multiplication process all by itself. For this multiplication, one can use either Algorithm 1 or Algorithm 2. However, an efficient scheme which does not need multiplication is presented in [18].

4.2   Cost and Comparison

In an effort to determine the cost of Algorithm 3, we give the dynamic instruction counts for its software implementation. We also consider the number of memory accesses to read the pre-computed values of $\Delta w_{j,k}$. For software implementation of the above algorithm, one would heavily rely on instructions, such as, XOR, AND and others which can be used to emulate cyclic shifts (in the C like programming language). XOR instructions are needed in lines 6 and 9, which are repeated v and $\sum_{j=1}^{v} h_j$ times, respectively. Since $v = \frac{m-1}{2}$ and $\sum_{j=1}^{v} h_j = \frac{C_N - 1}{2}$ [10], the total number of XOR instructions is $\frac{1}{2}(C_N + m - 1)\lceil \frac{m}{\omega} \rceil$. Because of the $\odot$ operations in lines 1 and 5, one can also see that the above algorithm requires $m \lceil \frac{m}{\omega} \rceil$ AND instructions. We assume that each i-fold cyclic shift, $1 \leq i \leq m - 1$, in lines 2, 4 and 8 needs $\rho \lceil \frac{m}{\omega} \rceil$ instructions where ρ is as defined earlier. In Algorithm 3, the number of cyclic shifts in lines 2, 4 and 8 are 1, $2v$ and $\sum_{j=1}^{v} h_j$, respectively. Thus, the total number of cyclic shifts in this algorithm is $1 + 2v + \sum_{j=1}^{v} h_j = \frac{1}{2}(C_N + 2m - 1)$ and so the total number of instructions to emulate cyclic shifts used in Algorithm 3 is $\frac{\rho}{2}(C_N + 2m - 1)\lceil \frac{m}{\omega} \rceil$. Based on the above discussion, we have the following theorem.

Theorem 3. *The dynamic instruction count for Algorithm 3 is given by*

$$\#Instructions \approx \left(\frac{1+\rho}{2} C_N + \frac{3+2\rho}{2} m - \frac{2+\rho}{2} \right) \left\lceil \frac{m}{\omega} \right\rceil.$$

For software implementation of Algorithm 3, if the loops are not unrolled and the values of $\Delta w_{j,k}$'s are not hard-coded, one needs to store all these $\Delta w_{j,k}$, $1 \leq j \leq v$, $1 \leq k \leq h_j$. Since the total number of $\Delta w_{j,k}$'s is $\sum_{j=1}^{v} h_j$ and each $\Delta w_{j,k} \in [0, m-1]$ needs $\lceil \log_2 m \rceil$ bits of memory, a total of about $\frac{C_N - 1}{2} \lceil \log_2 m \rceil$ bits of memory is needed to store the pre-computed $\Delta w_{j,k}$'s.

Table 2. Comparison of multiplication algorithms in terms of number of instructions and memory requirements.

Algorithms	# Instructions			Memory	
	XOR	AND	Others	Size in bits	# Accesses
Alg. 1	$\frac{1}{4} m (tm-1)$	$m (tm-1)$	$2\rho m \lceil \frac{m}{\omega} \rceil$	$(tm-1) \lceil \log_2 m \rceil$	$2m(tm-1)$
Alg. 2	$(tm-1) \lceil \frac{m}{\omega} \rceil$	$(tm-1) \lceil \frac{m}{\omega} \rceil$	$2\rho (tm-1) \lceil \frac{m}{\omega} \rceil$	$\frac{tm}{2} \lceil \log_2 m \rceil$	tm
Alg. 3	$\frac{1}{2} (C_N+m-2) \lceil \frac{m}{\omega} \rceil$	$m \lceil \frac{m}{\omega} \rceil$	$\frac{\rho}{2} (C_N+2m-1) \lceil \frac{m}{\omega} \rceil$	$\frac{C_N-1}{2} \lceil \log_2 m \rceil$	$\frac{C_N-1}{2}$
Ratio of Alg. 2 to Alg. 3	$\approx \frac{2t}{t+1}$	$\approx t$	$\approx \frac{4t}{t+1}$	≈ 1	≈ 2

Table 2 compares the number of dynamic instructions of the three algorithms we have described so far. This table also gives memory sizes and numbers of memory accesses of these algorithms. As it can be seen in Table 2, both our proposed schemes (i.e., Algorithms 2 and 3) are superior to the conventional bit-level multiplication scheme (i.e., Algorithm 1). The final row of Table 2 gives approximate improvement factors of Algorithm 3 to Algorithm 2. A more detailed comparison of these two algorithms are given in Table 3 for the five binary fields recommended by NIST for ECDSA (elliptic curve digital signature algorithm) [13]. We have also coded these algorithms in software using the C programming language. Table 3 also shows timing (in μs) for these codes executed on Pentium III 533 MHz PC[2]. Our codes are parameterized in the sense that they can be used for various m and t without major modifications. For high speed implementation, the codes can be optimized for special values of m and t.

Agnew et. al. in [1] have proposed a bit-serial architecture for the NB multiplication. Although their work has been targeted to hardware implementation, the main idea can be used for software implementation similar to the vector level method proposed here. For such a software implementation of [1], one would require $(C_N - 1) \left\lceil \frac{m}{\omega} \right\rceil$ XOR instructions, $m \left\lceil \frac{m}{\omega} \right\rceil$ AND instructions, and $\rho (C_N + m - 1) \left\lceil \frac{m}{\omega} \right\rceil$ other instructions. Thus, the dynamic instruction count would be $(\rho + 1)(C_N + m - 1) \left\lceil \frac{m}{\omega} \right\rceil$ which is about twice of that in Algorithm 3 (see Theorem 3). In [19], one can find software implementation of the NB multiplication for two special cases, namely, two optimal normal bases. The method used in [19] is similar to that of the NB multiplication of [1].

Some of the recently proposed *polynomial basis* multiplication algorithms, for example [6], [9], create a look-up table on the fly based on one of the inputs (say B) and yield significant speed-ups by processing a group of bits of the other input (i.e., A) at a time. At this point, it is not clear whether such a group-level processing of A can be incorporated into our Algorithm 3. However, if m is a composite number, then one can essentially achieve similar kind of group-level processing by performing computations in the sub-fields. This idea is explored in the following section.

Table 3. Comparison of the proposed algorithms for binary fields recommended by NIST for ECDSA applications ($\omega = 32$).

Parameters			Algorithm 2, Algorithm 3						
			# Instructions				Memory		Timing
m	t	C_N	XOR	AND	Others/ρ	Total ($\rho = 4$)	Size in bits	# Accesses	in μs	Ratio
163	4	645	3906, 2418	3906, 978	7812, 2910	39060, 15036	2608, 2576	652, 322	307, 99	3.1:1
233	2	465	3720, 2784	3720, 1864	7440, 3720	37200, 19528	1864, 1856	466, 232	346, 126	2.75:1
283	6	1677	15273, 8811	15273, 2547	30546, 10089	152730, 51714	7641, 7542	1698, 838	1005, 318	3.16:1
409	4	1629	21255, 13234	21255, 5317	42510, 15899	212550, 82147	7362, 7326	1636, 814	1466, 473	3.1:1
571	10	5637	102762, 55854	102762, 10278	205524, 61002	1027620, 310140	28550, 28180	5710, 2818	8423, 2949	2.86:1

---

[2] The PC has 64 M bytes of RAM, 32 K bytes of L1 cache and 512 K bytes of L2 cache.

5   Efficient Composite Field NB Multiplication Algorithm

In this section, we consider multiplications in the finite field $GF(2^m)$ where m is a composite number. These fields are referred to as composite fields and have been used in the recent past to develop efficient multiplication schemes [14], [15]. When these fields are to be used for elliptic curve crypto-systems, one must choose m such that its factor are large enough to resist the attack described by Galbraith and Smart [4].

Lemma 3. *[11] Let* $\gcd(m_1, m_2) = 1$. *Let* $N_1 = \{\beta_1^{2^j} \mid 0 \le j \le m_1 - 1\}$ *be a normal basis of* $GF(2^{m_1})$ *over* $GF(2)$. *Then* N_1 *is also a normal basis of* $GF(2^{m_1 m_2})$ *over* $GF(2^{m_2})$.

Here, we consider composite fields with only two prime factors[3] (i.e., both m_1 and m_2 are prime). Thus, in the following we give all equations and algorithm for odd degrees (i.e., m_1 and m_2). The reader can easily extend it for even degrees using the results of the previous section. Also, the parameters, namely δ_j, h_j, v, β, and $\Delta w_{j,k}$ of the previous section are used here in the context of the sub-fields $GF(2^{m_1})$ and $GF(2^{m_2})$ by putting an extra sub/superscript for example $\delta_j^{(1)}$ for $GF(2^{m_1})$ and $\delta_j^{(2)}$ for $GF(2^{m_2})$.

Let A and B be two elements of $GF(2^{m_1})$ over $GF(2)$ and C be their product. Then we have the following from [17].

$$C = \sum_{i=0}^{m_1-1} a_i b_i \beta_1^{2^i} + \sum_{j=1}^{v_1} \sum_{k=1}^{h_j^{(1)}} \left(\sum_{i=0}^{m_1-1} y_{i,j} \beta_1^{2^{i+w_{j,k}^{(1)}}} \right), \text{ for } m_1 \text{ odd} \qquad (14)$$

where

$$y_{i,j} = (a_i + a_{i+j})(b_i + b_{i+j}), \ 1 \le j \le v_1, \ 0 \le i \le m_1 - 1,$$

$$v_1 = \frac{m_1 - 1}{2}, \quad \beta_1^{2^j+1} = \sum_{k=1}^{h_j^{(1)}} \beta_1^{2^{w_{j,k}^{(1)}}}.$$

By combining Lemma 3 with (14), the following is obtained.

Lemma 4. *Let* $A = (A_0, A_1, \cdots, A_{m_1-1})$ *and* $B = (B_0, B_1, \cdots, B_{m_1-1})$ *be two elements of* $GF(2^{m_1 m_2})$ *over* $GF(2^{m_2})$ *and* C *be their product. Then*

$$C = \sum_{i=0}^{m_1-1} A_i B_i \beta_1^{2^i} + \sum_{j=1}^{v_1} \sum_{k=1}^{h_j^{(1)}} \left(\sum_{i=0}^{m_1-1} Y_{i,j} \beta_1^{2^{i+w_{j,k}^{(1)}}} \right), \text{ for } m_1 \text{ odd} \qquad (15)$$

---

[3] This is important for elliptic curve crypto-systems. For such systems in today's security applications, the values of m appear to be in the range of 160 to several hundreds only (571 as given in [13]). To avoid the attack of [4], one however may like to choose m such that it has no small factors such as 2, 3, 5, 7, 11. This basically makes one to choose m as the product of two primes.

where

$$Y_{i,j} = (A_i + A_{i+j})(B_i + B_{i+j}),\ 1 \le j \le v_1,\ 0 \le i \le m_1 - 1, \qquad (16)$$

and $A_i = (a_{i,0}, a_{i,1}, \cdots, a_{i,m_2-1})$, $B_i = (b_{i,0}, b_{i,1}, \cdots, b_{i,m_2-1}) \in GF(2^{m_2})$ *are sub-field coordinates of A and B.*

Lemma 4 leads to an algorithm for multiplication in composite fields using normal basis. The algorithm is stated below.

Algorithm 4 (ECFNB Multiplication of $GF(2^{m_1 m_2})$ over $GF(2^{m_2})$)

Input: $A, B \in GF(2^m)$, $\Delta w_{j,k}^{(1)} \in [0, m_1 - 1]$, $1 \le j \le v_1$, $v_1 = \frac{m_1 - 1}{2}$, $1 \le k \le h_j^{(1)}$

Output: $C = AB$

1.   Initialize $C := A \otimes B$, $S_A := A$, $S_B := B$
2.   For $j = 1$ to v_1 {
3. $S_A \ll m_2$, $S_B \ll m_2$
4. $T_A := A + S_A$, $T_B := B + S_B$
5. $R := T_A \otimes T_B$
6.   For $k = 1$ to $h_j^{(1)}$ {
7. $R \gg m_2 \Delta w_{j,k}^{(1)}$
8. $C := C + R$
9.   }
10.   }

In lines 1 and 5 of Algorithm 4, $A \otimes B = (A_0 B_0, A_1 B_1, \cdots, A_{m_1-1} B_{m_1-1})$ denotes parallel sub-field multiplications of A and B. This sub-field multiplication can be implemented with an extension of Algorithm 3 such that it produces m_1 sub-field multiplications over $GF(2^{m_2})$. This is shown in Algorithm 5 where $A \triangleright i$ (resp. $A \triangleleft i$) $0 \le i \le m_2 - 1$, denotes an i-fold right (resp. left) sub-field cyclic shift of all sub-field elements of A, i.e., $A_0, A_1, \cdots, A_{m_1-1}$, respectively.

Algorithm 5 (Parallel Sub-Field Multiplication over $GF(2^{m_2})$)

Input: $A, B \in GF(2^m)$, $\Delta w_{j,k}^{(2)} \in [0, m_2 - 1]$, $1 \le j \le v_2$, $1 \le k \le h_j^{(2)}$, $v_2 = \frac{m_2 - 1}{2}$

Output: $C = A \otimes B$

1.   Initialize $C := A \odot B$, $S_A := A$, $S_B := B$
2. $C \triangleright 1$
3.   For $j = 1$ to v_2 {
4. $S_A \triangleleft 1$, $S_B \triangleleft 1$
5. $T_A := A \odot S_B$, $T_B := B \odot S_A$
6. $R := T_A + T_B$
7.   For $k = 1$ to $h_j^{(2)}$ {
8. $R \triangleright \Delta w_{j,k}^{(2)}$
9. $C := C + R$
10.   }
11.   }

In order to obtain the cost of Algorithm 4, we need to evaluate the cost of Algorithm 5 which is called $1 + v_1 = \frac{m_1+1}{2}$ times by the former. Like Algorithm 3, one can determine the dynamic instruction counts of Algorithm 5 to be $\frac{1}{2}(C_2 + m_2 - 2)$ XOR, m_2 AND and $\frac{1}{2}(C_2 + 2m_2 - 1)$ others to emulate cyclic shifts. The total cost of Algorithm 4 also depends on how sub-field elements, each of m_2 bits, are stored in registers. For the sake of simplicity we assume that an element of $GF(2^{m_2})$ is stored in one w-bit register (for software implementation of elliptic curve crypto-systems with both m_1 and m_2 being prime, most general purpose processors would have w bit registers where $w \geq m_2$). For $w = 24$ and 32, the best values of m_2 are those which have ONBs, i.e., 23 and 29, respectively. Thus, each element of $GF(2^m)$ needs m_1 registers and the cyclic shifts in lines 3 and 7 of Algorithm 4 are almost free of cost (or at best register renaming). Based on this assumption, we give the dynamic instruction counts of Algorithm 4 in Table 4. In this table, μ is the number of instructions needed for one sub-field cyclic shift in each register and it is 4 in the C programming language.

Table 4. Cost of Algorithm 4.

	XOR	$\frac{m_1}{2}\left[(C_1 + 2m_1 - 3) + \frac{(m_1+1)}{2}(C_2 + m_2 - 2)\right]$
# Instructions	AND	$\frac{m_1 m_2(m_1+1)}{2}$
	Others	$\frac{\mu m_1}{4}(m_1 + 1)(C_2 + 2m_2 - 1)$
Memory	Size in bits	$\frac{C_1-1}{2}\lceil \log_2 m_1 \rceil + \frac{C_2-1}{2}\lceil \log_2 m_2 \rceil$
	# Accesses	$\frac{C_1-1}{2} + \frac{(m_1+1)(C_2-1)}{4}$

Table 5 shows the number of instructions and memory requirements of Algorithm 4 for six different composite fields. These six fields are obtained by combining three m_1's and two m_2's. Algorithm 4 is also coded for these composite fields using the C programming language. The actual timing (in μs) of Algorithm 4 executed on Pentium III 533 MHz PC are also shown in Table 5.

Table 5. Cost of Algorithm 4 for certain composite fields ($\mu = 4$).

Parameters					# Instructions				Memory		Actual timing
m	m_1	m_2	C_1	C_2	XOR	AND	Others	Total	Size in bits	# Accesses	(in μs)
299	13	23	45	45	3445	2093	16380	21918	198	176	114
377	"	29	"	57	4264	2639	20748	27651	228	218	150
391	17	23	81	45	6001	3519	27540	37060	310	238	188
437	19	"	117	"	7714	4370	34200	46284	400	278	249
493	17	29	81	57	7378	4437	34884	46699	340	292	242
551	19	"	117	"	9424	5510	43320	58254	430	338	309

6   Conclusions

In this article, we have presented a number of software algorithms for normal basis multiplication over $GF(2^m)$. Both Algorithms 2 and 3 make maximal use of the full width of the data-path of the processor on which the software is to be executed and they provide significant speed-ups compared to the conventional bit-level multiplication scheme (i.e., Algorithm 1). Algorithms 2 and 3 are particularly suitable if m is a prime. Such values of m are of importance, especially for designing high speed crypto-systems based on Koblitz curves and for protecting elliptic curve crypto-systems against the attack of Galbraith and Smart [4]. Both Algorithms 2 and 3 have been coded for software implementation using C, and our timing results show that Algorithm 3 is about 200% faster that Algorithm 2. These results are for those five Gaussian normal bases over the binary fields which NIST has described in their ECDSA document [13]. For the purpose of using NIST parameters, although we have presented our results for Gaussian normal bases, our algorithms are quite generic and can be used for any normal bases of $GF(2^m)$ over $GF(2)$.

We have also considered composite fields with $m = m_1 \cdot m_2$. To avoid the attack of [4] on elliptic curve crypto-systems defined over these composite fields, we choose both m_1 and m_2 to be prime. We have presented an algorithm (i.e., Algorithm 4) for normal basis multiplication for $GF(2^m)$ over $GF(2^{m_2})$. Our results show that for similar values of m, Algorithm 4 can be much more efficient than Algorithm 3. For example, the actual timing of Algorithm 3 is 318 micro-seconds for $GF(2^{283})$ whereas the timing of Algorithm 4 is 114 micro-seconds for $GF(2^{299})$. Composite fields also provide an added flexibility to hardware-software co-design of finite field processors. For example, Algorithm 5 which is *called* by Algorithm 4 a total of $\frac{m_1+1}{2}$ times, can be implemented in hardware for small values of m_2, and Algorithm 4 can be embedded in a micro-controller which would give us a high speed, yet quite flexible, normal basis multiplier over very large fields.

Acknowledgment

The authors would like to thank Z. Zhang for his help with implementing the algorithms and getting their timing results.

References

1. G. B. Agnew, R. C. Mullin, I. M. Onyszchuk, and S. A. Vanstone. "An Implementation for a Fast Public-Key Cryptosystem". *Journal of Cryptology*, 3:63–79, 1991.
2. G. B. Agnew, R. C. Mullin, and S. A. Vanstone. "An Implementation of Elliptic Curve Cryptosystems Over $F_{2^{155}}$". *IEEE J. Selected Areas in Communications*, 11(5):804–813, June 1993.
3. D. W. Ash, I. F. Blake, and S. A. Vanstone. "Low Complexity Normal Bases". *Discrete Applied Mathematics*, 25:191–210, 1989.

4. S. D. Galbraith and N. Smart. A Cryptographic Application of Weil Descent. In *Proceedings of the Seventh IMA Conf. on Cryptography and Coding, LNCS 1764*, pages 191–200. Springer-Verlag, 1999.
5. S. Gao and Jr. H. W. Lenstra. "Optimal Normal Bases". *Designs, Codes and Cryptography*, 2:315–323, 1992.
6. M. A. Hasan. Look-up Table-Based Large Finite Field Multiplication in Memory Constrained Cryptosystems. *IEEE Transactions on Computers*, 49:749–758, July 2000.
7. M. A. Hasan, M. Z. Wang, and V. K. Bhargava. "A Modified Massey-Omura Parallel Multiplier for a Class of Finite Fields". *IEEE Transactions on Computers*, 42(10):1278–1280, Oct. 1993.
8. D. Johnson and A. Menezes. "The Elliptic Curve Digital Signature Algorithm (ECDSA)". *Technical Report CORR 99-34, Dept. of C & O, University of Waterloo, Canada*, August 23 1999. Updated: Feb. 24, 2000.
9. J. Lopez and R. Dahab. High Speed Software Multiplication in F_{2^m}. In *Proceedings of Indocrypt 2000*, pages 203–212. LNCS 1977, Springer, 2000.
10. Chung-Chin Lu. "A Search of Minimal Key Functions for Normal Basis Multipliers". *IEEE Transactions on Computers*, 46(5):588–592, May 1997.
11. A. J. Menezes, I. F. Blake, X. Gao, R. C. Mullin, S. A. Vanstone, and T. Yaghoobian. *Applications of Finite Fields*. Kluwer Academic Publishers, 1993.
12. R. C. Mullin, I. M. Onyszchuk, S. A. Vanstone, and R. M. Wilson. "Optimal Normal Bases in $GF(p^n)$". *Discrete Applied Mathematics*, 22:149–161, 1988/89.
13. National Institute of Standards and Technology. *Digital Signature Standard*. FIPS Publication 186-2, February 2000.
14. S. Oh, C. H. Kim, J. Lim, and D. H. Cheon. "Efficient Normal Basis Multipliers in Composite Fields". *IEEE Transactions on Computers*, 49(10):1133–1138, Oct. 2000.
15. C. Paar, P. Fleishmann, and P. Soria-Rodriguez. "Fast Arithmetic for Public-Key Algorithms in Galois Fields with Composite Exponents". *IEEE Transactions on Computers*, 48(10):1025–1034, Oct. 1999.
16. A. Reyhani-Masoleh and M. A. Hasan. "A Reduced Redundancy Massey-Omura Parallel Multiplier over $GF(2^m)$". In 20th *Biennial Symposium on Communications*, pages 308–312, Kingston, Ontario, Canada, May 2000.
17. A. Reyhani-Masoleh and M. A. Hasan. "On Efficient Normal Basis Multiplication". In *LNCS 1977 as Proceedings of Indocrypt 2000*, pages 213–224, Calcutta, India, December 2000. Springer Verlag.
18. A. Reyhani-Masoleh and M. A. Hasan. "Fast Normal Basis Multiplication Using General Purpose Processors". *Technical Report CORR 2001-25, Dept. of C & O, University of Waterloo, Canada*, April 2001.
19. M. Rosing. *Implementing Elliptic Curve Cryptography*. Manning Publications Company, 1999.
20. B. Sunar and C. K. Koc. "An Efficient Optimal Normal Basis Type II Multiplier". *IEEE Transactions on Computers*, 50(1):83–88, Jan. 2001.

Fast Multiplication of Integers
for Public-Key Applications

Gurgen H. Khachatrian, Melsik K. Kuregian,
Karen R. Ispiryan, and James L. Massey

Cylink Corporation, 3131 Jay Street, Santa Clara, CA 95056, USA
JamesMassey@compuserve.com

Abstract. A new method for multiplication of large integers and designed for efficient software implementation is presented and compared
with the well-known "schoolbook" method that is currently used for
both software and hardware implementations of public-key cryptographic
techniques. The comparison for the software-efficient method is made in
terms of the required number of basic operations on small integers. It
is shown that a significant performance gain is achieved by the new
software-efficient method for integers from 192 to 1024 bits in length,
which is the range of interest for all current public-key implementations. For 1024-bit integer multiplication, the savings over the schoolbook
method is conservatively estimated to be about 33%. A new method for
multiplication of large integers, which is analogous to the new software-
efficient method but is designed for efficient hardware implementation,
is also presented and compared to the schoolbook method in terms of
the number of processor clock cycles required.

1   Introduction

Multiplication of large integers plays a decisive role in the efficient implementation of all existing public-key cryptographic techniques such as the Diffie-
Hellman and the elliptic-curve key-agreement protocols and the Rivest-Shamir-
Adelman (RSA) cryptosystem. The standard "schoolbook" method of multiplication is today the most used method for integer multiplication in practical
public-key systems, cf. pp. 630-631 in [1]. For very large integers beyond the range
of practical interest in current cryptographic systems, more efficient methods of
multiplication are known, cf. [2]. One of these methods that has some practical
significance is that due to Karatsuba and Ofman [3], which reduces the asymptotic complexity of multiplying two N-bit integers to $0(N^{1.585})$ bit operations
compared to $0(N^2)$ bit operations for the schoolbook method.

The main contribution of this paper is a new software-efficient method of multiplication that improves on the schoolbook method when used in any current
public-key cryptographic application. Section 2 provides a brief description of
the schoolbook and the Karatsuba-Ofman methods. In Section 3 we introduce
the new software-efficient method of multiplication and compare its complexity with that of the schoolbook method. We also provide explicit performance

S. Vaudenay and A. Youssef (Eds.): SAC 2001, LNCS 2259, pp. 245–254, 2001.
© Springer-Verlag Berlin Heidelberg 2001

figures for the new soft ware-efficient method and for the schoolbook method
for integers in the range from 192 to 1024 bits in length, which is the range
of interest for all current public-key techniques. In Section 4 we introduce a
new hardware-efficient method of multiplication, which is analogous to the new
software-efficient method, and we compare its complexity in hardware with that
of the schoolbook method.

2   Schoolbook and Karatsuba-Ofman Methods

Let $\beta = 2^w$ be the radix in which integers are represented for calculation. Normally, w is the word size in bits of the processor on which the algorithm is
implemented. By an *n-symbol integer*, we will mean an integer between 0 and
$\beta^n - 1$ inclusive, i.e., an integer that can be written as an n-place radix-β integer.
Note that a *symbol* is a w-bit integer and that an n-symbol integer is an N-bit
integer where $N = nw$.

Let $A = (a_{n-1}, a_{n-2}, ..a_0)$ and $B = (b_{n-1}, b_{n-2}, ..b_0)$, where a_i and b_i are
w-bit integers, be two n-symbol integers. The result of their multiplication is
the $2n$-symbol integer $A \cdot B$ where

$$A \cdot B = \sum_{i=0}^{n-1} a_i \beta^i \sum_{j=0}^{n-1} b_j \beta^j$$

or, equivalently,

$$A \cdot B = \sum_{i=0}^{n-1} \sum_{j=0}^{n-1} a_i \cdot b_j \beta^{i+j}. \tag{1}$$

The *schoolbook method* of multiplication computes $A \cdot B$ essentially by carrying out the n^2 multiplications of w-bit integers in (1), one for each of the n^2
terms, and adding coefficients of like powers of β. The schoolbook method thus
requires n^2 multiplications of w-bit integers to calculate the product of two n-
symbol integers. The precise order in which the multiplications and additions
are carried out will not concern us here, but this order affects the "overhea d"
in implementing the schoolbook method.

For counting the number of additions required by the schoolbook method, it
is convenient first to write the $2w$-bit integer $a_i \cdot b_j$ in (1) as $c_{i,j}\beta + d_{i,j}$ where
$c_{i,j}$ and $d_{i,j}$ are w-bit integers. Then (1) can be written as

$$A \cdot B = \sum_{i=0}^{n-1} \sum_{j=0}^{n-1} (c_{i,j}\beta + d_{i,j})\beta^{i+j}$$

$$= \sum_{i=0}^{n-1} \sum_{j=0}^{n-1} c_{i,j}\beta^{i+j+1} + \sum_{i=0}^{n-1} \sum_{j=0}^{n-1} d_{i,j}\beta^{i+j} \tag{2}$$

We note that there are only $2n$ distinct powers of β among the $2n^2$ terms in (2).
Because each addition of coefficients of some power of β reduces the number of

terms by one, it follows that exactly $2n^2 - 2n = 2n(n-1)$ additions of w-bit integers are required to add the coefficients of like powers of β in (2). Thus, the schoolbook method requires $2n(n-1)$ additions of w-bit integers to calculate the product of two n-symbol integers. The additions of the terms in (2) with coefficients $c_{i,j}$ are called "carry additions" because these terms originate from the "overflow" into the next higher w-bits when two w-bit integers are multiplied. Of the $2n(n-1)$ additions of w-bit integers required by the schoolbook method, exactly half are such carry additions. Finally, we note that each addition of w-bit integers can result in a bit carry to another w-bit integer, which increments this latter integer by 1. The schoolbook method requires a maximum of $2n(n-1)$ such carry-bit additions.

The Karatsuba-Ofman method [3] is a divide-and-conquer technique for computing the components of $C = A \cdot B$ based on the following observation. Suppose that A and B are n-symbol integers where $n = 2^t$. Let $A = \beta^{2^{t-1}} A_1 + A_0$ and $B = \beta^{2^{t-1}} B_1 + B_0$ where A_0, A_1, B_0 and B_1 are 2^{t-1}-symbol integers. Then $A \cdot B = C_2 \beta^{2^t} + C_1 \beta^{2^{t-1}} + C_0$, where $C_0 = A_0 \cdot B_0$, $C_2 = A_1 \cdot B_1$, and $C_1 = (A_0 + A_1) \cdot (B_0 + B_1) - C_0 - C_2$. It follows that $C = A \cdot B$ can be computed by performing three multiplications of 2^{t-1}-symbol integers together with two additions and two subtractions of such integers. This procedure is iterated conceptually t times, i.e., until the integers reach the size of one symbol (w-bits), at which point the multiplications and additions are actually performed. This algorithm requires only $3^t \approx n^{1.585}$ multiplications of w-bit integers, compared to n^2 such multiplications for the schoolbook method. Combining Karatsuba-Ofman algorithm with schoolbook multiplication may have some practical significance. However, the recursive nature of the Karatsuba-Ofman algorithm results in such a significant overhead that its direct application to integers of the size used in current public-key cryptography is not efficient, cf. pp. 630-631 in [1].

3 A Software-Efficient Multiplication Method

3.1 The Underlying Idea

Our new software-efficient multiplication method is based on the formula

$$A \cdot B = \sum_{u=1}^{n-1} \sum_{v=0}^{u-1} (a_u + a_v) \cdot (b_u + b_v) \beta^{u+v} + 2 \sum_{u=0}^{n-1} a_u \cdot b_u \beta^{2u} - \sum_{v=0}^{n-1} \beta^v \sum_{u=0}^{n-1} a_u \cdot b_u \beta^u. \quad (3)$$

We will use the same notation here as we used for the schoolbook method except that we will write the radix as $\beta = 2^W$ rather than as $\beta = 2^w$ for a reason that will become apparent in Subsection 3.2.

It is easy to check by multiplying out and combining terms that (3) gives the correct result for multiplication. We note here for future use that $n(n-1)/2$ additions of W-bit integers are required to form the coefficients $a_u + a_v$ in (3) and another $n(n-1)/2$ such additions are required to form the coefficients $b_u + b_v$. To facilitate the counting of multiplications and further additions of W-bit integers needed to implement the multiplication formula (3), it is convenient to write

$$a_u + a_v = a_{u,v}^{\mathrm{cb}}\beta + a_{u,v}^{\mathrm{sum}}$$

where $a_{u,v}^{\mathrm{cb}}$ is the carry bit and $a_{u,v}^{\mathrm{sum}}$ is the least significant W-bits of the sum of the W-bit integers a_u and a_v. Using analogous notation for the sum of the W-bit integers b_u and b_v, we can write (3) in the manner

$$A \cdot B = \sum_{u=1}^{n-1}\sum_{v=0}^{u-1}(a_{u,v}^{\mathrm{cb}}\beta + a_{u,v}^{\mathrm{sum}}) \cdot (b_{u,v}^{\mathrm{cb}}\beta + b_{u,v}^{\mathrm{sum}}) \cdot \beta^{u+v}$$

$$+2\sum_{u=0}^{n-1} a_u \cdot b_u \beta^{2u} - \sum_{v=0}^{n-1}\beta^v \sum_{u=0}^{n-1} a_u \cdot b_u \beta^u$$

or, equivalently,

$$A \cdot B = \sum_{u=1}^{n-1}\sum_{v=0}^{u-1} a_{u,v}^{\mathrm{cb}} b_{u,v}^{\mathrm{cb}} \beta^{u+v+2}$$

$$+ \sum_{u=1}^{n-1}\sum_{v=0}^{u-1}(b_{u,v}^{\mathrm{cb}} a_{u,v}^{\mathrm{sum}} + a_{u,v}^{\mathrm{cb}} b_{u,v}^{\mathrm{sum}})\beta^{u+v+1}$$

$$+ \sum_{u=1}^{n-1}\sum_{v=0}^{u-1} a_{u,v}^{\mathrm{sum}} \cdot b_{u,v}^{\mathrm{sum}} \beta^{u+v}$$

$$+2\sum_{u=0}^{n-1} a_u \cdot b_u \beta^{2u}$$

$$- \sum_{v=0}^{n-1}\beta^v \sum_{u=0}^{n-1} a_u \cdot b_u \beta^u. \tag{4}$$

The only multiplications of W-bit integers occur within the third, fourth and fifth lines in (4). Each of the $\binom{n}{2} = \frac{n(n-1)}{2}$ terms in the third line requires one such multiplication. Each of the n terms within the sum on u in the fourth line also requires one such multiplication and these are the same products as are required in the fifth line. Thus, to implement the multiplication formula (3) requires a total of $\frac{n(n-1)}{2} + n = \frac{n(n+1)}{2}$ multiplications of W-bit integers, which we note is about half that required by the schoolbook method when we choose $W = w$ as is required for a direct comparison.

In counting additions of W-bit integers, we consider the worst case where all the carry bits $a_{u,v}^{\mathrm{cb}}$ and $b_{u,v}^{\mathrm{cb}}$ are equal to 1. It is again convenient to write the $2W$-bit integer $a_u \cdot b_u$ in (4) as $c_u\beta + d_u$ where c_u and d_u are W-bit integers, and to write $a_{u,v}^{\mathrm{sum}} \cdot b_{u,v}^{\mathrm{sum}}$ in (4) as $c_{u,v}^{\mathrm{sum}}\beta + d_{u,v}^{\mathrm{sum}}$ where $c_{u,v}^{\mathrm{sum}}$ and $d_{u,v}^{\mathrm{sum}}$ are W-bit integers. We can then rewrite (4) for this worst case as

$$A \cdot B = \sum_{u=1}^{n-1}\sum_{v=0}^{u-1} \beta^{u+v+2}$$

$$+ \sum_{u=1}^{n-1} \sum_{v=0}^{u-1} a_{u,v}^{\mathrm{sum}} \beta^{u+v+1} + \sum_{u=1}^{n-1} \sum_{v=0}^{u-1} b_{u,v}^{\mathrm{sum}} \beta^{u+v+1}$$

$$+ \sum_{u=1}^{n-1} \sum_{v=0}^{u-1} c_{u,v}^{\mathrm{sum}} \beta^{u+v+1} + \sum_{u=1}^{n-1} \sum_{v=0}^{u-1} d_{u,v}^{\mathrm{sum}} \beta^{u+v}$$

$$+ \sum_{u=0}^{n-1} (c_u + c_u) \beta^{2u+1} + \sum_{u=0}^{n-1} (d_u + d_u) \beta^{2u}$$

$$- \sum_{u=0}^{n} \sum_{v=0}^{n-1} (c_{u-1} + d_u) \beta^{u+v}$$

with the convention that $c_j = d_j = 0$ for $j < 0$ and for $j \geq n$. Upon setting $e_u = c_{u-1} + d_u$ and then $u = i - v$ in the last line, we can rewrite this equivalently as

$$A \cdot B = \sum_{u=1}^{n-1} \sum_{v=0}^{u-1} \beta^{u+v+2}$$

$$+ \sum_{u=1}^{n-1} \sum_{v=0}^{u-1} a_{u,v}^{\mathrm{sum}} \beta^{u+v+1} + \sum_{u=1}^{n-1} \sum_{v=0}^{u-1} b_{u,v}^{\mathrm{sum}} \beta^{u+v+1}$$

$$+ \sum_{u=1}^{n-1} \sum_{v=0}^{u-1} c_{u,v}^{\mathrm{sum}} \beta^{u+v+1} + \sum_{u=1}^{n-1} \sum_{v=0}^{u-1} d_{u,v}^{\mathrm{sum}} \beta^{u+v}$$

$$+ \sum_{u=0}^{n-1} (c_u + c_u) \beta^{2u+1} + \sum_{u=0}^{n-1} (d_u + d_u) \beta^{2u}$$

$$- \sum_{i=0}^{2n-1} [\sum_{v=0}^{n-1} e_{i-v}] \beta^i. \tag{5}$$

The terms $e_i = c_{i-1} + d_i$ for $i = 0, 1, \ldots, n$ require $n - 1$ additions of W-bit integers for their formation because the terms for $i = 0$ and $i = n$, namely d_0 and c_{n-1} respectively, require no additions. We next consider the number of additions of W-bit integers required to form the coefficients

$$s_i = \sum_{v=0}^{n-1} e_{i-v} \quad \text{for } i = 0, 1, \ldots, 2n - 1$$

that appear in the fifth line of (5). We observe that we can rewrite this sum separately over two ranges of the index as

$$s_i = \sum_{j=0}^{i} e_i = s_{i-1} + e_i \quad \text{for } i = 0, 1, \ldots n - 1 \tag{6}$$

and

$$s_{2n-1-i} = \sum_{j=0}^{i} e_{n-j} = s_{2n-i} + e_{n-i} \quad \text{for } i = 0, 1, \ldots n - 1. \tag{7}$$

where we have taken $s_0 = s_{2n} = 0$. We see that $n - 2$ additions of W-bit integers are required to form the nested sums in (6) and another such $n - 2$ additions are required to form the nested sums in (7). Hence a total of $3(n - 1)$ additions of W-bit integers are required to form all the coefficients in the fifth line of (5).

The summation in the first line of (5) concerns only carry bits, which we will consider later. There are $n(n - 1)$ terms in the second line, another $n(n - 1)$ terms in the third line, $4n$ terms in the fourth line, and $2n$ terms in the fifth line—a total of $2n^2 + 4n$ terms. But there are only $2n$ distinct powers of β in (5) so that $2n^2 + 4n - 2n = 2n^2 + 2n$ additions of W-bit numbers are required to combine the like powers of β. To this, we must add the $n(n - 1)$ additions of W-bit integers required to form the coefficients $a_u + a_v$ and $b_u + b_v$ in (3) as well as the $3(n - 1)$ additions required to form the coefficients in the last line of (5). This gives a total of $3n^2 + 4n - 3$ additions of W-bit integers required to implement the multiplication formula (3). We note that this is greater by a factor of about $\frac{3}{2}$ than the $2n(n - 1)$ additions required by the schoolbook method when we choose $W = w$ as is required for a direct comparison.

Finally, we note that the first line of (5) specifies $\frac{1}{2}n(n-1)$ additions of carry bits (in this worst case) and each of the $3n^2 + 4n - 3$ additions of W-bit numbers can also result in a carry bit. Thus, to implement the multiplication formula (3) requires a maximum of $\frac{7}{2}n(n + 1) - 3$ carry-bit additions.

3.2 Achieving Efficiency

As we have just seen, the direct implementation of formula (3) for multiplication of nW-bit integers requires only about half as many multiplications, but about 50% more additions, of W-bit integers compared to the schoolbook method when we take $W = w$. To convert the multiplication formula (3) into an efficient method for multiplication of N-bit integers on a w-bit processor, we first set $N = nsw$ and then split the problem of multiplication into (1) the problem of multiplying N-bit integers using a virtual processor with word size $W = sw$, followed by (2) the problem of implementing the necessary multiplications and additions of W-bit integers using the actual processor with word size w. We solve the first problem by implementing the multiplication formula (3), after which we solve the second problem by implementing the necessary multiplications of sw-bit integers by the schoolbook method. We now count the number of multiplications and additions of w-bit integers required by this "hybrid method" for multiplying N-bit integers.

As was shown in Subsection 3.1, the multiplication of N-bit integers, where $N = nW$, according to the multiplication formula (3) requires $\frac{n(n+1)}{2}$ multiplications of W-bit integers where $W = sw$. Each such multiplication when performed by the schoolbook method requires s^2 multiplications and $2s(s - 1)$ additions of w-bit numbers, as well as $2s(s - 1)$ carry-bit additions. Thus, *the multiplications performed in the first step of the hybrid method* result intal numbers of w-bit integer operations shown in the following table: The multiplication of N-bit integers, where $N = nW$, according to the multiplication formula (3) requires $3n^2 + 4n - 3$ additions of W-bit integers where $W = sw$. Each such addition is

multiplications	additions	carry-bit additions
$\frac{1}{2}s^2 n(n+1)$	$s(s-1)n(n+1)$	$s(s-1)n(n+1)$

equivalent to s additions of w-bit integers and s carry-bit additions. Thus, the *the additions performed in the first step of the hybrid method* result in the total numbers of w-bit integer operations shown in the following table: Finally, the

multiplications	additions	carry-bit additions
0	$(3n^2 + 4n - 3)s$	$(3n^2 + 4n - 3)s$

multiplication of N-bit integers, where $N = nW$, according to the multiplication formula (3) requires in the worst case $\frac{7}{2}n(n+1) - 3$ carry-bit additions for W-bit integers where $W = sw$. Each such carry-bit addition for sw-bit integers is equivalent to a single carry-bit addition for w-bit integers so that *the carry-bit additions performed in the first step of the hybrid method* result in the total numbers of w-bit integer operations shown in the following table:

multiplications	additions	carry-bit additions
0	0	$\frac{7}{2}n(n+1) - 3$

Tallying the counts in the three previous tables gives the figures shown in the following table:

For ease of comparison, we include here the table of counts for schoolbook method as calculated in Section 2.

3.3   Numerical Examples

Example 1: Consider the multiplication of 1024-bit integers [where we note that 1024 is a length commonly used for current implementations of the RSA cryptosystem and of the Diffie-Hellman key agreement protocol] on a processor with word size $w = 16$ bits. As a basis for comparison, we assume that one 16-bit addition constitutes 1 unit of computation as also does one carry-bit addition, but that one 16-bit multiplication constitutes 2 units of computation.

The specifications $N = nsw = 1024$ and $w = 16$ give $ns = 64$ and hence the allowed values of (n, s) are $(1, 64)$, $(2, 32)$, $(4, 16)$, $(8, 8)$, $(16, 4)$, $(32, 8)$ and $(64, 1)$. Calculating the cost for each of these choices with the aid of the values in Table 1 shows that the choice $n = s = 8$ yields the minimum cost of 16,457 computational units for the new software-efficient method, but the choice $n = 4$ and $s = 16$ is nearly as good with a cost of 16,739 units. For the choice $n = s = 8$, the number of multiplications, additions and carry-bit additions are 2304, 5800 and 6049, respectively. By comparison, we calculate from Table 2 that the

Table 1. Total counts of w-bit integer operations for the software-efficient multiplication method for nsw-bit integers

multiplications	additions	carry-bit additions
$\frac{1}{2}s^2n(n+1)$	$s[(s+2)n^2 + (s+3)n - 3]$	$s[(s+2)n^2 + (s+3)n - 3]$ $+\frac{7}{2}n(n+1) - 3$

Table 2. Total counts of w-bit integer operations for the schoolbook multiplication method for nsw-bit integers

multiplications	additions	carry-bit additions
$(sn)^2$	$2sn(sn - 1)$	$2sn(sn - 1)$

schoolbook method has a cost of 24,320 computational units arising from the 4096 multiplications, 8064 additions, and 8064 carry-bit additions that must be performed. In this example, the new software-efficient multiplication method uses about one-third less computation than does the schoolbook method.

Example 2: Consider the multiplication of 192-bit integers [which is one of the lengths for the Elliptic curve system recommended for the FIPS 186-2 standard] on a processor with word size $w = 8$ bits. Again we assume that one 8-bit addition or one carry-bit addition constitutes 1 unit of compution, but that one 8-bit multiplication constitutes 2 units of computation.

The specifications $N = nsw = 192$ and $w = 8$ give $ns = 24$ and hence the allowed values of (n, s) are $(1, 24)$, $(2, 12)$, $(3, 8)$, $(4, 6)$, $(6, 4)$, $(8, 3)$, $(12, 2)$ and $(24, 1)$. Calculating the cost for each of these choices with the aid of the values in Table 1 shows that the choice $n = 4$ and $s = 6$ yields the minimum cost of 2719 computational units for the new software-efficient method, but the choice $n = 3$ and $s = 8$ is virtually as good with a cost of 2727 units. For the choice $n = 4$ and $s = 6$, the number of multiplications, additions and carry-bit additions are 360, 966 and 1033, respectively. By comparison, we calculate from Table 2 that the schoolbook method has a cost of 3360 computational units arising from the 576 multiplications, 1104 additions, and 1104 carry-bit additions that must be performed. In this example, the new software-efficient multiplication method uses about 19% less computation than does the schoolbook method.

It should be pointed out that *actual performance results for the new software-efficient multiplication method may well be substantially better than predicted by our analysis*, which was made using worst-case assumptions. For instance, our 8-bit implementation of the new software-efficient multiplication method on a Pentium 2 processor for the parameters of Example 2 actually used about 40% less computation than did the schoolbook method, rather than only 19% less as our analysis had predicted.

4   A Hardware-Efficient Multiplication Method

The following formula, analogous to (3), is the basis for our new hardware-efficient method of multiplication:

$$A \cdot B = (\sum_{v=0}^{n-1} \beta^v)(\sum_{u=0}^{n-1} a_u \cdot b_u \beta^u) + \sum_{u=1}^{n-1}\sum_{v=0}^{u-1} (a_u - a_v) \cdot (b_v - b_u)\beta^{u+v}. \qquad (8)$$

The complexity of implementing multiplication according to (8) is comparable to that for implementing multiplication according to (3). Which method is superior depends on the computational environment. For example, using (8) will give fewer carry-bit additions but will require sign checks. In general the use of (8) is better suited to hardware implementations and therefore we now analyze the use of (8) in a hardware implementation by estimating the number of clock cycles needed to multiply two N-bit numbers A and B. We will compare this performance to a hardware implementation of the schoolbook method using the shift-and-add technique, which requires N clock cycles when all N bits can be processed in parallel.

Let $N = nW$ and consider the multiplication formula (8) where a_i and b_i are W-bit numbers. Calculating all $n(n-1)$ required differences $(a_i - a_j)$ and $(b_i - b_j)$ in the second double summation of (8) requires $\frac{n(n-1)}{n} = n - 1$ clock cycles if the same resources as for the schoolbook method are used. Formula (8) requires $\frac{n(n+1)}{2}$ multiplications of W-bit numbers. Because n such multiplications can be performed in parallel, another $W\lceil\frac{(n+1)}{2}\rceil$ clock cycles are needed. Summing the results of these multiplications requires in the worst case an additional $\frac{n(n+1)}{2}$ clock cycles. The total number of clock cycles required for the multiplication $A \cdot B$ is thus

$$W(\lceil\frac{(n+1)}{2}\rceil) + \frac{n(n+1)}{2} + n - 1, \qquad (9)$$

which is about half that required by the schoolbook method for large $N = nW$.

Example 3: Suppose that A and B are 1024 bit numbers and consider the choice $W = 128$ and $n = 8$. Multiplication according to (9) requires at most 683 clock cycles compared to 1024 clock cyles for the schoolbook method using the shift-and-add technique, a reduction of 33%.

5   Conclusion

The analyses of the new software-efficient multiplication method and of the new hardware-efficient multiplication method both show that a significant performance improvement over the schoolbook method can be obtained for all current applications in public key cryptography. Moreover, our complexity estimates for the new methods are conservative–actual gains can exceed those predicted, as was pointed out in Example 2.

References

1. A. Menezes, P. van Oorschot, S. Vanstone: Handbook of Applied Cryptography. CRC press, Boca Raton and New York (1997).
2. D. E. Knuth: The Art of Computer Programming, Vol. 2, 2nd Ed. Addison-Wesley, Reading, Mass. (1981).
3. A. Karatsuba, Yu. Ofman: Multiplication of Multidigit Numbers on Automata. Soviet Physics Doclady, **7**, (1963) 595-596.

Fast Simultaneous Scalar Multiplication
on Elliptic Curve with Montgomery Form

Toru Akishita

Sony Corporation, 6-7-35 Kitashinagawa Shinagawa-ku, Tokyo, 141-0001, Japan
akishita@pal.arch.sony.co.jp

Abstract. We propose a new method to compute x-coordinate of $kP +$ lQ simultaneously on the elliptic curve with Montgomery form over $\mathbb{F}_p$ without precomputed points. To compute x-coordinate of $kP + lQ$ is required in ECDSA signature verification. The proposed method is about 25% faster than the method using scalar multiplication and the recovery of Y-coordinate of kP and lQ on the elliptic curve with Montgomery form over $\mathbb{F}_p$, and also slightly faster than the simultaneous scalar multiplication on the elliptic curve with Weierstrass form over $\mathbb{F}_p$ using NAF and mixed coordinates. Furthermore, our method is applicable to Montgomery method on elliptic curves over $\mathbb{F}_{2^n}$.

1 Introduction

Elliptic curve cryptography was first proposed by Koblitz [10] and Miller [15]. In recent years, efficient algorithms and implementation techniques of elliptic curves over $\mathbb{F}_p$ [3,4], $\mathbb{F}_{2^n}$ [7,13] and $\mathbb{F}_{p^n}$ [2,9,12] has been investigated. In particular, the scalar multiplication on the elliptic curve with Montgomery form over $\mathbb{F}_p$ can be computed efficiently without precomputed points [16], and is immune to timing attacks [11,17]. This method is extended to elliptic curves over $\mathbb{F}_{2^n}$ [14].

We need to compute $kP + lQ$, where P and Q are points on the elliptic curve and k, l are integers less than the order of the base point, in Elliptic Curve Digital Signature Algorithm (ECDSA) signature verification [1]. On the elliptic curve with Weierstrass form, $kP + lQ$ can be efficiently computed by a simultaneous multiple point multiplication [3,7,19], which we call a *simultaneous scalar multiplication*. On the other hand, the simultaneous scalar multiplication on the elliptic curve with Montgomery form has not been proposed yet. Then we propose it and call it *Montgomery simultaneous scalar multiplication*. This method is about 25% faster than the method using Montgomery scalar multiplication and the recovery of Y-coordinate of kP and lQ, and about 1% faster than Weierstrass simultaneous scalar multiplication over $\mathbb{F}_p$ using NAF [8] and mixed coordinates [4]. Moreover, our method is applicable to elliptic curves over $\mathbb{F}_{2^n}$.

This paper is described as follows. Section 2 presents preliminaries including arithmetic over the elliptic curve with Montgomery form and Weierstrass form. In Section 3, we describe the new method, Montgomery simultaneous scalar

S. Vaudenay and A. Youssef (Eds.): SAC 2001, LNCS 2259, pp. 255–267, 2001.
© Springer-Verlag Berlin Heidelberg 2001

multiplication. Section 4 presents comparison of our method with others and Section 5 presents implementation results. We then apply our method to elliptic curves over $\mathbb{F}_{2^n}$ in Section 6 and conclude in Section 7.

2   Preliminaries

2.1   Elliptic Curve with Montgomery Form

Montgomery introduced the new form of elliptic curve over $\mathbb{F}_p$ [16]. For $A, B \in \mathbb{F}_p$, the elliptic curve with Montgomery form E_M is represented by

$$E_M : By^2 = x^3 + Ax^2 + x \quad ((A^2 - 4)B \neq 0). \tag{1}$$

We remark that the order of any elliptic curve with Montgomery form is always divisible by 4.

In affine coordinates (x, y), the x-coordinate of the sum of the two points on E_M can be computed without the y-coordinates of these points if the difference between these points is known. Affine coordinates (x, y) can be transformed into projective coordinates (X, Y, Z) by $x = X/Z, y = Y/Z$. Equation (1) can also be transformed as

$$E_M : BY^2 Z = X^3 + AX^2 Z + XZ^2.$$

Let $P_0 = (X_0, Y_0, Z_0)$ and $P_1 = (X_1, Y_1, Z_1)$ be points on E_M and $P_2 = (X_2, Y_2, Z_2) = P_1 + P_0$, $P_3 = (X_3, Y_3, Z_3) = P_1 - P_0$. Addition formulas and doubling formulas are described as follows.

Addition formulas $P_2 = P_1 + P_0$ $(P_1 \neq P_0)$

$$X_2 = Z_3((X_0 - Z_0)(X_1 + Z_1) + (X_0 + Z_0)(X_1 - Z_1))^2$$
$$Z_2 = X_3((X_0 - Z_0)(X_1 + Z_1) - (X_0 + Z_0)(X_1 - Z_1))^2 \tag{2}$$

Doubling formulas $P_2 = 2P_0$

$$4X_0 Z_0 = (X_0 + Z_0)^2 - (X_0 - Z_0)^2$$
$$X_2 = (X_0 + Z_0)^2 (X_0 - Z_0)^2$$
$$Z_2 = (4X_0 Z_0)((X_0 - Z_0)^2 + ((A + 2)/4)(4X_0 Z_0))$$

$P_2 = (X_2, Z_2)$ can be computed without Y-coordinate. Since the computational cost of a field addition and subtraction is much lower than that of a field multiplication and squaring, we can ignore it. The computational cost of the addition formulas is $4M + 2S$, where M and S respectively denote that of a field multiplication and squaring. If $Z_3 = 1$, the computational cost of the addition formulas is $3M + 2S$. If $(A + 2)/4$ is precomputed, the computational cost of the doubling formulas is also $3M + 2S$.

Let $(k_t \cdots k_1 k_0)_2$ be the binary representation of k with $k_t = 1$. To compute the scalar multiplication kP from $P = (x, y)$, we hold $\{m_i P, (m_i + 1)P\}$ for $m_i = (k_t \cdots k_i)_2$. If $k_i = 0$, $m_i P = 2m_{i+1}P$ and $(m_i + 1)P = (m_{i+1} + 1)P + m_{i+1}P$.

Otherwise, $m_i P = (m_{i+1}+1)P + m_{i+1}P$ and $(m_i+1)P = 2(m_{i+1}+1)P$. We can compute $\{kP, (k+1)P\}$ from $\{P, 2P\}$. Montgomery scalar multiplication requires the addition formulas $t - 1$ times and the doubling formulas t times. Since the difference between $(m_{i+1} + 1)P$ and $m_{i+1}P$ is P, we can assume $(X_3, Z_3) = (x, 1)$ in addition formulas (2). The computational cost of Montgomery scalar multiplication kP is $(6|k| - 3)M + (4|k| - 2)S$, where $|k|$ is the bit length of k.

In ECDSA signature verification, we need to compute x-coordinate of $kP + lQ$, where P, Q are points on the elliptic curve and k, l are integers less than the order of the base point. kP and lQ can be computed using Montgomery scalar multiplication, but $kP + lQ$ cannot be computed from kP and lQ using formulas (2) because the difference between kP and lQ is unknown. Therefore, the recovery of Y-coordinate of kP and lQ is required to compute $kP + lQ$ from kP and lQ using other addition formulas. The method of recovering Y-coordinate is described in [18]. If $kP = (X_0, Z_0)$, $(k + 1)P = (X_1, Z_1)$ and $P = (x, y)$, we can recover Y-coordinate of $kP = (X, Y, Z)$ as:

$$X = 2ByZ_0Z_1X_0$$
$$Y = Z_1((X_0 + xZ_0 + 2AZ_0)(X_0x + Z_0) - 2AZ_0{}^2) - (X_0 - xZ_0)^2X_1$$
$$Z = 2ByZ_0Z_1Z_0$$

The computational cost of recovering Y-coordinate is $12M + S$.

To compute x-coordinate of $kP + lQ$, we require these 6 steps.

Step1 Compute kP using Montgomery scalar multiplication
Step2 Recover Y-coordinate of kP
Step3 Compute lQ using Montgomery scalar multiplication
Step4 Recover Y-coordinate of lQ
Step5 Compute $kP + lQ$ from kP and lQ in projective coordinates
Step6 Compute x-coordinate of $kP + lQ$ using $x = X/Z$

The computational cost of Step5 is $10M + 2S$ and that of Step6 is $M + I$, where I denotes that of a field inversion. We can assume $|k| = |l|$ without loss of generality. The computational cost of x-coordinate of $kP + lQ$ is $(12|k| + 29)M + 8|k|S + I$.

2.2 Simultaneous Scalar Multiplication on Elliptic Curve with Weierstrass Form

For $a, b \in \mathbb{F}_p$, the elliptic curve with Weierstrass form E_W is represented by

$$E_W : y^2 = x^3 + ax + b \ (4a^2 + 27b^3 \neq 0).$$

We remark that all elliptic curves with Montgomery form can be transformed into Weierstrass form, but not all elliptic curves with Weierstrass form can be transformed into Montgomery form.

$kP + lQ$ can be computed simultaneously on the elliptic curve with Weierstrass form without precomputed points [19]. This method is known as Shamir's trick [5]. On the elliptic curve with Weierstrass form over $\mathbb{F}_p$, the most effective method computing $kP + lQ$ without precomputed points is the simultaneous scalar multiplication using non-adjacent form (NAF) $(k'_{t'} \cdots k'_1 k'_0)$, where $k'_i \in \{0, \pm 1\}$ $(0 \le i \le t')$, and mixed coordinates [4]. In [3], Weierstrass simultaneous scalar multiplication using window method and mixed coordinates is described. This method is faster than Weierstrass simultaneous scalar multiplication using NAF, but requires much more memories where the points are stored. That is why we pick Weierstrass simultaneous scalar multiplication using NAF and mixed coordinates in this section.

NAF has the property that no two consecutive coefficients k'_i are non-zero and the average density of non-zero coefficients is approximately $1/3$. In mixed coordinates, we use the addition formulas of $J^m \leftarrow J + A$ for the additions, the doubling formulas of $J \leftarrow 2J^m$ for the doublings ahead of addition, and the doubling formulas of $J^m \leftarrow 2J^m$ for the doublings ahead of doubling, where J, A, and J^m respectively denote Jacobian coordinate, affine coordinate, and modified Jacobian coordinate. This method is described as follows.

Algorithm 1: Weierstrass Simultaneous Scalar Multiplication using NAF and mixed coordinates

---

Input: $k = (k_t \cdots k_1 k_0)_2$, $l = (l_t \cdots l_1 l_0)_2$, $P, Q \in E_W$ (k_t or $l_t = 1$).
Output: x-coordinate of $W = kP + lQ$.

1. Compute $P + Q$, $P - Q$.
2. Let $(k'_{t'} \cdots k'_1 k'_0)$ and $(l'_{t'} \cdots l'_1 l'_0)$ be NAF of k and l ($k'_{t'}$ or $l'_{t'} = 1$).
3. $W \leftarrow k'_{t'} P + l'_{t'} Q$.
4. For i from $t' - 1$ downto 0 do
    4.1 if $(k'_i, l'_i) = (0, 0)$ then
        $W \leftarrow 2W$ $(J^m \leftarrow 2J^m)$;
    4.2 else then
        $W \leftarrow 2W$ $(J \leftarrow 2J^m)$,
        $W \leftarrow W + (k'_i P + l'_i Q)$ $(J^m \leftarrow J + A)$.
5. Compute x-coordinate of W.

At step 1, $P + Q$, $P - Q$ are computed in affine coordinates and their computational cost is $4M + 2S + I$. In mixed coordinates, the computational cost of the addition formulas of $J^m \leftarrow J + A$, the doubling formulas of $J \leftarrow 2J^m$, and the doubling formulas of $J^m \leftarrow 2J^m$ are respectively $9M + 5S$, $3M + 4S$, and $4M + 4S$. Since the probability that $(k'_i, l'_i) = (0, 0)$ is $(1 - 1/3)^2 = 4/9$, we repeat step 4.1 $4|k|/9$ times and step 4.2 $5|k|/9$ times if $t' = t + 1$. The computational cost of step 4.1 is $(4|k|/9) \cdot (4M + 4S)$ and that of step 4.2 is $(5|k|/9) \cdot (12M + 9S)$. This shows that the computational cost of step 4 is $76|k|/9 \cdot M + 61|k|/9 \cdot S$. The computational cost of step 5 is $M + S + I$ by $x = X/Z^2$. Therefore, the computational cost of x-coordinate of $kP + lQ$ is $(76|k| + 45)/9 \cdot M + (61|k| + 27)/9 \cdot S + 2I$.

3   Proposed Method — Montgomery Simultaneous Scalar Multiplication

Now we propose the new method to compute $kP + lQ$ simultaneously on the elliptic curve with Montgomery form over $\mathbb{F}_p$.

At first, we define a set of four points G_i,

$$G_i = \left\{ \begin{array}{c} m_iP + n_iQ, \\ m_iP + (n_i + 1)Q, \\ (m_i + 1)P + n_iQ, \\ (m_i + 1)P + (n_i + 1)Q \end{array} \right\}, \tag{3}$$

for $m_i = (k_t \cdots k_i)_2$, $n_i = (l_t \cdots l_i)_2$. Now, we present how to compute G_i from G_{i+1} in every case of (k_i, l_i).

1. $(k_i, l_i) = (0, 0)$
   Since $m_i = 2m_{i+1}$ and $n_i = 2n_{i+1}$, we can compute all elements of G_i from G_{i+1} as:

   $$m_iP + n_iQ = 2(m_{i+1}P + n_{i+1}Q)$$
   $$m_iP + (n_i + 1)Q = (m_{i+1}P + (n_{i+1} + 1)Q) + (m_{i+1}P + n_{i+1}Q)$$
   $$(m_i + 1)P + n_iQ = ((m_{i+1} + 1)P + n_{i+1}Q) + (m_{i+1}P + n_{i+1}Q)$$
   $$(m_i + 1)P + (n_i + 1)Q = ((m_{i+1} + 1)P + n_{i+1}Q) + (m_{i+1}P + (n_{i+1} + 1)Q)$$

   All elements of G_i can be computed without $(m_{i+1} + 1)P + (n_{i+1} + 1)Q \in G_{i+1}$.

2. $(k_i, l_i) = (0, 1)$
   Since $m_i = 2m_{i+1}$ and $n_i = 2n_{i+1} + 1$, we can compute all elements of G_i from G_{i+1} as:

   $$m_iP + n_iQ = (m_{i+1}P + (n_{i+1} + 1)Q)) + (m_{i+1}P + n_{i+1}Q)$$
   $$m_iP + (n_i + 1)Q = 2(m_{i+1}P + (n_{i+1} + 1)Q)$$
   $$(m_i + 1)P + n_iQ = ((m_{i+1} + 1)P + (n_{i+1} + 1)Q) + (m_{i+1}P + n_{i+1}Q)$$
   $$(m_i + 1)P + (n_i + 1)Q = ((m_{i+1} + 1)P + (n_{i+1} + 1)Q)$$
   $$+ (m_{i+1}P + (n_{i+1} + 1)Q)$$

   All elements of G_i can be computed without $(m_{i+1} + 1)P + n_{i+1}Q \in G_{i+1}$.

3. $(k_i, l_i) = (1, 0)$
   Since $m_i = 2m_{i+1} + 1$ and $n_i = 2n_{i+1}$, we can compute all elements of G_i from G_{i+1} as:

   $$m_iP + n_iQ = ((m_{i+1} + 1)P + n_{i+1}Q) + (m_{i+1}P + n_{i+1}Q)$$
   $$m_iP + (n_i + 1)Q = ((m_{i+1} + 1)P + (n_{i+1} + 1)Q) + (m_{i+1}P + n_{i+1}Q)$$
   $$(m_i + 1)P + n_iQ = 2((m_{i+1} + 1)P + n_{i+1}Q)$$
   $$(m_i + 1)P + (n_i + 1)Q = ((m_{i+1} + 1)P + (n_{i+1} + 1)Q)$$
   $$+ ((m_{i+1} + 1)P + n_{i+1}Q)$$

All elements of G_i can be computed without $m_{i+1}P + (n_{i+1} + 1)Q \in G_{i+1}$.

4. $(k_i, l_i) = (1, 1)$

Since $m_i = 2m_{i+1} + 1$ and $n_i = 2n_{i+1} + 1$, we can compute all elements of G_i from G_{i+1} as:

$$
\begin{aligned}
m_iP + n_iQ &= ((m_{i+1} + 1)P + n_{i+1}Q)) \\
&\quad + (m_{i+1}P + (n_{i+1} + 1)Q) \\
m_iP + (n_i + 1)Q &= ((m_{i+1} + 1)P + (n_{i+1} + 1)Q) \\
&\quad + (m_{i+1}P + (n_{i+1} + 1)Q) \\
(m_i + 1)P + n_iQ &= ((m_{i+1} + 1)P + (n_{i+1} + 1)Q) \\
&\quad + ((m_{i+1} + 1)P + n_{i+1}Q) \\
(m_i + 1)P + (n_i + 1)Q &= 2((m_{i+1} + 1)P + (n_{i+1} + 1)Q)
\end{aligned}
$$

All elements of G_i can be computed without $m_{i+1}P + n_{i+1}Q \in G_{i+1}$.

In every case, all elements of G_i can be computed from G_{i+1} without $(m_{i+1} + 1 - k_i)P + (n_{i+1} + 1 - l_i)Q \in G_{i+1}$. When we define a set of three points G'_i,

$$G'_i = G_i - \{(m_i + 1 - k_{i-1})P + (n_i + 1 - l_{i-1})Q\}, \tag{4}$$

all elements of G_i can be computed from G'_{i+1}. Therefore, we can compute G'_i from G'_{i+1}. The way to compute G'_i fro m G'_{i+1} depends on $(k_i, l_i, k_{i-1}, l_{i-1})$, since computing G_i from G'_{i+1} depends on (k_i, l_i) while extracting G'_i from G_i depends on (k_{i-1}, l_{i-1}).

Example 1 $(k_i, l_i, k_{i-1}, l_{i-1}) = (0, 0, 0, 0)$

m_i, n_i, G'_{i+1} and G'_i would be described as: $m_i = 2m_{i+1}$, $n_i = 2n_{i+1}$,

$$
G'_{i+1} = \left\{ \begin{array}{l} m_{i+1}P + n_{i+1}Q, \\ m_{i+1}P + (n_{i+1} + 1)Q, \\ (m_{i+1} + 1)P + n_{i+1}Q \end{array} \right\}, G'_i = \left\{ \begin{array}{l} m_iP + n_iQ, \\ m_iP + (n_i + 1)Q, \\ (m_i + 1)P + n_iQ \end{array} \right\}
$$

Therefore, we can compute G'_i from G'_{i+1} as:

$$
\begin{aligned}
m_iP + n_iQ &= 2(m_{i+1}P + n_{i+1}Q) \\
m_iP + (n_i + 1)Q &= (m_{i+1}P + (n_{i+1} + 1)Q) + (m_{i+1}P + n_{i+1}Q) \quad (5) \\
(m_i + 1)P + n_iQ &= ((m_{i+1} + 1)P + n_{i+1}Q) + (m_{i+1}P + n_{i+1}Q)
\end{aligned}
$$

If we define $G'_i = \{T_0[i], T_1[i], T_2[i]\}$, equations (5) can be described as:

$$
\begin{aligned}
T_0[i] &= 2T_0[i + 1] \\
T_1[i] &= T_1[i + 1] + T_0[i + 1] \ (T_1[i + 1] - T_0[i + 1] = Q) \\
T_2[i] &= T_2[i + 1] + T_0[i + 1] \ (T_2[i + 1] - T_0[i + 1] = P).
\end{aligned}
$$

Example 2 $(k_i, l_i, k_{i-1}, l_{i-1}) = (0, 1, 1, 0)$

m_i, n_i, G'_{i+1} and G'_i would be described as: $m_i = 2m_{i+1}$, $n_i = 2n_{i+1} + 1$,

$$G'_{i+1} = \left\{ \begin{array}{l} m_{i+1}P + n_{i+1}Q, \\ m_{i+1}P + (n_{i+1} + 1)Q, \\ (m_{i+1} + 1)P + (n_{i+1} + 1)Q \end{array} \right\}, G'_i = \left\{ \begin{array}{l} m_iP + n_iQ, \\ (m_i + 1)P + n_iQ, \\ (m_i + 1)P + (n_i + 1)Q \end{array} \right\}$$

Therefore, we can compute G'_i from G'_{i+1} as:

$$m_iP + n_iQ = (m_{i+1}P + (n_{i+1} + 1)Q) + (m_{i+1}P + n_{i+1}Q)$$
$$(m_i + 1)P + n_iQ = ((m_{i+1} + 1)P + (n_{i+1} + 1)Q)$$
$$+ (m_{i+1}P + n_{i+1}Q) \qquad (6)$$
$$(m_i + 1)P + (n_i + 1)Q = ((m_{i+1} + 1)P + (n_{i+1} + 1)Q)$$
$$+ (m_{i+1}P + (n_{i+1} + 1)Q)$$

Equations (6) can be described as:

$$T_0[i] = T_1[i+1] + T_0[i+1] \ (T_1[i+1] - T_0[i+1] = Q)$$
$$T_1[i] = T_2[i+1] + T_0[i+1] \ (T_2[i+1] - T_0[i+1] = P + Q)$$
$$T_2[i] = T_2[i+1] + T_1[i+1] \ (T_2[i+1] - T_1[i+1] = P)$$

From equations (3) and (4), we can define G'_{t+1} as an initial set of G'_i as:

$$G_{t+1} = \{\mathcal{O}, Q, P, P + Q\}$$
$$G'_{t+1} = G_{t+1} - \{(1 - k_t)P + (1 - l_t)Q\},$$

where $\mathcal{O}$ is the point at infinity. By calculating G'_i from G'_{i+1} repeatedly, we can compute G'_1 from G'_{t+1} whereas $kP + lQ$ will be computed from G'_1. Our method to compute x-coordinate of $kP + lQ$ can be described as next page. $T_i + T_j \ (P)$ means that the difference between T_i and T_j is P.

At step1, $P - Q$ must be computed because the difference between $(m_i + 1)P + n_iQ$ and $m_iP + (n_i + 1)Q$ is $P - Q$.

We consider about the computational cost of the proposed method. At step1, we compute $P + Q, P - Q$ in affine coordinates and their computation cost is $4M + 2S + I$. At step2, 3 and 4, we use projective coordinates in the same way as Section 2.1. We assume $|k| = |l|$ as referred in the previous section. At step3, we require the addition formulas twice and the doubling formulas once, or the addition formulas three times per bit of k. In either case, since the computational cost per bit of k is $9M + 6S$, the computational cost of step3 is $9(|k| - 1)M + 6(|k| - 1)S$. The computational cost of step4 is $3M + 2S$ and that of step5 is $M + I$. Therefore, the computational cost of the proposed method is $(9|k| - 1)M + (6|k| - 2)S + 2I$.

Algorithm 2: Montgomery Simultaneous Scalar Multiplication

Input: $k = (k_t \cdots k_1 k_0)_2$, $l = (l_t \cdots l_1 l_0)_2$, $P, Q \in E_M$ (k_t or $l_t = 1$).
Output: x-coordinate of $W = kP + lQ$.

1. Compute $P + Q$, $P - Q$.
2. If $(k_t, l_t) = (0, 1)$ then: $T_0 \leftarrow \mathcal{O}$, $T_1 \leftarrow Q$, $T_2 \leftarrow P + Q$;
   else if $(k_t, l_t) = (1, 0)$ then: $T_0 \leftarrow \mathcal{O}$, $T_1 \leftarrow P$, $T_2 \leftarrow P + Q$;
   else then: $T_0 \leftarrow Q$, $T_1 \leftarrow P$, $T_2 \leftarrow P + Q$.
3. For i from t downto 1 do
   3.1. If $(k_i, l_i, k_{i-1}, l_{i-1}) = (0, 0, 0, 0)$ then:
        $T_2 \leftarrow T_2 + T_0$ (P), $T_1 \leftarrow T_1 + T_0$ (Q), $T_0 \leftarrow 2T_0$;
   3.2. else if $(k_i, l_i, k_{i-1}, l_{i-1}) = (0, 0, 0, 1)$ then:
        $T_2 \leftarrow T_2 + T_1$ $(P - Q)$, $T_1 \leftarrow T_1 + T_0$ (Q), $T_0 \leftarrow 2T_0$;
   3.3. else if $(k_i, l_i, k_{i-1}, l_{i-1}) = (0, 0, 1, 0)$ then:
        $T \leftarrow T_1$, $T_1 \leftarrow T_2 + T_0$ (P), $T_0 \leftarrow 2T_0$, $T_2 \leftarrow T_2 + T$ $(P - Q)$;
   3.4. else if $(k_i, l_i, k_{i-1}, l_{i-1}) = (0, 0, 1, 1)$ then:
        $T \leftarrow T_1$, $T_1 \leftarrow T_2 + T_0$ (P), $T_0 \leftarrow T + T_0$ (Q), $T_2 \leftarrow T_2 + T$ $(P - Q)$;
   3.5. else if $(k_i, l_i, k_{i-1}, l_{i-1}) = (0, 1, 0, 0)$ then:
        $T_2 \leftarrow T_2 + T_0$ $(P + Q)$, $T_0 \leftarrow T_1 + T_0$ (Q), $T_1 \leftarrow 2T_1$;
   3.6. else if $(k_i, l_i, k_{i-1}, l_{i-1}) = (0, 1, 0, 1)$ then:
        $T_2 \leftarrow T_2 + T_1$ (P), $T_0 \leftarrow T_1 + T_0$ (Q), $T_1 \leftarrow 2T_1$;
   3.7. else if $(k_i, l_i, k_{i-1}, l_{i-1}) = (0, 1, 1, 0)$ then:
        $T \leftarrow T_1$, $T_1 \leftarrow T_2 + T_0$ $(P + Q)$, $T_0 \leftarrow T + T_0$ (Q), $T_2 \leftarrow T_2 + T$ (P);
   3.8. else if $(k_i, l_i, k_{i-1}, l_{i-1}) = (0, 1, 1, 1)$ then:
        $T \leftarrow T_1$, $T_1 \leftarrow T_2 + T_0$ $(P + Q)$, $T_0 \leftarrow 2T$, $T_2 \leftarrow T_2 + T$ (P);
   3.9. else if $(k_i, l_i, k_{i-1}, l_{i-1}) = (1, 0, 0, 0)$ then:
        $T \leftarrow T_1$, $T_1 \leftarrow T_2 + T_0$ $(P + Q)$, $T_0 \leftarrow T + T_0$ (P), $T_2 \leftarrow 2T$;
   3.10. else if $(k_i, l_i, k_{i-1}, l_{i-1}) = (1, 0, 0, 1)$ then:
        $T \leftarrow T_1$, $T_1 \leftarrow T_2 + T_0$ $(P + Q)$, $T_0 \leftarrow T + T_0$ (P), $T_2 \leftarrow T_2 + T$ (Q);
   3.11. else if $(k_i, l_i, k_{i-1}, l_{i-1}) = (1, 0, 1, 0)$ then:
        $T_0 \leftarrow T_1 + T_0$ (P), $T_2 \leftarrow T_2 + T_1$ (Q), $T_1 \leftarrow 2T_1$;
   3.12. else if $(k_i, l_i, k_{i-1}, l_{i-1}) = (1, 0, 1, 1)$ then:
        $T_0 \leftarrow T_2 + T_0$ $(P + Q)$, $T_2 \leftarrow T_2 + T_1$ (Q), $T_1 \leftarrow 2T_1$;
   3.13. else if $(k_i, l_i, k_{i-1}, l_{i-1}) = (1, 1, 0, 0)$ then:
        $T \leftarrow T_1$, $T_1 \leftarrow T_2 + T_0$ (P), $T_0 \leftarrow T + T_0$ $(P - Q)$, $T_2 \leftarrow T_2 + T$ (Q);
   3.14. else if $(k_i, l_i, k_{i-1}, l_{i-1}) = (1, 1, 0, 1)$ then:
        $T \leftarrow T_1$, $T_1 \leftarrow T_2 + T_0$ (P), $T_0 \leftarrow T + T_0$ $(P - Q)$, $T_2 \leftarrow 2T_2$;
   3.15. else if $(k_i, l_i, k_{i-1}, l_{i-1}) = (1, 1, 1, 0)$ then:
        $T_0 \leftarrow T_1 + T_0$ $(P - Q)$, $T_1 \leftarrow T_2 + T_1$ (Q), $T_2 \leftarrow 2T_2$;
   3.16. else then:
        $T_0 \leftarrow T_2 + T_0$ (P), $T_1 \leftarrow T_2 + T_1$ (Q), $T_2 \leftarrow 2T_2$.
4. If $(k_0, l_0) = (0, 0)$ then $W \leftarrow 2T_0$;
   else if $(k_0, l_0) = (0, 1)$ then $W \leftarrow T_1 + T_0$ (Q);
   else if $(k_0, l_0) = (1, 0)$ then $W \leftarrow T_1 + T_0$ (P);
   else then $W \leftarrow T_1 + T_0$ $(P - Q)$.
5. Compute x-coordinate of W by $x = X/Z$.

---

4   Comparison

Now we compare the computational cost of the proposed method to that of both methods in Section 2. In addition, we compare this to the computational cost of the method described in IEEE P1363 Draft [8], which is based on the scalar multiplication using NAF on the elliptic curve with Weierstrass form. This is a fair comparison because all four methods require no precomputed point. Table 1 shows the computational cost of each method to compute x-coordinate of $kP + lQ$. M, S and I respectively denote the computational costs of a field multiplication, squaring and inversion.

Table 1. The computational costs of every method

Method	M	S	I
Weierstrass NAF [8]	$(40|k| - 1)/3$	$14|k| - 9$	1
Montgomery	$12|k| + 29$	$8|k|$	1
Weierstrass Simultaneous +NAF (Algorithm 1)	$(76|k| + 45)/9$	$(61|k| + 27)/9$	2
Montgomery Simultaneous (Algorithm 2)	$9|k| - 1$	$6|k| - 2$	2

Table 2 shows the computational cost of each method for $|k| = 160$ and the total cost when we assume $S/M = 0.8$ and $I/M = 30$ [12]. We also compare the computational cost of our method to that of Weierstrass simultaneous scalar multiplication using window method and mixed coordinates, which requires memories for 13 points [3]. The proposed method, Montgomery simultaneous scalar multiplication, is about 45% faster than the method described in IEEE P1363 Draft, and about 25% faster than the method using Montgomery scalar multiplication and the recovery of Y-coordinate. Moreover, the proposed method is about 1% faster than Weierstrass simultaneous scalar multiplication using NAF. Our method is about 2% slower than Weierstrass simultaneous scalar multiplication using window method, but requires much less memories.

5   Running Times

Here we present the running times of each method described in Section 4. To calculate arbitrary precision arithmetic over $\mathbb{F}_p$, we used the GNU MP library GMP [6]. The running times were obtained on a Pentium II 300 MHz machine. We used the following elliptic curve over $\mathbb{F}_p$, where $|p| = 162$ and the order of the base point r was 160 bits. $\#E$ means the order of this elliptic curve.

Table 2. The computational cost of each method for $|k| = 160$ and $S/M = 0.8, I/M = 30$.

Method	M	S	I	M $(S/M = 0.8, I/M = 30)$
Weierstrass NAF	2133	2231	1	3938
Montgomery	1949	1280	1	3003
Weierstrass Simultaneous + NAF	1356	1087	2	2286
Montgomery Simultaneous	1439	958	2	2265
Weierstrass Simultaneous + Window	1281	1018	4	2215

$$p = \quad 2 \ \texttt{0aa6fc4d} \ \texttt{8396f3ac} \ \texttt{06200db7} \ \texttt{3e819694} \ \texttt{067a0e7b}$$
$$a = \quad 1 \ \texttt{5fed3282} \ \texttt{429907d6} \ \texttt{03b41b7a} \ \texttt{309abf87} \ \texttt{bed9bd83}$$
$$b = \quad 1 \ \texttt{74019686} \ \texttt{9a423134} \ \texttt{f3cdf013} \ \texttt{b13564d0} \ \texttt{ba3999e8}$$
$$\#E = 4 * \texttt{82a9bf13} \ \texttt{60e5bceb} \ \texttt{01878167} \ \texttt{1d478cea} \ \texttt{881e1d1d}$$
$$A = \quad 1 \ \texttt{8be6a098} \ \texttt{c28d6bc0} \ \texttt{3286dc51} \ \texttt{e7e3f705} \ \texttt{8a5b9d98}$$
$$B = \quad 0 \ \texttt{120c2550} \ \texttt{f6ff7a01} \ \texttt{440d78d1} \ \texttt{122fa3ac} \ \texttt{aa70fd53}$$

We obtained average running times to compute x-coordinate of $kP + lQ$ by randomly choosing 100 points P, Q over this elliptic curve and 100 integers $k, l < r$. Table 3 shows the average time of each method to compute x-coordinate of $kP + lQ$. From Table 3, we notice the proposed method, Montgomery simultaneous scalar multiplication, is about 44% faster than the method described in IEEE P1363 Draft, and about 25% faster than the method using Montgomery scalar multiplication and the recovery of Y-coordinate. Moreover, the proposed method is about 3% faster than Weierstrass simultaneous scalar multiplication using NAF. This shows that the theoretical advantage of our method is actually observed.

Table 3. The average time of each method

Method	Average time (ms)
Weierstrass NAF	36.2
Montgomery	26.9
Weierstrass Simultaneous + NAF	20.8
Montgomery Simultaneous	20.1

6   Elliptic Curve over $\mathbb{F}_{2^n}$

A non-supersingular elliptic curve E over $\mathbb{F}_{2^n}$ is represented by $E : x^2 + xy = x^3 + ax^2 + b$, where $a, b \in \mathbb{F}_{2^n}, b \neq 0$. We can apply the proposed method to Montgomery method on elliptic curves over $\mathbb{F}_{2^n}$ [14]. The advantage of Montgomery method on elliptic curves over $\mathbb{F}_{2^n}$ is that we need not consider transformability from Weierstrass form to Montgomery form. Since the computational cost of a field squaring over $\mathbb{F}_{2^n}$ is much lower than that of a field multiplication over $\mathbb{F}_{2^n}$, we can ignore it.

In Montgomery method, the computational cost of addition formulas and doubling formulas are respectively $4M$ and $2M$ in projective coordinates. If we compute $kP + lQ$ using Montgomery scalar multiplication, we requires addition formulas twice and doubling formulas twice per bit of k. Therefore, the computational cost per bit of k is estimated to be about $12M$.

On the other hand, if we compute $kP + lQ$ using Montgomery simultaneous scalar multiplication, we require addition formulas twice and doubling formulas once at probability of $3/4$, and addition formulas three times at probability of $1/4$ per bit of k, as described in Algorithm 2. Since we require addition formulas $9/4$ times and doubling formulas $3/4$ times per bit of k, the computational cost per bit of k is estimated to be about $21/2 \cdot M$.

In Weierstrass simultaneous scalar multiplication over $\mathbb{F}_{2^n}$ using NAF [7,13], the computational cost per bit of k is estimated to be about $9M$ if $a = 0, 1$.

Therefore, the proposed method is only 13% faster than the method using Montgomery scalar multiplication and 17% slower than Weierstrass simultaneous scalar multiplication using NAF. This shows that the proposed method on elliptic curves over $\mathbb{F}_{2^n}$ is not so efficient as that on elliptic curves with Montgomery form over $\mathbb{F}_p$.

7   Conclusion

We proposed the new method to compute x-coordinate of $kP + lQ$ simultaneously on the elliptic curve with Montgomery form over $\mathbb{F}_p$ without precomputed points. To compute x-coordinate of $kP + lQ$ is required in ECDSA signature verification. Our method is about 25% faster than the method using Montgomery scalar multiplication and the recovery of Y-coordinate over $\mathbb{F}_p$, and slightly faster than Weierstrass simultaneous scalar multiplication over $\mathbb{F}_p$ using NAF and mixed coordinates. Our method is considered to be particularly useful in case that ECDSA signature generation is performed using Montgomery scalar multiplication on the elliptic curve over $\mathbb{F}_p$ because of its efficiency of computation and its immunity to timing attacks, since all arithmetic on the elliptic curve can be computed with Montgomery form and we don't require transformation to the elliptic curve with Weierstrass form. Furthermore, we showed that our method was applicable to Montgomery method on elliptic curves over $\mathbb{F}_{2^n}$.

Acknowledgements

We would like to thank Taizo Shirai for his careful reading and valuable suggestions. We also thank the referees for their numerous helpful comments.

References

1. ANSI X9.62-1998, *Public Key Cryptography for the Financial Services Industry:
   The Elliptic Curve Digital Signature Algorithm (ECDSA)*, Working Draft, September 1998.
2. D. V. Bailey and C. Paar, "Optimal Extension Field for Fast Arithmetic in Public
   Key Algorithms", *Advances in Cryptography - CRYPTO '98*, LNCS 1462, pp. 472-
   485, 1998.
3. M. Brown, D. Hankerson, J. Lopez, and A. Menezes, "Software Implementation
   of the NIST Elliptic Curves Over Prime Fields", *Topics in Cryptology – CT-RSA
   2001*, LNCS 2020, pp. 250-265, 2001.
4. H. Cohen, A. Miyaji, and T. Ono, "Efficient Elliptic Curve Exponentiation Using
   Mixed Coordinates", *Advances in Cryptography - ASIACRYPT '98*, LNCS 1514,
   pp. 51-65, 1998.
5. T. ElGamal, "A Public Key Cryptosystem and a Signature Scheme Based on Discrete Logarithms", *IEEE Transaction on Information Theory*, Vol. 31, pp. 469-472,
   1985.
6. GNU MP Library GMP, version 3.1.1, October 2000. http://www.swox.com/gmp/
7. D. Hankerson, J. Lopéz, and A. Menezes, "Software Implementation of Elliptic
   Curve Cryptography over Binary Fields", *Cryptographic Hardware and Embedded
   Systems - CHES 2000*, LNCS 1965, pp. 3-24, 2000.
8. IEEE P1363, *Standard Specifications for Public Key Cryptography*, Draft Version
   13, November 1999. http://grouper.ieee.org/groups/1363/
9. T. Kobayashi, H. Morita, K. Kobayashi, and F. Hoshino, "Fast Elliptic Curve
   Algorithm Combining Frobenius Map and Table Reference to Adapt to Higher
   Characteristic", *Advances in Cryptography - EUROCRYPT '99*, LNCS 1592, pp.
   176-189, 1999.
10. N. Koblitz, "Elliptic Curve Cryptosystems", *Mathematics of Computation*, vol. 48,
    pp. 203-209, 1987.
11. P. C. Kocher, "Timing Attacks on Implementations of Diffie-Hellman, RSA, DSS,
    and Other Systems", *Advances in Cryptology – CRYPTO '96*, LNCS 1109, pp.
    104-113, 1996.
12. C. H. Lim and H. S. Hwang, "Fast Implementation of Elliptic Arithmetic in
    $GF(p^n)$", *Public Key Cryptography - PKC 2000*, LNCS 1751, pp. 405-421, 2000.
13. J. Lopéz and R. Dahab, "Improved Algorithms for Elliptic Curve Arithmetic in
    $GF(2^n)$", *Selected Areas in Cryptology - SAC '98*, LNCS 1556, pp. 201-212, 1999.
14. J. Lopéz and R. Dahab, "Fast Multiplication on Elliptic Curves over $GF(2^m)$
    without Precomputation", *Cryptographic Hardware and Embedded Systems - CHES
    '99*, LNCS 1717, pp. 316-327, 1999.
15. V. Miller, "Uses of Elliptic Curves in Cryptography", *Advances in Cryptography -
    CRYPTO '85*, LNCS 218, pp. 417-426, 1986.
16. P. L. Montgomery, "Speeding the Pollard and Elliptic Curve Methods of Factor-
    ization", *Mathematics of Computation*, vol. 48, pp. 243-264, 1987.

17. K. Okeya, H. Kurumatani, and K. Sakurai, "Elliptic Curves with the Montgomery-Form and Their Cryptographic Applications", *Public Key Cryptography - PKC 2000*, LNCS 1751, pp. 238-257, 2000.

18. K. Okeya and K. Sakurai, "Efficient Elliptic Curve Cryptosystems from a Scalar Multiplication Algorithm with Recovery of the y-coordinate on a Montgomery-Form Elliptic Curve", *Preproceedings of Cryptographic Hardware and Embedded Systems - CHES 2001*, pp. 129-144, 2001.

19. E. D. Win, S. Mister, B. Preneel, and M. Wiener, "On the Performance of Signature Schemes Based on Elliptic Curves", *Algorithm Number Theory - ANTS III*, LNCS 1423, pp. 252-266, 1998.

On the Power of Multidoubling in Speeding Up Elliptic Scalar Multiplication

Yasuyuki Sakai[1] and Kouichi Sakurai[2]

[1] Mitsubishi Electric Corporation,
5-1-1 Ofuna, Kamakura, Kanagawa 247-8501, Japan
ysakai@iss.isl.melco.co.jp
[2] Kyushu University,
6-10-1 Hakozaki, Higashi-ku, Fukuoka 812-8581, Japan
sakurai@csce.kyushu-u.ac.jp

Abstract. We discuss multidoubling methods for efficient elliptic scalar multiplication. The methods allows computation of $2^k P$ directly from P without computing the intermediate points, where P denotes a randomly selected point on an elliptic curve. We introduce algorithms for elliptic curves with Montgomery form and Weierstrass form defined over finite fields with characteristic greater than 3 in terms of affine coordinates. These algorithms are faster than k repeated doublings. Moreover, we apply the algorithms to scalar multiplication on elliptic curves and analyze computational complexity. As a result of our implementation with respect to the Montgomery and Weierstrass forms in terms of affine coordinates, we achieved running time reduced by 28% and 31%, respectively, in the scalar multiplication of an elliptic curve of size 160-bit over finite fields with characteristic greater than 3.

Keywords. Elliptic curve cryptosystems, Scalar multiplication, Montgomery form, Multidoubling, Fast implementation

1 Introduction

Elliptic curve cryptosystems, which were suggested by Miller [Mi85] and Koblitz [Ko87], are now widely used in various security services. IEEE and other standardizing bodies such as ANSI and ISO are in the process of standardizing elliptic curve cryptosystems. Therefore, it is very attractive to provide algorithms that allow efficient implementation. Encryption/decryption or signature generation/verification schemes require computation of scalar multiplication. The computational performance of cryptographic protocols with elliptic curves strongly depends on the efficiency of the scalar multiplication. Thus, fast scalar multiplication is essential for elliptic curve cryptosystems.

One method to increase doubling speed involves the "*multidoubling*", which computes $2^k P$ directly from $P \in E(\mathbb{F}_q)$, without computing the intermediate points $2P, 2^2 P, \cdots, 2^{k-1} P$. The concept of multidoubling was first suggested by Guajardo and Paar in [GP97]. They formulated algorithms for the multidoubling

S. Vaudenay and A. Youssef (Eds.): SAC 2001, LNCS 2259, pp. 268–283, 2001.

of $4P$, $8P$ and $16P$ on elliptic curves over $\mathbb{F}_{2^n}$ in terms of affine coordinates. Recent related results include a formula for computing $4P$ on elliptic curves over $\mathbb{F}_p$ in affine coordinates by Müller [Mu97] and a formula for computing $4P$ on elliptic curves over $\mathbb{F}_p$ in projective coordinates by Miyaji, Ono and Cohen [MOC97a]. These formulae are more efficient than repeated doublings. All of the previous works were on the subject of elliptic curves with Weierstrass form. Another model of an elliptic curve that is useful for cryptosystems is the Montgomery form. Montgomery introduced the equation to speed up integer factorization with elliptic curves [Mo87]. The elliptic curve method of factoring was proposed by H.W.Lenstra [Le87]. In recent years, several authors have proposed elliptic curve cryptosystems using the Montgomery model [Iz99,LD99,OKS00].

In this paper, we propose efficient algorithms for speeding up elliptic curve cryptosystems with Montgomery elliptic curves in terms of affine coordinates. We construct efficient formulae that compute $2^k P$ directly for $\forall k \geq 2$. In the case of an elliptic curve with Montgomery form, our formulae have computational complexity $(8k+4)\mathcal{M}+(4k-1)\mathcal{S}+\mathcal{I}$, where $\mathcal{M}$, $\mathcal{S}$, and $\mathcal{I}$ denote multiplication, squaring and inversion in $\mathbb{F}_p$, respectively. This is more efficient than k repeated doublings, which require k inversions. When implementing our multidoubling method, experimental results show that computing $16P$ achieved running time reduced by 40% over 4 doublings in affine coordinates. Moreover we introduce formulae that compute $2^k P$ directly for $\forall k \geq 2$, for Weierstrass elliptic curves in terms of affine coordinates. Our formulae have computational complexity $(4k + 1)\mathcal{M} + (4k + 1)\mathcal{S} + \mathcal{I}$. The formulae have slightly simple form compared to the formulae described in [SS01] and have computational advantage, due to one field multiplication, over the formulae proposed in [SS00]

As a results of our implementation with respect to Montgomery and Weierstrass forms in terms of affine coordinates, we achieved running time reduced by 28% and 31%, respectively, in the scalar multiplication of an elliptic curve of size 160-bit. We also discuss the computational complexity of scalar multiplication using multidoubling. The proposed algorithm improve the performance of scalar multiplication with the binary method, as well as the window method. Therefore, they are effective in restricted environments where resources are limited, such as smart cards.

2   Previous Work

In this section, we summarize the multidoubling, the direct computation and arithmetic for an elliptic curve with Montgomery form.

2.1   Multidoubling and Direct Computation

The concept of using multidoubling and direct computation of $2^k P$ to efficiently implement elliptic scalar multiplication was first proposed by Guajardo and Paar in [GP97]. They formulated algorithms for computing $4P$, $8P$, and $16P$ on elliptic curves over $\mathbb{F}_{2^n}$ in terms of affine coordinates. In recent years, several authors have reported methods that compute $2^k P$ directly, but some of them are limited to small k. The following section summarizes the previous work on multidoubling and direct computation.

1. Guajardo and Paar [GP97] proposed formulae for computing $4P$, $8P$, and $16P$ on elliptic curves over $\mathbb{F}_{2^n}$ in terms of affine coordinates.
2. Müller [Mu97] proposed formulae for computing $4P$ on elliptic curves over $\mathbb{F}_p$ in terms of affine coordinates.
3. Miyaji, Ono, and Cohen [MOC97a] proposed formulae for computing $4P$ on elliptic curves over $\mathbb{F}_p$ in terms of projective coordinates.
4. Han and Tan [HT99] proposed formulae for computing $3P$, $5P$, $6P$, $7P$, etc, on elliptic curves over $\mathbb{F}_{2^n}$ in terms of affine coordinates.
5. Sakai and Sakurai [SS00,SS01] proposed formulae for computing $2^k P$ ($\forall k \geq 1$) on elliptic curves over $\mathbb{F}_p$ in terms of affine coordinates.

We should remark that the algorithm proposed by Cohen, Miyaji and Ono in [CMO98] can be efficiently used for the direct computation of several doublings. The authors call their algorithm a *"modified jacobian"* coordinate system. The coordinate system uses (redundant) mixed representation such as (X, Y, Z, aZ^4). Doubling in terms of the modified jacobian coordinates has computational advantages over weighted projective (jacobian) coordinates. Itoh et al. also gave a similar method for doubling [ITTTK99].

All of these works dealt with computations on Weierstrass elliptic curves. In later sections, we will formulate algorithms that work on Montgomery elliptic curves in terms of affine coordinates, and analyze their computational complexity.

2.2   Elliptic Curves with Montgomery Model

Let $a, b \in \mathbb{F}_p$, $4a^3 + 27b^2 \neq 0$, $p > 3$, and p be a prime number. An elliptic curve defined over $\mathbb{F}_p$ for Weierstrass model is defined by the following equation (1). Elliptic curve cryptosystems using curves with Weierstrass form are in the process of being standardized, e.g., [IEEE], and are widely used in various security services.

$$E : y^2 = x^3 + ax + b \tag{1}$$

H. W. Lenstra proposed the elliptic curve method of factoring [Le87]. Montgomery introduced the following equation to speed up integer factorization with elliptic curves [Mo87]. In recent years, several authors have proposed cryptosystems using an elliptic curve with Montgomery form [Iz99,LD99,OKS00]. Let $A, B \in \mathbb{F}_p$, $(A^2 - 4)B \neq 0$. An elliptic curve of Montgomery model is defined by the following equation (2).

$$E_m : Bv^2 = u^3 + Au^2 + u \tag{2}$$

The formulae for transforming Montgomery and Weierstrass forms are given by the following (See [Iz99] for details).

By the transformation $u = \frac{3x-AB}{3B}$ and $v = \frac{y}{B^2}$, we obtain $y^2 = x^3 + \frac{B^2(3-A^2)}{3}x + \frac{AB^3(2A^2-9)}{27}$. Therefore, by the relationship $a = \frac{B^2(3-A^2)}{3}$ and $b = \frac{AB^3(2A^2-9)}{27}$, we can transform a Montgomery form into a Weierstrass form.

The above linear transformation clearly converts any elliptic curve with Montgomery form into a curve with Weierstrass form. However, the inverse transformation, from Weierstrass form to Montgomery form, works only if there exists a particular curve. Based on the above relationship between (a, b) and (A, B), we eliminate B, then we obtain $A^6 - 9A^4 - 27(r-1)A^2 + 27(4r-1) = 0$, where $r = \frac{a^3}{4a^4+27b^2}$. Let we consider the equation $f(t) = t^3 - 9t^2 - 27(r-1)t + 27(4r-1) = 0$, where $t = A^2$. If $f(t)$ has a solution $t = \alpha$ such as a quadratic residue in $\mathbb{F}_p$, a Weierstrass form can be transform into a Montgomery form by the following relation. Let β be a square root of α, we obtain $A = \beta$ and $B = \frac{9b(3-\beta^2)}{\alpha\beta(2\beta^2-9)}$. Then the relation $x = \frac{3Bu+AB}{3}$ and $y = B^2v$ are derived.

A detailed analysis on the case which can transform a Weierstrass form into a Montgomery form was given by Izu [Iz99]. He concluded that approximately 40% of curves with Weierstrass form can be transformed into a curve with Montgomery form.

2.3 Group Operation for Elliptic Curves with Montgomery Form

We describe algorithms for group operation in an elliptic curve with Montgomery form. When we estimate a computational efficiency, we will ignore the cost of a field addition, as well as the cost of a multiplication by small constants.

Affine Coordinates. Suppose $P_3(u_3, v_3) = P_1(u_1, v_1) + P_2(u_2, v_2) \in E_m(\mathbb{F}_p)$, and $P_1 \neq P_2$. The addition formulae are given by the following.

$$
\begin{aligned}
u_3 &= B\lambda^2 - A - u_1 - u_2 \\
v_3 &= \lambda(u_1 - u_3) - v_1 \\
\lambda &= \frac{v_1 - v_2}{u_1 - u_2}
\end{aligned}
\tag{3}
$$

The computational complexity for an addition involves $3\mathcal{M} + \mathcal{S} + \mathcal{I}$.

Suppose $P_3(u_3, v_3) = 2P_1(u_1, v_1) \in E_m(\mathbb{F}_p)$. Point doubling can be accomplished by the following.

$$
\begin{aligned}
u_3 &= B\lambda^2 - A - 2u_1 \\
v_3 &= \lambda(u_1 - u_3) - v_1 \\
\lambda &= \frac{3u_1^2 + 2Au_1 + 1}{2Bv_1}
\end{aligned}
\tag{4}
$$

The computational complexity of a doubling involves $5\mathcal{M} + 2\mathcal{S} + \mathcal{I}$.

Projective Coordinates. Next, we describe formulae for the group operations in projective coordinates. Let $u = \frac{U}{W}$, $v = \frac{V}{W}$. Suppose $P_2(U_2, W_2) =$

$P_1(U_1, W_1) + P(u, v, 1)$. The point $P_3(U_3, W_3) = P_1(U_1, W_1) + P_2(U_2, W_2)$ can be computed by the following.

$$\begin{aligned}
U_3 &= (U_1 U_2 - W_1 W_2)^2 \\
&= (Sub_1 Add_2 + Add_1 Sub_2)^2 \\
W_3 &= u (U_1 W_2 - W_1 U_2)^2 \\
&= u (Sub_1 Add_2 - Add_1 Sub_2)^2
\end{aligned}$$

where, $Add_1 = U_1 + W_1$, $Sub_1 = U_1 - W_1$, $Add_2 = U_2 + W_2$ and $Sub_2 = U_2 - W_2$. The computational complexity for an addition involves $3\mathcal{M} + 2\mathcal{S}$.

Note that V-coordinate does not enter into any of the formulae. An addition can be accomplished without computation of the V-coordinate if the difference between the two given points is known [Mo87]. Point doubling can be accomplished by the following.

$$\begin{aligned}
U_3 &= \left(U_1^2 - W_1^2\right)^2 = Add_1^2 Sub_1^2 \\
W_3 &= 4 U_1 W_1 \left(U_1^2 + A U_1 W_1 + W_1^2\right) \\
&= \left(Add_1^2 - Sub_1^2\right) \left\{ Sub_1^2 + C \left(Add_1^2 - Sub_1^2\right) \right\}
\end{aligned}$$

where, $C = \frac{A+2}{4}$. For a given curve, C can be pre-computed. Therefore above formulae have computational complexity $3\mathcal{M} + 2\mathcal{S}$.

The basic well known method for elliptic scalar multiplication on curves with Weierstrass form is the "*double-and-add*" (or binary) method. There are several methods which have computational advantage over the binary method such as the window method. However, in the case of elliptic curves with Montgomery form in terms of projective coordinates, we can not apply such methods, because the difference between the two given points, i.e., the U-coordinate of $P_2 - P_1$, must be known when adding the two points. To compute kP, we compute $2P$ and then repeatedly compute two points $(2mP, (2m+1)P)$ or $((2m+1)P, (2m+2)P)$, depending on whether the corresponding bit in the binary representation of k is a 0 or a 1 [AMV93,Mo87,MV93]. This method maintains the invariant relationship such that the difference of the two points always P.

3   The Proposed Algorithms

In this section, we describe new algorithms for elliptic curves with Montgomery form, which compute $2^k P$ directly from a given point $P \in E_m(\mathbb{F}_p)$ without computing the intermediate points $2P, 2^2 P, \cdots, 2^{k-1} P$. We will begin by constructing formulae for small k, then we will construct an algorithm for general k ($k \geq 2$). We will also show an algorithm that compute $2^k P$ directly for elliptic curves with Weierstrass form. This is an improved version of the algorithm proposed in [SS00,SS01].

3.1   Montgomery Form

As an example, we give an algorithm that compute $8P$ directly from $P \in E_m(\mathbb{F}_p)$.

Computing 8P. Let $P_8(u_8, v_8) = 8P(u_1, v_1) \in E_m(\mathbb{F}_p)$. For an elliptic curve with Montgomery form in terms of affine coordinates, P_8 can be computed in the following way. The derivation is based on repeated substitution of the point doubling formulae, such that only one field inversion needs to be calculated. First we compute C, D_i, E_i, F_i, for $1 \leq i \leq 3$ as follows.

$$C = AB$$
$$D_1 = u_1 B$$
$$E_1 = v_1$$
$$F_1 = 1 + u_1(2A + 3u_1)$$
$$D_2 = -2^2 C E_1^2 - 8D_1 E_1^2 + F_1^2$$
$$E_2 = -8B^2 E_1^4 + F_1(4D_1 E_1^2 - D_2)$$
$$F_2 = 2^4 B^2 E_1^4 + D_2(2^3 C E_1^2 + 3D_2)$$
$$D_3 = -2^4 C(E_1 E_2)^2 - 8D_2 E_2^2 + F_2^2$$
$$E_3 = -8E_2^4 + F_2(4D_2 E_2^2 - D_3)$$
$$F_3 = 2^8 B^2 (E_1 E_2)^4 + D_3(2^5 C(E_1 E_2)^2 + 3D_3)$$

Then we compute u_8 and v_8 as follows.

$$u_8 = \frac{-8E_3^2(2^3 C(E_1 E_2)^2 + D_3) + F_3^2}{2^6 B(E_1 E_2 E_3)^2}$$
$$v_8 = \frac{F_3((2^6 C(E_1 E_2)^2 + 12D_3)E_3^2 - F_3^2) - 8E_3^4}{2^9 B^2(E_1 E_2 E_3)^3}$$

Note that C and B^2 can be pre-computed, and that although the denominator of u_8 differs from that of v_8, the above formulae require only one inversion if we multiply the numerator of u_8 by $2^3 B E_1 E_2 E_3$. The above formulae have computational complexity $25\mathcal{M} + 11\mathcal{S} + \mathcal{I}$.

Multidoubling. From the formulae that compute $2^k P$ for small k, given in the previous subsection, we can easily obtain general formulae that allow direct doubling $P \mapsto 2^k P$ for $k \geq 2$. The figure shown below describes the formulae, and their computational complexity is given as Theorem 1.

Algorithm 1: Direct computation of $2^k P$ in affine coordinates on an elliptic curve with Montgomery form, where $k \geq 2$ and $P \in E_m(\mathbb{F}_p)$.

INPUT: $P_1 = (u_1, v_1) \in E_m(\mathbb{F}_p)$
OUTPUT: $P_{2^k} = 2^k P_1 = (u_{2^k}, v_{2^k}) \in E_m(\mathbb{F}_p)$

Pre Computations

$$C = AB$$
$$B_2 = B^2$$

Step 1. Compute D_0, E_0 and F_0

$$D_0 = u_1 B$$
$$E_0 = v_1$$
$$F_0 = 1 + u_1(2A + 3u_1)$$

Step 2. For i from 1 to k compute D_i and E_i, for i from 1 to $k-1$ compute F_i

$$D_i = -2^{2i}C \left(\prod_{j=0}^{i-1} E_j \right)^2 - 8D_{i-1}E_{i-1}^2 + F_{i-1}^2$$

if $i = 1$ $E_1 = -8B_2E_0^4 + F_0(4D_0E_0^2 - D_1)$

else $E_i = -8E_{i-1}^4 + F_{i-1}(4D_{i-1}E_{i-1}^2 - D_i)$

$$F_i = \left(\prod_{j=0}^{i-1} E_j \right)^2 \left(2^{4i}B_2 \left(\prod_{j=0}^{i-1} E_j \right)^2 + 2^{2i+1}CD_i \right) + 3D_i^2$$

Step 3. Compute u_{2^k} and v_{2^k}

$$u_{2^k} = \frac{D_k}{B \left(2^k \prod_{i=0}^{k-1} E_i \right)^2}$$

$$v_{2^k} = \frac{E_k}{B_2 \left(2^k \prod_{i=0}^{k-1} E_i \right)^3}$$

Theorem 1. *For an elliptic curve with Montgomery form in terms of affine coordinates, there exists an algorithm that computes $2^k P$, with $k \geq 2$, in at most $8k + 4$ field multiplication, $4k - 1$ field squaring and one inversion in $\mathbb{F}_p$ for any point $P \in E_m(\mathbb{F}_p)$, excluding precomputation.*

The proof is outlined in Appendix A.1.

3.2   Weierstrass Form

The multidoubling for Weierstrass elliptic curves in terms of affine coordinates is given below. Their computational complexity have, given as Theorem 2, $(4k+1)\mathcal{M}+(4k+1)\mathcal{S}+\mathcal{I}$. The complexity has a one field multiplication computational advantage over the formulae proposed in [SS00]. Moreover, the formulae have slightly simple form compared to the formulae described in [SS01]

Algorithm 2: Direct computation of $2^k P$ in affine coordinates on an elliptic curve with Weierstrass form, where $k \geq 1$ and $P \in E(\mathbb{F}_p)$.

INPUT: $P_1 = (x_1, y_1) \in E(\mathbb{F}_p)$
OUTPUT: $P_{2^k} = 2^k P_1 = (x_{2^k}, y_{2^k}) \in E(\mathbb{F}_p)$

Step 1. Compute A_0, B_0 and C_0

$$A_0 = x_1$$
$$B_0 = y_1$$
$$C_0 = 3x_1^2 + a$$

Step 2. For i from 1 to k compute A_i and B_i, for i from 1 to $k-1$ compute C_i

$$A_i = C_{i-1}^2 - 8A_{i-1}B_{i-1}^2$$
$$B_i = 8B_{i-1}^4 + C_{i-1}(A_i - 4A_{i-1}B_{i-1}^2)$$
$$C_i = 3A_i^2 + 16^{i-1}a\left(\prod_{j=1}^{i-1} B_j\right)^4$$

Step 3. Compute x_{2^k} and y_{2^k}

$$x_{2^k} = \frac{A_k}{\left(2^k\prod_{i=1}^k B_i\right)^2}$$
$$y_{2^k} = \frac{B_k}{\left(2^k\prod_{i=1}^k B_i\right)^3}$$

Theorem 2. *For an elliptic curve with Weierstrass form in terms of affine coordinates, there exists an algorithm that computes $2^k P$ in at most $4k + 1$ multiplications, $4k + 1$ squarings, and one inversion in $\mathbb{F}_p$ for any point $P \in E(\mathbb{F}_p)$.*

The proof is outlined in Appendix A.2.

3.3 Complexity Comparison on Direct Computation

In this subsection, we compare the computational complexity of the multidoubling given in the previous subsection with the complexity of k separate repeated doublings. The complexity of a doubling is estimated from the algorithm given by the formulae (4) or as shown in [IEEE]. Tables 1 and 2 show the number of multiplications $\mathcal{M}$, squarings $\mathcal{S}$, and inversions $\mathcal{I}$ in the base field $\mathbb{F}_p$. Note that our method reduces inversions at the cost of multiplications. Therefore, the performance of the new formulae depends on the cost factor of one inversion relative to one multiplication. For this purpose, we introduce the notation of a "*break-even point*", as used in [GP97]. It is possible to express the time that it takes to perform one inversion in terms of the equivalent number of multiplications needed per inversion. In this comparison, we assume that one squaring has complexity $\mathcal{S} = 0.8\mathcal{M}$, and that the costs of field addition and multiplication by small constants can be ignored.

As we can see from Table 1, if a field inversion has complexity $\mathcal{I} > 10.4\mathcal{M}$, one quadrupling will be more efficient than two separate doublings. If $\mathbb{F}_p$ has size 160-bit or larger, it is likely that $\mathcal{I} > 10.4\mathcal{M}$ in many implementations (e.g., see [WMPW98]). In addition, if $k > 2$, our direct computation method is more efficient than individual doublings in most implementations. For Weierstrass form, shown in Table 2, our direct computation method is more efficient than individual doublings in most implementations.

Table 1. Complexity comparison on direct computation : Montgomery form

Calculation	Method	Complexity			Break-Even
		$\mathcal{M}$	$\mathcal{S}$	$\mathcal{I}$	Point
$4P$	Direct computation	18	7	1	$10.4\mathcal{M} < \mathcal{I}$
	Separate 2 doublings	10	4	2	
$8P$	Direct computation	25	11	1	$7.0\mathcal{M} < \mathcal{I}$
	Separate 3 doublings	15	6	3	
$16P$	Direct computation	32	15	1	$5.9\mathcal{M} < \mathcal{I}$
	Separate 4 doublings	20	8	4	
$2^k P$	Direct computation	$8k+4$	$4k-1$	1	$(4.6 + \frac{7.8}{k-1})\mathcal{M} < \mathcal{I}$
	Separate k doublings	$5k$	$2k$	k	

Table 2. Complexity comparison on direct computation : Weierstrass form

Calculation	Method	Complexity			Break-Even Point
		$\mathcal{M}$	$\mathcal{S}$	$\mathcal{I}$	
$4P$	Direct computation	9	9	1	$8.6\mathcal{M} < \mathcal{I}$
	Separate 2 doublings	4	4	2	
$8P$	Direct computation	13	13	1	$6.3\mathcal{M} < \mathcal{I}$
	Separate 3 doublings	6	6	3	
$16P$	Direct computation	17	17	1	$5.4\mathcal{M} < \mathcal{I}$
	Separate 4 doublings	8	8	4	
$2^k P$	Direct computation	$4k+1$	$4k+1$	1	$(3.6 + \frac{5.4}{k-1})\mathcal{M} < \mathcal{I}$
	Separate k doublings	$2k$	$2k$	k	

4   Scalar Multiplication with Direct Computation

4.1   The Algorithm

Using our previous formulae for direct computation of $2^k P$, we can improve elliptic scalar multiplication with the sliding signed binary window method [Go98,KT92]. For example, we apply our new formulae to the window method with windows of length 4. We represent a scalar m in $P \mapsto mP$ with a *nonadjacent form* (NAF) [1]. For example, $m = (1101110111)_2$ will be represented as $m' = (100\bar{1}000\bar{1}00\bar{1})_{NAF}$, where $\bar{1}$ denotes -1.

---

[1]   Koyama and Tsuruoka pointed out that an NAF is not necessarily the optimal representation to use [Go98,KT92]. Although it has minimal weight, allowing a few adjacent nonzeros may increase the length of zero-runs, which, in turn, would reduces the total number of additions. Their method may be useful for our scalar multiplication with direct computations of $2^k P$.

Algorithm 3 describes scalar multiplication on elliptic curves using our direct computations of $2^k P$ for the case k up to 4.

Algorithm 3: Elliptic scalar multiplication combining our direct computation of $2^k P$ with the window method, and window size $k = 4$

INPUT: $P \in E_m(\mathbb{F}_p)$ or $E(\mathbb{F}_p)$, $m \in \mathbb{Z}$
OUTPUT: $mP \in E_m(\mathbb{F}_p)$ or $E(\mathbb{F}_p)$

Step 1. Construct NAF representation
$$m = (e_t e_{t-1} \cdots e_1 e_0)_{NAF}, \quad e_i \in \{-1, 0, 1\}$$
Step 2. Precomputation
  2.1 $P_6 \leftarrow 6P$
  2.2 For i from 7 to 10 do: $P_i \leftarrow P_{i-1} + P$
Step 3. $P_m \leftarrow \mathcal{O}, \quad i \leftarrow t$
Step 4. While $i \geq 3$ do the following:
  4.1 If $e_i = 0$ then:
        find the longest bitstring $e_i e_{i-1} \cdots e_l$ such that $e_i = e_{i-1} = \cdots e_l = 0$, and do the following
        $$P_m \leftarrow 2^{i-l+1} P_m$$
        $$i \leftarrow l - 1$$
  4.2 else ($e_i \neq 0$):
        If $(e_i e_{i-1} e_{i-2} e_{i-3})_{NAF} > 0$ then:
        $$P_m \leftarrow 16 P_m + P_{(e_i e_{i-1} e_{i-2} e_{i-3})_{NAF}}$$
        else:
        $$P_m \leftarrow 16 P_m - P_{|(e_i e_{i-1} e_{i-2} e_{i-3})_{NAF}|}$$
        $$i \leftarrow i - 4$$
Step 5. $P_m \leftarrow (e_i \cdots e_0)_{NAF} P_m$ using the traditional double-and-add method
Step 6. Return P_m

In Algorithm 3, we compute $16P$ directly from P in each window rather than using 4 separate doublings. In **Step 4.1** with strings of zero-runs in the scalar m_{NAF}, we should choose computations $16P$, $8P$, $4P$ or $2P$ optimally. This can be done with rules such as: 1) If a length of zero equals to 4, we compute $16P$. 2) If a length of zero equals to 3, we compute $8P$, and so on. Note that the computation for **Step 5** is inexpensive if m is large.

Using our algorithms for scalar multiplication, many of the doublings in the traditional window method will be replaced by the direct computation of $16P$. Therefore, if one computation of $16P$ is relatively faster than four doublings, scalar multiplication with our method will be significantly improved. We will examine this improvement by real implementation in the next section.

5   Complexity Comparison on Scalar Multiplication

In this section, we discuss the computational complexity of scalar multiplication using our direct computation.

Table 3. Number of computations of $2^k P$, where $1 \leq k \leq 4$, and addition in the sliding signed binary window method with window length of 4

Curves	Add	2P	4P	8P	16P
160-bit	36.82	14.93	4.99	2.58	31.75
192-bit	37.63	15.29	5.06	2.64	39.54
224-bit	37.77	15.31	5.05	2.69	47.49
256-bit	41.77	15.31	5.06	2.70	55.53
384-bit	48.76	17.13	5.10	3.59	86.26
521-bit	90.39	31.34	14.51	6.94	109.87

5.1   Number of $2^k P$ Computations in the Window Method

Table 3 shows the number of required computations of $2^k P$ and additions in the sliding signed binary window method based on Algorithm 3. The window size shown in the table is 4 as an example. The numbers were counted by our implementation such that We randomly generated 10000 exponents and counted the number of operations. The averages of the numbers are shown in Table 3. In the case of a window of length 4, direct computations of $4P$, $8P$, and $16P$ can be used.

From the table, we can see that with direct computations of up to $16P$, the computational efficiency of $16P$ significantly affects scalar multiplication.

5.2   Break-Even Point

Based on the number of computations of $2^k P$ in scalar multiplication, given in Table 3, we compared the computational complexity of scalar multiplication. For example, in the case of a 160-bit scalar, the complexity of scalar multiplication using direct computation with $k = 4$ can be evaluated as: $C = 36.82A + 14.93D_2 + 4.99D_4 + 2.58D_8 + 31.75D_{16}$, where A, D_2, D_4, D_8, and D_{16} denote the complexity of the computation for point addition, doubling, $4P$, $8P$, and $16P$, respectively. The complexity of those point operations can be evaluated using the algorithms given in the previous sections. For the proposed scalar multiplication with the window method, we used Algorithm 3, which is based on the sliding window method with NAF representation for a scalar.

The complexity comparisons in the case of 160-bit are described in Tables 4 and 5. By the "Traditional method", we mean a scalar multiplication using the double-and-add method in terms of affine coordinates. Again, we assume that one squaring has complexity $S = 0.8M$. For larger sizes, the comparison can be obtained in the same way.

Table 4. Break-even point in scalar multiplication on a 160-bit elliptic curve with Montgomery form

	Method	Complexity	Break-Even Point
Binary	Traditional	$1360\mathcal{M} + 240\mathcal{I}$	$9.3\mathcal{M} < \mathcal{I}$
	Proposed	$1686\mathcal{M} + 205\mathcal{I}$	
NAF	Traditional	$1259\mathcal{M} + 213\mathcal{I}$	$6.6\mathcal{M} < \mathcal{I}$
	Proposed	$1551\mathcal{M} + 169\mathcal{I}$	
Window with NAF	Traditional	$1195\mathcal{M} + 197\mathcal{I}$	$6.1\mathcal{M} < \mathcal{I}$
	Proposed	$1840\mathcal{M} + 91\mathcal{I}$	

Table 5. Break-even point in scalar multiplication on a 160-bit elliptic curve with Weierstrass form

	Method	Complexity	Break-Even Point
Binary	Traditional	$800\mathcal{M} + 240\mathcal{I}$	$6.6\mathcal{M} < \mathcal{I}$
	Proposed	$1030\mathcal{M} + 205\mathcal{I}$	
NAF	Traditional	$724\mathcal{M} + 213\mathcal{I}$	$7.1\mathcal{M} < \mathcal{I}$
	Proposed	$1038\mathcal{M} + 169\mathcal{I}$	
Window with NAF	Traditional	$679\mathcal{M} + 197\mathcal{I}$	$5.6\mathcal{M} < \mathcal{I}$
	Proposed	$1269\mathcal{M} + 91\mathcal{I}$	

6   Running Time

In this section, we present the running times that we obtained with our software implementation of the proposed algorithms.

The platform consisted of a 600MHz Pentium III, which has 32-bit word, using Windows 2000, Visual C++ 6.0, and MASM 6.15. The programs were written in assembly language for multi-precision integer operations, which may be time-critical in our implementation, or in ANSI C language for other operations.

We used the following domain parameters for an elliptic curve with Montgomery form.

$$p = 800000000000000000000000000000000000000012b$$
$$A = 49cb474d172aadfd987191a490ae0671674fe5a9$$
$$B = 17240aee6e1c8c00a7ec1df1b8721d3f90437803$$
$$G_u = 31c0186c5389ec1c81d85f4e1449390c954f7f39$$
$$G_v = 534a718a33d4e2c2089ac68e48c8f6eb101ec46d$$
$$\sharp E_m(\mathbb{F}_p) = 80000000000000000000005b4c33272e33dfe2cb9c$$

where $(G_u, G_v) \in E_m(\mathbb{F}_p)$, and $\sharp E_m(\mathbb{F}_p)$ denotes the number of points on E_m.

Table 6. Running time of elliptic curve and field operations in msec

Curve	Elliptic (160-bit)					Field (160-bit)		
	Add	2P	4P	8P	16P	multiply	square	inversion
Montgomery	0.11	0.12	0.19	0.25	0.29	$1.92 \cdot 10^{-3}$	$1.63 \cdot 10^{-3}$	$56.0 \cdot 10^{-3}$
Weierstrass	0.093	0.094	0.13	0.16	0.20			

Table 7. Running time of scalar multiplication of a randomly selected point in msec

Curve	Binary		NAF		Window with NAF	
(160-bit)	Traditional	Proposed	Traditional	Proposed	Traditional	Proposed
Montgomery	27.4	25.7	25.0	22.7	23.3	16.7
Weierstrass	22.5	20.2	20.0	17.1	17.9	12.3

Table 8. Improvement of the performance of scalar multiplications in %

Curve (160-bit)	Binary	NAF	Window with NAF
Montgomery	6	9	28
Weierstrass	10	14	31

Table 6 shows the running times of elliptic curve and definition field operations. [2] Table 7 shows the running times of scalar multiplications.

We achieved running time reduction as shown in Table 8. As a result of our implementation with respect to Montgomery and Weierstrass form in terms of affine coordinates, we achieved running time reduced by 28% and 31%, respectively, in the scalar multiplication of the elliptic curve of size 160-bit. The proposed algorithms improved the performance of a scalar multiplication with the binary method, as well as the window method. Therefore they are effective in an restricted environment where resources are limited, such as with a smart card.

References

[AMV93]   G. B. Agnew, R. C. Mullin, S. A. Vanstone, "An Implementation of Elliptic Curve Cryptosystems Over F_2^{155}", *IEEE journal on selected areas in communications*, **11**, No.5 (1993)

[CMO98]   H. Cohen, A. Miyaji, T. Ono, "Efficient Elliptic Curve Exponentiation Using Mixed Coordinates", *Advances in Cryptology – ASIACRYPT'98*, LNCS, **1514** (1998), Springer-Verlag, 51–65.

[Go98]   D. M. Gordon, "A survey of fast exponentiation methods", *Journal of Algorithms*, **27** (1998), 129–146.

---

[2]   We used classical euclidean method for field inversion. Although the prime number p has a special form which provide first implementation, we applied general method for modular reduction. Therefore further optimization will derive faster implementation.

[GP97]   J. Guajardo, C. Paar, "Efficient Algorithms for Elliptic Curve Cryptosystems", *Advances in Cryptology – CRYPTO'97*, LNCS, **1294** (1997), Springer-Verlag, 342–356.

[HT99]   Y. Han, P. C. Tan, "Direct Computation for Elliptic Curve Cryptosystems", Pre-proceedings of CHES'99, (1999), Springer- Verlag, 328–340.

[IEEE]   IEEE P1363-2000, (2000), http://grouper.ieee.org/ groups/1363/

[ITTTK99] K. Itoh, M. Takenaka, N. Torii, S. Temma, Y. Kurihara, "Fast Implementation of Public-key Cryptography on a DSP TMS320C6201", *Cryptography Hardware and Embedded Systems*, LNCS, **1717** (1999), Springer-Verlag, 61–72.

[Iz99]   T. Izu, "Elliptic Curve Exponentiation for Cryptosystem", *The 1999 Symposium on Cryptography and Information Security*, (1999), 275–280.

[Ko87]   N. Koblitz, "Elliptic curve cryptosystems", *Mathematics of Computation*, **48** (1987), 203–209.

[KT92]   K. Koyama, Y. Tsuruoka, "Speeding up Elliptic Cryptosystems by Using a Signed Binary Window Method", *Advances in Cryptology – CRYPTO'92*, LNCS, **740** (1993), Springer-Verlag, 345–357.

[LD99]   J. Lopez, R. Dahab, "Fast Multiplication on Elliptic Curves over $GF(2^m)$ without Precomputation", *CHES'99*, LNCS, **1717** (1999), Springer-Verlag, 316–327.

[Le87]   H. W. Lenstra, Jr., "Factoring integers with elliptic curves", *Ann. of Math*, **126** (1987), 649–673.

[Mi85]   V. Miller, "Uses of elliptic curves in cryptography", *Advances in Cryptology – CRYPTO'85*, LNCS, **218** (1986), Springer-Verlag, 417–426.

[MV93]   A. Menezes, A. Vanstone, "Elliptic Curve Cryptosystems and Their Implementation", *J. Cryptology*, **6** (1993), Springer-Verlag, 209–224.

[Mo87]   P. L. Montgomery, "Speeding the Pollard and Elliptic Curve Methods of Factorization", *Mathematics of Computation*, **48** (1987), 243–264.

[MOC97a]   A. Miyaji, T. Ono, H. Cohen, "Efficient Elliptic Curve Exponentiation (I)", *Technical Report of IEICE*, ISEC97-16, (1997)

[MOC97b]   A. Miyaji, T. Ono, H. Cohen, "Efficient Elliptic Curve Exponentiation", *Advances in Cryptology – ICICS'97*, LNCS, **1334** (1997), Springer-Verlag, 282–290.

[Mu97]   V. Müller, "Efficient Algorithms for Multiplication on Elliptic Curves", Proceedings of *GI - Arbeitskonferenz Chipkarten 1998*, TU München, (1998)

[OKS00]   K. Okeya, H. Kurumatani, K. Sakurai, "Elliptic Curves with the Montgomery Form and Their Cryptographic Applications", *Public Key Cryptography (PKC) 2000*, LNCS, **1751** (2000), Springer-Verlag, 238–257.

[SS00]   Y. Sakai, K. Sakurai, "Efficient Scalar Multiplications on Elliptic Curves without Repeated Doublings and Their Practical Performance". *Information Security and Privacy, ACISP 2000*, LNCS, **1841** (2000), Springer-Verlag, 59–63. The final version of this paper has been published [SS01].

[SS01]   Y. Sakai, K. Sakurai, "Efficient Scalar Multiplications on Elliptic Curves with Direct Computations of Several Doublings". *IEICE Trans. Fundamentals*, **E84-A** No.1 (2001), 120–129. Available at http://search.ieice.or.jp/2001/files/ e120a01.htm#e84-a,1,107

[WMPW98]   E. De Win, S. Mister, B. Preneel, M. Wiener, "On the Performance of Signature Schemes Based on Elliptic Curves", *Algorithmic Number Theory III*, LNCS, **1423** (1998), Springer-Verlag, 252–266.

A   Computational Complexity of Direct Computations

In this appendix, we give proofs of Theorems 1 and 2. In these proofs, we ignore the cost of field additions and a subtractions, as well as the cost of multiplications by small constants.

A.1   Proof of Theorem 1

In **Step 1** of Algorithm 1, two multiplications are performed to compute $u_1 B$ and $u_1(2A + 3u_1)$. The complexity of **Step 1** involves $2\mathcal{M}$.

In **Step 2**, the following computations are performed k times to compute D_i and E_i, and $k - 1$ times to compute F_i. We first perform 2 squarings to compute E_{i-1}^2 and F_{i-1}^2. If $i > 1$, we perform one multiplication to compute $(\prod_{j=0}^{i-1} E_j)^2$. Next we perform 2 multiplications for the computation of $D_{i-1} E_{i-1}^2$ and $C(\prod_{j=0}^{i-1} E_j)^2$. Note that $(\prod_{j=0}^{i-2} E_j)^2$ should be stored in the previous loop of the iteration. This gives D_i, and so the complexity of computing D_i involves $3\mathcal{M} + 2\mathcal{S}$ if $i > 1$ and $2\mathcal{M} + 2\mathcal{S}$ if $i = 1$. Next, we perform one squaring and one multiplication to compute E_{i-1}^4 and $F_{i-1}(D_{i-1} E_{i-1}^2 - D_{i-1})$. If $i = 1$, we perform one more multiplication to compute $B_2 E_0^4$. This gives E_i, and so the complexity of computing E_i involves $\mathcal{M} + \mathcal{S}$ if $i > 1$ and $2\mathcal{M} + \mathcal{S}$ if $i = 1$. Next, if $i \neq k$, we perform 3 multiplications and one squaring to compute CD_i, $B_2(\prod_{j=0}^{i-1} E_j)^2$, $(\prod_{j=0}^{i-1} E_j)^2(2^{4i} B_2(\prod_{j=0}^{i-1} E_j)^2 + 2^{2i+1} CD_i)$ and D_i^2. This gives F_i, and so the complexity of computing F_i involves $3\mathcal{M} + \mathcal{S}$.

The total complexity of **Step 2** involves $k(4\mathcal{M} + 3\mathcal{S}) + (k - 1)(3\mathcal{M} + \mathcal{S})$.

In **Step 3**, we first perform $k - 1$ multiplications to compute $(\prod_{i=0}^{k-1} E_i)$ and set the result to T_1. Next, two multiplications for $(\prod_{i=0}^{k-1} E_i)^3$ and $B_2(\prod_{i=0}^{k-1} E_i)^3$ are performed. Note that $(\prod_{i=0}^{k-1} E_i)^2$ has already been computed in **Step 2**. Then, we perform on inversion to compute $(2^{3k} B_2(\prod_{j=0}^{i-1} E_j)^3)^{-1}$ and set the result to T_2. Next, we perform one multiplication to compute $E_k T_2$. Then, we obtain v_{2^k}. The complexity of computing v_{2^k} involves $(k - 1)\mathcal{M} + 3\mathcal{M} + \mathcal{I}$.

To compute u_{2^k}, we perform 3 multiplications to compute $D_k B$, $D_k B T_1$, and $D_k B T_1 T_2$. Then, we obtain u_{2^k}. The complexity of computing u_{2^k} involves $3\mathcal{M}$.

According to above computation, the complexity of Algorithm 1 involves $(8k + 4)\mathcal{M} + (4k - 1)\mathcal{S} + \mathcal{I}$.

A.2   Proof of Theorem 2

In **Step 1** of Algorithm 2, one squaring is performed to compute x_1^2. The complexity of **Step 1** involves $\mathcal{S}$.

In **Step 2**, the following computations are performed k times to compute A_i and B_i, and $k - 1$ times to compute C_i. First, we perform 3 squarings to compute B_{i-1}^2, B_{i-1}^4, and C_{i-1}^2. Second, we perform one multiplication to compute $A_{i-1} B_{i-1}^2$. Then we obtain A_i. Third, we perform one multiplication to compute $C_{i-1}(A_i - 4A_{i-1} B_{i-1}^2)$. Then we obtain B_i. Next, we perform one squaring to compute A_i^2. If $i = 1$, we perform one multiplication to compute aB_1^4 and set the result to U, and if $i > 1$, we perform one multiplication to compute UB_{i-1}^4 and set the result to U. Then, U equals $a(\prod_{j=1}^{i-1} B_j)^4$. Then we obtain C_i. The complexity of **Step 2** involves $(2\mathcal{M} + 3\mathcal{S})k + (\mathcal{M} + \mathcal{S})(k - 1)$.

In **Step 3**, we first compute $\prod_{i=1}^{k} B_i$ which takes $k - 1$ multiplications. Second, we perform one inversion to compute $(2^k \prod_{i=1}^{k} B_i)^{-1}$ and set the result to T. Next, we perform one squaring to compute T^2. Next, we perform one multiplication to compute $A_k T^2$. Then, we obtain x_{2^k}. Finally, we perform 2 multiplications to compute $B_k T^2 T$. Then, we obtain y_{2^k}. The complexity of **Step 3** involves $(k-1)\mathcal{M} + 3\mathcal{M} + \mathcal{S} + \mathcal{I}$.

According to above computation, the complexity of Algorithm 2 involves $(4k+1)\mathcal{M} + (4k+1)\mathcal{S} + \mathcal{I}$.

The GH Public-Key Cryptosystem

Guang Gong[1], Lein Harn[2], and Huapeng Wu[3]

[1] Department of Electrical and Computer Engineering, University of Waterloo,
Waterloo, Ontario N2L 3G1, Canada
[2] Department of Computer Networking, University of Missouri-Kansas City,
Kansas City, Missouri 64110-2499, USA
[3] Department of Combinatorics & Optimization, University of Waterloo,
Waterloo, Ontario N2L 3G1, Canada

Abstract. This paper will propose an efficient algorithm that utilizes the signed-digit representation to compute the kth term of a characteristic sequence generated by a linear feedback shift register of order 3 over $GF(q)$. We will also propose an efficient algorithm to compute the $(h - dk)$th term of the characteristic sequence based on the knowledge of the kth term where k is unknown. Incorporating these results, we construct the ElGamal-like digital signature algorithm for the public-key cryptography based on the 3rd-order characteristic sequences which was proposed by Gong and Harn in 1999.

Key words. Public-key cryptosystem, digital signature, third-order linear feedback shift register sequences over finite fields.

1 Introduction

Gong and Harn have published papers on applying third-order linear feedback shift register (LFSR) sequences with some initial states to construct public-key cryptosystems (PKC) in the ChinaCrypt'98 [4] and in the IEEE Transactions on Information Theory [5], respectively. This type of LFSR sequences is called the characteristic sequence in the area of sequence study. The security of the PKC is based on the difficulty of solving the discrete logarithm (DL) in $GF(q^3)$; but all computations involved in the system are still performed in $GF(q)$.

In [5], Gong and Harn have proposed the Diffie-Hellman (DH) key agreement protocol [1] and the RSA-like encryption scheme [14] as examples of the applications of the GH public-key cryptosystem. Along this line, Lenstra and Verheul [7] have published their XTR public-key system at the Crypto'2000. In the XTR public-key system, they have also used the 3rd-order characteristic sequence; but with a special polynomial. They have proposed the XTR DH and the XTR Nyberg-Rueppel signature scheme as examples.

In this paper, we will review some fundamental properties of 3rd-order characteristic sequences and the original GH public-key cryptosystem and point it out that the XTR cryptosystem is constructed based on a special type of 3rd-order characteristic sequences as Gong and Harn have analyzed in [5]. Then,

S. Vaudenay and A. Youssef (Eds.): SAC 2001, LNCS 2259, pp. 284–300, 2001.

we will explore some useful properties of 3rd-order characteristic-sequences over $GF(q)$ and utilize these results in the construction of the GH ElGamal-like digital signature algorithm [2] [13].

The paper is organized as follows. In section 2, we introduce LFSR sequences and 3rd-order characteristic sequences over $GF(q)$. Then, we will review the original GH public-key cryptosystem and explain the relations of design approach between GH and XTR public-key cryptosystems. In Section 3, using the maximal-weight signed-digit representation, we propose a fast algorithm for evaluating the kth term of a pair of reciprocal characteristic sequences over $GF(q)$ and discuss the computational complexity for a special case when $q = p^2$. This algorithm is more efficient than the previously proposed one [5]. In Section 4, we will introduce the Duality Law of a pair of reciprocal characteristic sequences. Using this law, we show the property of redundancy in states of characteristic sequences over $GF(q)$. Lenstra and Verheul in [8] have also found this type of redundancy for a special case of characteristic sequences over $GF(p^2)$. However the technique used in [8] can not be extended to the general case of the characteristic sequences over either $GF(p^2)$ or $GF(q)$ for any arbitrary q. In Section 5, using the linear feedback shift register concept, we will propose an efficient algorithm to compute the $(h - dk)$th term of a characteristic sequence based on the knowledge of the kth term where k is unknown to the user. This mentioned property is required for digital signature verification. This algorithm can save the matrix computation needed in Algorithm 2.4.8 proposed by Lenstra and Verheul in [7]. Then we will apply these results to the design of the GH ElGamal-like digital signature algorithm as an example of digital signature schemes for the GH public-key cryptosystem.

2   Characteristic Sequences and the GH Public-Key Cryptosystem

In this section, we will briefly introduce LFSR sequences, characteristic sequences, and the GH Public-key Cryptosystem and explain that the sequences used to construct the XTR cryptosystem is just a special case used in the design of the GH cryptosystem. We will use the notation $K = GF(q)$ where $q = p^r$, p is a prime and r is a positive integer throughout this paper.

2.1   LFSR Sequences

Let
$$f(x) = x^n - c_{n-1}x^{n-1} - \cdots - c_1 x - c_0, c_i \in K$$
be a polynomial and
$$\underline{s} = \{s_i\} = s_0, s_1, s_2, \cdots, s_i \in K$$
be a sequence over K. If $\underline{s}$ satisfies the following linear recursive relation
$$s_{k+n} = \sum_{i=0}^{n-1} c_i s_{k+i}, k = 0, 1, \cdots.$$

then we say that $\underline{s}$ is an LFSR sequence of order n (generated by $f(x)$). $(s_0, s_1, \cdots, s_{n-1})$ is called an *initial state* of the sequence $\underline{s}$ or $f(x)$. A vector $(s_k, s_{k+1}, \cdots, s_{k+n-1})$ containing consecutive n terms of $\underline{s}$ is called a state of $\underline{s}$, or the kth state of $\underline{s}$, which is denoted by $\underline{s}_k$.

Example 1. Let $K = GF(5), n = 3$ and $f(x) = x^3 - x - 1$ which is an irreducible polynomial over K. An LFSR sequence generated by $f(x)$ is given below:

3	0	3	3	2	0	1	2	4	4
3	0	1	3	4	3	4	1	4	3
2	1	1	1	0	0	1	0	4	1
1	$\cdots$								

which has period $31 = 5^2 + 5 + 1$ and the initial state is $\underline{s}_0 = (3, 0, 3)$.

2.2 Irreducible Case and Trace Representation

If $f(x)$ is an irreducible polynomial over K, let α be a root of $f(x)$ in the extension $E = GF(q^3)$, then there exists some $\beta \in K$ such that

$$s_i = Tr(\beta \alpha^i), i = 0, 1, 2, \cdots,$$

where $Tr(x) = x + x^q + \cdots + x^{q^{n-1}}$ is the trace function from E to K. If $\beta = 1$, then $\underline{s}$ is called a *characteristic sequence* of $f(x)$, or a *char-sequence* for short.

The sequence given in Example 1 is the characteristic sequence of $f(x)$.

2.3 Period and Order

Let $f(x) \in K[x]$, we say that $f(x)$ has period t if t is the smallest integer such that $f(x)|x^t - 1$. We denote it as $per(f) = t$.

For $\beta \in E$, the order of β is the smallest integer t such that

$$\beta^t = 1.$$

We denote it as $ord(\beta) = r$. A proof of the following result can be found in several references on sequences, for example, in [9,10].

Lemma 1. *If $f(x) \in K[x]$ is irreducible over K and $\underline{s}$ is generated by $f(x)$, then*

$$per(\underline{s}) = per(f) = ord(\alpha)$$

where α is a root of $f(x)$ in the extension $GF(q^n)$.

2.4 Third-Order Characteristic Sequences

Let

$$f(x) = x^3 - ax^2 + bx - 1, a, b \in K \tag{1}$$

be an irreducible polynomial over K and α be a root of $f(x)$ in the extension field $GF(q^3)$. Let $\underline{s} = \{s_i\}$ be a characteristic sequence generated by f(x). Then the initial state of $\{s_i\}$ can be given by

$$s_0 = 3, s_1 = a, \text{ and } s_2 = a^2 - 2b.$$

Example 1 is a characteristic sequence of oder 3.

We list the following lemmas which appeared in [5].

Lemma 2. *With the same notation, we have*

- $per(\underline{s})|q^2 + q + 1$, *i.e., period of $\underline{s}$ is a factor of $q^2 + q + 1$.*
- $\underline{s}$ *has the following trace representation:*

$$s_k = Tr(\alpha^k) = \alpha^k + \alpha^{kq} + \alpha^{kq^2}, k = 0, 1, 2, \cdots.$$

If the order of α satisfies an additional condition, then we have the following result whose proof can be found in [7].

Lemma 3. *With the same notation, let $K = GF(p^2)$, if $ord(\alpha)|p^2 - p + 1$, then*

$$f(x) = x^3 - ax^2 + a^p x - 1, a \in K.$$

2.5   Fundamental Results on 3rd-Order Char-sequences

Here we summarize some results obtained previously. In [5], all results on 3rd-order char-sequences related to the GH Diffie-Hellman key agreement protocol are presented in the finite field $GF(p)$. However, all these results are also true in $K = GF(q)$. So we just list the following two lemmas and the proofs of these lemmas are exactly the same as that described in [5]. Similar results can also be found in corollary 2.3.5 in [7]. For the sake of simplicity, we write $s_k = s_k(a, b)$ or $s_k(f)$ to indicate the generating polynomial. Let $f^{-1}(x) = x^3 - bx^2 + ax - 1$, which is the reciprocal polynomial of $f(x)$. Let $\{s_k(b, a)\}$ be the char-sequence of $f^{-1}(x)$, called *the reciprocal sequence of* $\{s_k(a, b)\}_{k \geq 0}$. Then we have $s_{-k}(a, b) = s_k(b, a), k = 1, 2, \cdots$ (see [5]).

Lemma 4. *Let $f(x) = x^3 - ax^2 + bx - 1$ be an irreducible polynomial over K and $\underline{s}$ be the char-sequence of $f(x)$ and α be a root of $f(x)$ in $GF(q^3)$. Then*

1. *For all integers r and e,*

$$s_r(s_e(a, b), s_{-e}(a, b)) = s_{re}(a, b).$$

2. *For all integers n and m,*
   (a) $s_{2n} = s_n^2 - 2s_{-n}$, *and*
   (b) $s_n s_m - s_{n-m} s_{-m} = s_{n+m} = s_{n-2m}.$
3. *If $gcd(k, per(\underline{s})) = 1$, then $\alpha^{kq^i}, i = 0, 1, 2$ are three roots of $g(x) = x^3 - s_k x^2 + s_{-k} x - 1$ in $GF(q^3)$.*

Lemma 5. *Let* $k = k_0 k_1 \cdots k_r = \sum_{i=0}^{r} k_i 2^{r-i}$ *be the binary representation of* k. *Let* $T_0 = k_0 = 1$, *and* $T_j = k_j + 2T_{j-1}, 1 \leq j \leq r$. *So,* $T_r = k$. *If* $(s_{T_{j-1}-1}, s_{T_{j-1}}, s_{T_{j-1}+1})$ *is computed, then* $(s_{T_j-1}, s_{T_j}, s_{T_j+1})$ *can be computed according to the following formulas.*

For $k_j = 0$

$$s_{T_j-1} = s_{T_{j-1}} s_{T_{j-1}-1} - b s_{-T_{j-1}} + s_{-(T_{j-1}+1)} \tag{2}$$

$$s_{T_j} = s_{T_{j-1}}^2 - 2 s_{-T_{j-1}} \tag{3}$$

$$s_{T_j+1} = s_{T_{j-1}} s_{T_{j-1}+1} - a s_{-T_{j-1}} + s_{-(T_{j-1}-1)} \tag{4}$$

For $k_j = 1$

$$s_{T_j-1} = s_{T_{j-1}}^2 - 2 s_{-T_{j-1}} \tag{5}$$

$$s_{T_j} = s_{T_{j-1}} s_{T_{j-1}+1} - a s_{-T_{j-1}} + s_{-(T_{j-1}-1)} \tag{6}$$

$$s_{T_j+1} = s_{T_{j-1}+1}^2 - 2 s_{-(T_{j-1}+1)} \tag{7}$$

Thus, to calculate a pair of kth terms s_k *and* s_{-k} *of the sequence* $\underline{s}$ *needs* $9 \log k$ *multiplications in* $GF(q)$ *in average.*

2.6   The GH Diffie-Hellman Key Agreement Protocol

In this subsection, we will review the GH Diffie-Hellman (DH) key agreement protocol. (Note. In [5], the GH-DH was presented in $GF(p)$. As we have mentioned in the beginning of Section 2.5, all results can also be true in $GF(q)$, where q is a power of a prime.) In the following discussion, we will present the GH-DH in $GF(q)$, where $q = p^2$ in the same setting as in the XTR cryptosystem.

GH-DH Key Agreement Protocol (Gong and Harn, 1999) [5] :

System parameters: p is a prime number, $q = p^2$ and $f(x) = x^3 - ax^2 + bx - 1$ which is an irreducible polynomial over $GF(q)$ with period $Q = q^2 + q + 1$.

User Alice chooses $e, 0 < e < Q$, with $gcd(e, Q) = 1$ as her private key and computes (s_e, s_{-e}) as her public key. Similarly, user Bob has $r, 0 < r < Q$, with $gcd(r, Q) = 1$ as his private key and (s_r, s_{-r}) as his public key. In the key distribution phase, Alice uses Bob's public key to form a polynomial:

$$g(x) = x^3 - s_r x^2 + s_{-r} x - 1$$

and then computes the eth terms of a pair of reciprocal char-sequences generated by $g(x)$. I.e., Alice computes

$$s_e(s_r, s_{-r}) \text{ and } s_{-e}(s_r, s_{-r}).$$

Similarly, Bob computes

$$s_r(s_e, s_{-e}) \text{ and } s_{-r}(s_e, s_{-e}).$$

They share the common secret key as (s_{er}, s_{-er}).

Let $\mathbb{Z}_Q = \{0, 1, 2, \cdots, Q-1\}$, $\mathbb{Z}_Q^*$ contain all numbers in $\mathbb{Z}_Q$ and these numbers are coprime with Q, R_Q contains all numbers in $\mathbb{Z}_Q^*$ and these numbers are not conjugate modulo Q respect to q (i.e., any two numbers t and r are conjugate modulo Q if there exists some integer j such that $r \equiv tq^j \mod Q$).

The mathematical function used in the GH public-key cryptosystem is:

$$\begin{aligned} \mu : R_Q &\to K \times K \\ i &\mapsto (s_i, s_{-i}) \end{aligned} \tag{8}$$

where $\underline{s}$ is the 3rd-order char-sequence over $GF(q)$ generated by $f(x)$ which is an irreducible polynomial with a period of $Q = q^2 + q + 1$. In [5], it is shown that this is an injective map from R_Q to $K \times K$.

Remark 1. The XTR [7] is designed based on the char-sequences generated by the 3rd-order polynomial of $f(x) = x^3 - ax^2 + a^p x - 1$ which is irreducible over $GF(q)$ with period $Q | p^2 - p + 1$. The XTR only uses one char-sequence instead of a pair of reciprocal char-sequences. The mathematical function used in the XTR public-key system is:

$$\begin{aligned} \nu : R_Q &\to K \\ i &\mapsto s_i \end{aligned} \tag{9}$$

However, the GH system is based on the char-sequences generated by the 3rd-order polynomial of $a^3 - ax^2 + bx - 1$, where a and b are from $GF(p^2)$. Thus, the 3rd-order char-sequence used to construct the XTR cryptosystem is just a special case used in the design of the GH cryptosystem. Two schemes have the same efficiency when they are applied to the DH key agreement protocol, because the GH-DH computes a pair of elements over $GF(p^2)$ and shares a pair of elements over $GF(p^2)$, and the XTR-DH computes one element over $GF(p^2)$ and shares one element over $GF(p^2)$.

In the following sections, we will explore some useful properties of 3rd-order char-sequences over K and use these results to the design of a new GH digital signature algorithm. For additional results on LFSR sequences and finite fields, the reader can refer to [3,9,10].

3   Fast Computational Algorithm Based on the Signed-Digit Representation

3.1   A New Signed-Digit Number Representation

Definition 1. Let $A = a_{n-1}a_{n-2} \cdots a_0$, $a_i \in \{0, 1\}$ be a binary representation. Then $A = b_{n-1}b_{n-2} \cdots b_0$, $b_i \in \{-1, 0, 1\}$ is called the binary *maximal-weight* signed-digit (SD) representation of A, if there does not exist another binary SD representation of length n for A whose Hamming weight is higher than that of the maximal-weight SD representation.

An algorithm to obtain such an SD representation is given in Appendix A. The following lemma is obvious from Algorithm 4 in Appendix A.

Lemma 6. *Let $a_{n-1}a_{n-2}\cdots a_0$, $a_i \in \{0,1\}$ and $a_{n-1} = 1$ be the binary representation of integer A. Let $d, 0 \le d \le n-2$, be the smallest integer such that $a_d \ne 0$. Then the Hamming weight of the binary maximal-weight SD representation of A is $n-d$. Moreover, all the zeroes are associated with least significant bit positions.*

Some examples are given for the maximal-weight SD representations: $101100111 = 1\,1\,\bar{1}\,1\,1\,\bar{1}\,\bar{1}\,1\,1$ and $11001011000 = 1\,1\,1\,\bar{1}\,\bar{1}\,1\,\bar{1}\,1\,0\,0\,0$. When this SD representations is involved in computing exponentiation-like functions using square-and-multiply method, it is obvious that efficiency can only be achieved when the "squaring and multiplication" is less expensive than the "squaring". It appears to be the case when computing terms in a third-order recurrence sequence as it will be discussed in detail in the next subsections.

3.2 Fast Computational Algorithm of Recurrence Terms

Let k be given in its maximal-weight SD representation as $k = k_0 k_1 \cdots k_r = \sum_{i=0}^{r} k_i 2^{r-i}$, $k \in \{-1,0,1\}$. It can be proven that Lemma 5 still holds true if we add the following formulas.
For $k_j = -1$

$$s_{T_j-1} = s_{T_{j-1}-1}^2 - 2s_{-(T_{j-1}-1)} \tag{10}$$

$$s_{T_j} = s_{T_{j-1}} s_{T_{j-1}-1} - bs_{-T_{j-1}} + s_{-(T_{j-1}+1)} \tag{11}$$

$$s_{T_j+1} = s_{T_{j-1}}^2 - 2s_{-T_{j-1}} \tag{12}$$

With values of initial terms as $s_0 = 3$, $s_1 = a$, $s_2 = a^2 - 2b$, $s_{-1} = b$, and $s_{-2} = b^2 - 2a$, $T_0 = k_0 = 1$ and $T_j = k_j + 2T_{j-1}, 1 \le j \le r, T_r = k$, a pair of dual terms, s_k and s_{-k} for $k \ge 0$, can be computed based on the following algorithm.

Algorithm 1 Computing $s_{\pm k}$

1. Set up initial values: $s_{T_0-1} = s_{-T_0+1} = 3$; $s_{T_0} = a$; $s_{T_0+1} = a^2 - 2b$; $s_{-T_0} = b$; $s_{-T_0-1} = b^2 - 2a$;
2. IF $k_r = 0$ THAN find $h < r$, such that $k_h \ne 0$ and $k_{h+1} = k_{h+2} = \cdots = k_r = 0$, ELSE $h = r$;
3. IF $h > 1$ THEN FOR $i = 1$ TO $h - 1$
   (a) IF $k_i = 1$ THEN
       i. compute s_{T_i} and $s_{T_i \pm 1}$ using (5)-(7);
       ii. compute s_{-T_i} and $s_{-T_i \pm 1}$ using (10)-(12);
   (b) ELSE
       i. compute s_{T_i} and $s_{T_i \pm 1}$ using (10)-(12);
       ii. compute s_{-T_i} and $s_{-T_i \pm 1}$ using (5)-(7);
4. FOR $i = \text{Max}\{1, h\}$ TO r
   (a) compute $s_{\pm T_i}$ using (3);

The final value is $s_{T_r} = s_k$. Note that T_j and T_{j-1} should be respectively replaced by $-T_j$ and $-T_{j-1}$ in (2)-(12), when these formulas are used in computing s_{-T_i} and $s_{-T_i \pm 1}$ as shown in Steps 3(a)(ii), 3(b)(ii) and 4(a) in the above algorithm.

Since the implementation of (5)-(7) or (10)-(12) is less costly than that of (2)-(4), certain efficiency can be achieved by using the maximal-weight SD representations. It can be shown that an evaluation of (5)-(7) or (10)-(12) needs one multiplication, two squarings, and one constant multiplication in $GF(q)$; while an evaluation of (3) requires one squaring in $GF(q)$. Thus, Step 3 needs two multiplications, four squarings, and two constant multiplications in $GF(q)$; while Step 4 requires two squarings in $GF(q)$. With the assumptions that Step 3(a) or Step 3(b) has to be performed for $h - 1$ times and Step 4(a) has to be executed for $r - h + 1$ times, and also with the estimation of the average value of h as included in Appendix B, we have the following lemma:

Lemma 7. *Let k be given in its maximal-weight SD representation, with $\log_2 k \geq 10$, then, on the average case (which is also the worst case), to compute a dual pair $\{s_{-k}, s_k\}$ using Algorithm 1 needs $4 \log_2 k$ multiplications and $4 \log_2 k$ squarings in $GF(q)$. On the best case to compute both s_k and s_{-k} needs $2 \log_2 k$ multiplications in $GF(q)$.*

Note that half number of $4 \log_2 k$ multiplications are in fact contant multiplications if both a and b are contant.

3.3   Complexity of Computing Recurrence Terms When $q = p^2$

When -1 is a quadratic non-residue in $GF(p)$, the binomial $f(x) = x^2 + 1$ is irreducible over $GF(p)$. Let α be a root of $f(x)$. Under previous assumptions, $\{1, \alpha\}$ forms a polynomial basis in $GF(p^2)$ over $GF(p)$. Any two elements, x and $y \in GF(p^2)$, can be represented in the polynomial basis as $x = x_0 + x_1 \alpha$ and $y = y_0 + y_1 \alpha$, $x_0, x_1, y_0, y_1 \in GF(p)$. (It is worth to mention that, in XTR system [7], the irreducible trinomial $f(x) = x^2 + x + 1$ and the normal basis have been chosen.) A multiplication in $GF(p^2)$ can be represented by $xy = (x_0 + x_1\alpha)(y_0 + y_1\alpha) = (x_0y_0 - x_1y_1) + (x_0y_1 + x_1y_0)\alpha = [x_0(y_0 + y_1) - y_1(x_0 + x_1)] + [x_0(y_0 + y_1) + y_0(x_1 - x_0)]\alpha$. Thus, three multiplications in $GF(p)$ are needed. Since the squaring can be represented by $x^2 = (x_0 + x_1\alpha)^2 = (x_0^2 - x_1^2) + 2x_0x_1\alpha = (x_0 - x_1)(x_0 + x_1) + 2x_0x_1\alpha$, two multiplications in $GF(p)$ are required for a squaring in $GF(p^2)$. The constant multiplication is the case where the multiplicand is a fixed element. If the constant element can be chosen to be a number with a specific form, then the constant multiplication can be extremely efficient. For example, if both x_0 and x_1 can be chosen to be a small power of two, then it can be seen from $xy = (x_0 + x_1\alpha)(y_0 + y_1\alpha) = (x_0y_0 - x_1y_1) + (x_0y_1 + x_1y_0)\alpha$ that the multiplication of xy can be obtained for free. If we choose the constant element $x = x_0 + x_1\alpha$ such that one of the two coefficients x_0 and x_1 is a small power of 2, then only two multiplications in $GF(p)$ are needed to perform multiplication of xy in $GF(p^2)$. We summarize the above results in the following lemma:

Lemma 8.

1. *A multiplication in $GF(p^2)$ can be realized by performing three multiplications in $GF(p)$.*
2. *A squaring in $GF(p^2)$ needs two multiplications in $GF(p)$.*
3. *A constant multiplication in $GF(p^2)$*
   - *(a) can be realized for free if both the coefficients of the constant element can be chosen to be a small power of 2.*
   - *(b) can be realized by performing two multiplications in $GF(p)$, if one of the two coefficients of the constant element can be chosen as a small power of 2.*

¿From Lemmas 7 and 8, we can find the complexity of computing $s_{\pm k}$ using Algorithm 1 and we summarized this result in the following lemma:

Lemma 9. *Let $q = p^2$ and k be given in its maximal-weight SD representation, with $\log_2 k \geq 10$, then on the average case (which can also be the worst case) to compute a dual pair $\{s_{-k}, s_k\}$ using Algorithm 1 needs:*

1. *at most $20\log_2 k$ multiplications in $GF(p)$;*
2. *at most $18\log_2 k$ multiplications in $GF(p)$, if one of the two coefficients of the constant elements $a, b \in GF(p^2)$, can be chosen to be a small power of 2;*
3. *at most $16\log_2 k$ multiplications in $GF(p)$, if one constant element is chosen in such a way that one of the coefficients is a small power of 2, and the other constant element is chosen such that both coefficients are small powers of 2;*
4. *at most $14\log_2 k$ multiplications in $GF(p)$, if the both constant elements can be chosen such that all the coefficients are small powers of 2.*

On the best case to compute both s_k and s_{-k} needs $4\log_2 k$ multiplications in $GF(p)$.

4   Redundancy in States of the 3rd-Order Char-Sequences

In this section, we will introduce the duality law of a pair of reciprocal char-sequences. Under this duality law, we can address some redundancy in states of char-sequence over K.

4.1   Duality Law

Let $f(x) = x^3 - ax^2 + bx - 1$ be an irreducible polynomial over K and $\underline{s}$ be its char-sequence. We define a dual operator as given below:

$$D(s_k) = s_{-k}$$

$$D(s_k, s_{k+1}, \cdots, s_{k+t}) = (s_{-k}, s_{-(k+1)}, \cdots, s_{-(+t)}), k \in \mathbb{Z}, t \geq 0,$$

where $T = (s_k, s_{k+1}, \cdots, s_{k+t})$ is a segment of $\underline{s}$.

    We call $(s_k, s_{k+1}, \cdots, s_{k+t})$ and $D(s_k, s_{k+1}, \cdots, s_{k+t})$ a *dual segment* of $\underline{s}$ or $f(x)$. If $t = 0$, we call s_k and s_{-k} a *dual pair* of $\underline{s}$ or $f(x)$.

Let $h(x_1, x_2, \cdots, x_t) \in \mathbb{K}[x_1, x_2, \cdots, x_t]$, i.e., h is a multivariables polynomial over $\mathbb{K}$. We define

$$D(h(s_{i_1}, s_{i_2}, \cdots, s_{i_t})) = h(s_{-i_1}, s_{-i_2}, \cdots, s_{-i_t}), i_j \in Z.$$

Duality Law. Let $f(x) = x^3 - ax^2 + bx - 1$ be an irreducible polynomial over K, $\underline{s}$ be its char-sequence and D be the dual operator. Then $D(D(T)) = T$, $D(D(h)) = h$ and

$$h(s_{i_1}, s_{i_2}, \cdots, s_{i_t}) = 0 \leftrightarrow D(h(s_{i_1}, s_{i_2}, \cdots, s_{i_t})) = 0.$$

4.2   Property of Redundancy

In the following theorem, we will show that three elements in any state of the 3rd-order char-sequence are not independent. If we know any two consecutive elements, the third remaining one can be uniquely determined according to a formula.

Theorem 1. *Let $f(x) = x^3 - ax^2 + bx - 1$ be an irreducible polynomial over K and $\underline{s}$ be its char-sequence. For given the dual segment (s_k, s_{k+1}) and $(s_{-k}, s_{-(k+1)})$, we assume that $\Delta = s_{k+1}s_{-(k+1)} - s_1 s_{-1} \neq 0$. Then s_{k-1} and its dual $s_{-(k-1)}$ can be computed by the following formulas:*

$$s_{k-1} = \frac{es_{-(k+1)} - s_{-1}D(e)}{\Delta} \tag{13}$$

$$s_{-(k-1)} = \frac{D(e)s_{(k+1)} - s_1 e}{\Delta} \tag{14}$$

where

$$e = -s_{-1}D(c_1) + c_2, \quad where \ c_1 = s_1 s_{k+1} - s_{-1} s_k \ and \ c_2 = s_k^2 - 3s_{-k} + (b^2 - a)s_{-(k+1)}.$$

(Note. Here $s_1 = a$ and $s_{-1} = b$. In order to keep symmetric forms in the formulas, we keep on using s_1 and s_{-1}.)

Proof. A sketch to prove this theorem is given below. From $U = (s_{k-1}, s_k, s_{k+1}, s_{k+2})$ and its dual, we will form four linear equations in terms of four variables $s_{k+2}, s_{-(k+2)}, s_{k-1}, s_{-(k-1)}$. Then based on linear algebra, we can solve these equations and obtain (13) and (14).

We now start to construct the linear equations. Since U is a segment of $\underline{s}$ generated by $f(x)$, it satisfies the linear recurrent relation,

$$s_{k+2} = s_1 s_{k+1} - s_{-1} s_k + s_{k-1}.$$

Thus, we have

$$s_{k+2} - s_{k-1} = c_1 \ where \ c_1 = s_1 s_{k+1} - s_{-1} s_k. \tag{15}$$

Applying Duality Law to the above expression, we have

$$s_{-(k+2)} - s_{-(k-1)} = D(c_1). \tag{16}$$

Let $n = k - 1$, $m = k + 1$ in formula 2(b) in Lemma 4, we get

$$s_{k-1}s_{k+1} - s_{-2}s_{-(k+1)} = s_{2k} - s_{-(k+3)}. \tag{17}$$

Since $(s_{-k}, s_{-(k+1)}, s_{-(k+2)}, s_{-(k+3)})$ is a dual of U, it satisfies the following linear recursive relation

$$s_{-(k+3)} = s_{-1}s_{-(k+2)} - s_1 s_{-(k+1)} + s_{-k}.$$

Note that $s_{2k} = s_k^2 - 2s_{-k}$ from 2(a) in Lemma 4. Substituting $s_{-(k+3)}$ and s_{2k} in (17) respectively with the above two identities, it follows that

$$s_{k-1}s_{k+1} - s_{-2}s_{-(k+1)} = s_k^2 - 2s_{-k} - s_{-1}s_{-(k+2)} + s_1 s_{-(k+1)} - s_{-k}.$$

Since $s_{-2} + s_1 = b^2 - 2a + a = b^2 - a$, we have

$$s_{-1}s_{-(k+2)} + s_{k+1}s_{k-1} = c_2 \text{ where } c_2 = s_k^2 - 3s_{-k} + (b^2 - a)s_{-(k+1)}. \tag{18}$$

By Duality Law, we have

$$s_1 s_{(k+2)} + s_{-(k+1)}s_{-(k-1)} = D(c_2). \tag{19}$$

The equations (15), (16), (18), and (19) form four linear equations in terms of variables $s_{k+2}, s_{-(k+2)}, s_{k-1}, s_{-(k-1)}$. Let A be the matrix of the coefficients of this linear system, i.e.,

$$A = \begin{pmatrix} 1 & 0 & -1 & 0 \\ 0 & 1 & 0 & -1 \\ 0 & s_{-1} & s_{k+1} & 0 \\ s_1 & 0 & 0 & s_{-(k+1)} \end{pmatrix}$$

Therefore, this linear system can be written as

$$AS^T = C^T, \tag{20}$$

where $S = (s_{k+2}, s_{-(k+2)}, s_{k-1}, s_{-(k-1)})$, $C = (c_1, D(c_1), c_2, D(c_2))$ and X^T is the transpose of the vector X. Let $\tilde{A} = (A, C^T)$. Then the reduced row-echelon form of $\tilde{A}$ is given below:

$$\tilde{A} \sim \begin{pmatrix} 1 & 0 & -1 & 0 & c_1 \\ 0 & 1 & 0 & -1 & D(c_1) \\ 0 & 0 & s_{k+1} & s_{-1} & e \\ 0 & 0 & s_1 & s_{-(k+1)} & D(e) \end{pmatrix}$$

where $e = -s_{-1}D(c_1) + c_2$. Thus (20) has a unique solution if and only if $\det(B) \neq 0$, where

$$B = \begin{pmatrix} s_{k+1} & s_{-1} \\ s_1 & s_{-(k+1)} \end{pmatrix}.$$

Since $det(B) = \Delta \neq 0$, then s_{k-1} is given by

$$s_{k-1} = \frac{det\begin{pmatrix} e & s_{-1} \\ D(e) & s_{-(k+1)} \end{pmatrix}}{\Delta},$$

which yields (13). The validity of (14) follows from the Duality Law.

Corollary 1. *With the same notation as used in Theorem 1, the dual pair s_{k+2} and $s_{-(k+2)}$ are given by*

$$s_{k+2} = s_{k-1} + c_1 \ \text{and} \ s_{-(k+2)} = s_{-(k-1)} + D(c_1).$$

Remark 2. If $\Delta \neq 0$, then three elements in a state (s_{k-1}, s_k, s_{k+1}) and their duals are dependent. With the knowledge of any two consecutive elements and their duals, the third one and its dual can be uniquely determined by Theorem 1. If $\Delta = 0$, then the third element in a state of $\underline{s}$ may have more than one solution. For the case of knowing (s_{k-1}, s_k) and its dual to compute s_{k+1} and its dual, it is similar to the previous case that we have discussed. We will not include the discussion here.

Remark 3. In [8], Lenstra et. al. have also given a formula to compute s_{k-1} (or s_{k+1}) with the knowledge of (s_k, s_{k+1}) (or (s_{k-1}, s_k)) for a special case of $K = GF(p^2)$ and $f(x) = x^3 - ax^2 + a^px - 1$. Here, we have discussed more general cases and proposed a simpler proof. The formulas between these two approaches are different. The technique used in [8] can not be extended to the general case of the char-sequences.

5   The GH Digital Signature Algorithm

In this section, we explain the method to evaluate $s_{c(h-dk)}$ and its dual with the knowledge of s_k and its dual; but without knowing k. Then, we apply this result together with Theorem 1 in Section 4 and Algorithm 1 in Section 3 to the design of GH ElGamal-like digital signature algorithm (GH-DSA).

5.1   Computation of a Mixed Term $S_{c(h-dk)}$

The following lemma is a direct result from the definition of LFSR sequences.

Lemma 10. *With the same notation of $f(x)$, $\underline{s}$, let (s_{k-1}, s_k, s_{k+1}) be a state of $\underline{s}$ and $\underline{u}$ be a sequence generated by $f(x)$ with (s_{k-1}, s_k, s_{k+1}) as an initial state. I.e.,*

$$u_0 = s_{k-1}, u_1 = s_k, \ \text{and} \ u_2 = s_{k+1}.$$

Then, $\underline{u}_{v-1} = (u_{v-1}, u_v, u_{v+1})$, the $(v-1)$th state of $\underline{u}$, is equal to the $(v-1+k)$th state of $\underline{s}$. In other words, we have

$$(u_{v-1}, u_v, u_{v+1}) = (s_{v-1+k}, s_{v+k}, s_{v+1+k}).$$

For simplicity, we denote $((s_{k-1}, s_k, s_{k+1}), f(x))$ as a sequence generated by $f(x)$ with an initial state (s_{k-1}, s_k, s_{k+1}).

Algorithm 2 *Assume that $f(x)$,(s_k, s_{k+1}) and its dual are given. Let $Q = per(\underline{s})$. Assume that c, h and d are given integers with $\gcd(d, Q) = 1$. Then $s_{c(h-dk)}$ and its dual can be computed according to the following procedures:*

1. *Compute $v = -hd^{-1} \bmod Q$ and $u = -cd \bmod Q$.*
2. *Compute the s_{k-1} and its dual according to Theorem 1.*
3. *Compute $(v - 1)$th state of a sequence generated by $((s_{k-1}, s_k, s_{k+1}), f(x))$ according to Algorithm 1. This step gives s_{v+k} and its dual.*
4. *Construct $g(x) = x^3 - s_{v+k}x^2 + s_{-(v+k)}x - 1$ and compute $s_u(g), s_{-u}(g)$ according to Algorithm 1.*

Here, we have $s_u(g) = s_{c(h-dk)}$ and $s_{-u}(g) = s_{-(c(h-dk))}$.

Note. All results that we have discussed so far are true for general q and Q.

Lemma 11. *With the same notation as used in Algorithm 2, to compute $s_{c(h-dk)}$ and its dual needs $2 \cdot 4(logv + logu)$ multiplications and $2 \cdot 4(logv + logu)$ squarings in $GF(q)$ in average. In particular, if $q = p^2$, to compute $s_{c(h-dk)}$ and its dual needs $2 \cdot 20(logv + logu)$ multiplications in $GF(p)$ in average.*

Proof. In Algorithm 2, the computational cost depends only on how many times Algorithm 1 is invoked. Since Algorithm 2 invoked Algorithm 1 twice, applying Lemma 7, it needs 16 multiplications in $GF(q)$. According to Lemma 9, for invoking Algorithm 1 each time, it needs $20(logv + logu)$ multiplications in $GF(p)$. In total, it needs $2 \cdot 20(logv + logu)$ multiplications in $GF(p)$.

Remark 4. When we apply Algorithm 2 to the char-sequences used in the XTR, it can save the matrix computation as given in Algorithm 2.4.8 [7]. So, this algorithm is more efficient than the algorithm given in [7].

5.2   The GH Digital Signature Algorithm

We are now ready to present the GH ElGamal-like digital signature algorithm. Note that the GH signature scheme can also be modified into variants of generalized ElGamal-like signature schemes as listed in [6].

Algorithm 3 *(GH-DSA)*
   System public parameters: *p is a prime, $q = p^2$, and $f(x) = x^3 - ax^2 + bx - 1$ which is an irreducible polynomial over $GF(q)$ with period Q, where Q satisfies the condition that $Q = P_1 P_2$, P_1 is a prime divisor of $p^2 + p + 1$ and P_2 is a prime divisor of $p^2 - p + 1$. Let $GF(p^2)$ be defined by the irreducible polynomial of $x^2 + 1$ (see Section 3) and γ be its root in $GF(p^2)$.*

   Alice: *Choose x, with $0 < x < Q$ and $\gcd(x, Q) = 1$ as her private key and compute (s_x, s_{-x}) as her public key. For a message m that Alice needs to sign, she follows the procedures:*

1. *Randomly choose k, with $0 < k < Q$, and $gcd(k, Q) = 1$, and use Algorithm 1 to compute (s_{k-1}, s_k, s_{k+1}) and its dual such that $r = s_{k,0} + s_{k,1}p$ is coprime with Q, where $s_k = s_{k,0} + s_{k,1}\gamma$. (Here, we adopt the similar approach as used in the Elliptic Curve digital signature algorithm [12] to form an integer r in digital signing process.)*
2. *Compute $h = h(m)$, where h is a hash function.*
3. *Compute $t = k^{-1}(h - xr) \mod Q$ (i.e., the signing equation is: $h \equiv xr + kt \mod Q$.)*

Then (r, t) is a digital signature of the message m. Alice sends Bob (m, r, t) together with (s_k, s_{k+1}) and its dual.

Bob: *Performing the following verifying process*

Check if $gcd(t, Q) = 1$.

Case 1. $gcd(t, Q) = 1$.

1. *Compute $v = tr^{(-1)} \mod Q$ and $u = hr^{(-1)} \mod Q$.*
2. *Compute s_{u-vk} and its dual according to Algorithm 2.*
3. *Check if both $s_{u-vk} = s_x$ and $s_{-(u-vk)} = s_{-x}$. If so, Bob accepts it as a valid signature. Otherwise, Bob rejects it.*

Case 2. $gcd(t, Q) > 1$.

1. *Compute s_{h-rx} and its dual according to Algorithm 2.*
2. *Form $g(x) = x^3 - s_k x^2 + s_{-k}x - 1$ and compute $s_t(g)$ and its dual according to Algorithm 1.*
3. *Check if both $s_{h-rx} = s_t(g)$ and $s_{-(h-rx)} = s_{-t}(g)$. If so, Bob accepts it as a valid signature. Otherwise, Bob rejects it.*

Lemma 12. *The security of the GH-DSA is based on the difficulty of solving the discrete logarithm in $GF(q^3) = GF(p^6)$. The signing and verifying processes need respectively $20logQ$ multiplications and $2 \cdot 20logQ$ multiplications in $GF(p)$ in average.*

Proof. Since $f(x)$ is an irreducible polynomial over $GF(q^3)$ and the period of $f(x)$ is $Q = P_1 P_2$, where $P_1 | p^2 + p + 1$ and $P_2 | p^2 - p + 1$, a root of $f(x)$ is in $GF(p^6) - (GF(p^3) \cup GF(p^2))$. Similarly, as we have proved in [5], the problem of solving for x from (s_x, s_{-x}) or solving for k from (s_k, s_{-k}) is equivalent to compute DL in $GF(p^6)$. Thus, the first assertion is established.

Note that the probability of any number less than Q which is not coprime with Q is given by

$$Prob\{gcd(z, Q) > 1 : 0 < z < Q\} = \frac{P_1 + P_2 - 1}{P_1 P_2}. \tag{21}$$

Thus, in the signing process, we only need to estimate the computational cost for invoking Algorithm 1 at Step 1, which is $20logQ$ multiplications in $GF(p)$ in average. For case 1 in the verifying process, it can be seen that from Lemma 9 invoking Algorithm 2 at Step 2 needs $2 \cdot 20logQ$ multiplications in $GF(p)$ in average. In Case 3, it invokes Algorithm 2 at Step 2 and Algorithm 1 at Step 3. Thus, it needs $3 \cdot 20logQ$ multiplications in $GF(p)$ in average. Combined with (21), the verifying process needs $2 \cdot 20logQ$ multiplications in $GF(p)$ in average.

6   Conclusion

In this paper, we discuss an efficient algorithm that utilizes the signed-digit representation to compute the k term of a characteristic sequence generated by a linear feedback shift register of order 3 over $GF(q)$. Then we propose an efficient algorithm to compute the $(h-dk)$th term of the characteristic sequence based on the knowledge of the kth term where k is unknown. By using these new results on the characteristic sequences, the GH-DSA (Digital Signature Algorithm) is developed.

Remark 5. The GH cryptosystem, just like the elliptic curve public-key cryptosystem, enjoys the benefit of using a shorter key to achieve high security. Also, the GH cryptosystem can be resistant to power analysis attack and timer analysis attack without increasing cost of computation. This is due to their evaluation formulas as given in Lemma 5 of Section 2.

References

1. W. Diffie and M. E. Hellman, "New directions in cryptography", *IEEE Trans. on Infor. Theory* vol. IT-22, no. 6, pp 644-654, Nov.1976.
2. T. ElGamal, "A public key cryptosystem and a signature scheme based on discrete logarithms," *IEEE Trans. on Inform. Theory*, vol. IT- 31, no. 4, July 1985, pp.469-472.
3. S. Golomb, *Shift Register Sequences,* Holden-Day, Inc., 1967. Revised edition, Aegean Park Press, 1982.
4. G. Gong and L. Harn, "A new approach on public-key distribution", *ChinaCRYPT '98*, pp 50-55, May, 1998, China.
5. G. Gong and L. Harn, "Public-key cryptosystems based on cubic finitie field extensions", *IEEE IT* vol 45, no 7, pp 2601-2605, Nov. 1999.
6. L. Harn and Y. Xu, "On the design of generalized ElGamal type digital signature schemes based on the discrete logarithm", *Electronics Letters* vol. 30, no. 24, pp 2025-2026, Nov. 1994.
7. A. K. Lenstra and E. R. Verheul, "The XTR public key system", *Advances in Cryptology, Proceedings of Crypto'2000*, pp. 1-19, Lecture Notes in Computer Science, Vol. 1880, Springer-Verlag, 2000.
8. A. K. Lenstra and E. R. Verheul, "Key improvements to XTR", *the Proceedings of Asiacrypt'2000*, Lecture Notes in Computer Science, vol. 1976, 2000.
9. R. Lidl and H. Niederreiter, *Finite Fields*, Addison-Wesley Publishing Company, Reading, MA, 1983.
10. R.J. McEliece, *Finite Fields for Computer Scientists and Engineers*, Kluwer Academic Publishers, Boston, 1987.

11. A. J. Menezes, *Elliptic Curve Public Key Cryptosystems*, Kluwer Academic Publishers, Boston, 1993.
12. A. J. Menezes and P. C. van Oorschot and S. A. Vanstone, *Handbook of Applied Cryptography*, CRC Press, 1996.
13. NIST, FIPS PUB 186-2, Digital Signature Standard (DSS), Jan. 2000, http://csrc.nist.gov/publications/fips/fips186-2/fips186-2.pdf
14. R. Rivest, A. Shamir and L. Adleman, "A method for obtaining digital signatures and public-key cryptosystems", in *Comm. Of the ACM*, vol. 21, no. 2, pp 120-126, 1978.

Appendix

A   An Algorithm to Obtain Maximal-Weight SD Representation

When the binary representation of an integer is given, its binary maximal-weight SD representation can be generated with the following algorithm.

Algorithm 4 Maximal-weight signed-digit recoding
Input: the binary representation of A: $a_{n-1}a_{n-2}\cdots a_0$, $a_i \in \{0,1\}$
         and $a_{n-1} = 1$;
Output: the binary maximal-weight representation of A:
         $b_{n-1}b_{n-2}\cdots b_0$, $b_i \in \{-1,0,1\}$;
1. initialize the flag: $t = 0$;
2. FOR $i = 0$ TO $n - 2$
   (a) IF $t = 0$ THEN
       i. IF $a_i = 0$ THEN $b_i = 0$;
       ii. ELSE $\{t = 1$; IF $a_{i+1} = 0$ THEN $b_i = -1$; ELSE $b_i = 1;\}$
   (b) ELSE
       i. IF $(a_i = 1$ AND $a_{i+1} = 0)$ THEN $b_i = -1$;
       ii. IF $(a_i = 1$ AND $a_{i+1} = 1)$ THEN $b_i = 1$;
       iii. IF $(a_i = 0$ AND $a_{i+1} = 0)$ THEN $b_i = -1$;
       iv. IF $(a_i = 0$ AND $a_{i+1} = 1)$ THEN $b_i = 1$;
3. $b_{n-1} = a_{n-1}$;

The correctness of this algorithm can be proved. It is worth to point out that the maximal-weight SD representation always has the same length as the binary form. If the maximal-weight SD representation of a negative integer $(-A)$ is required, it can be obtained by negating each bit in the maximal-weight SD representation of A. It can be seen that the Hamming weight of the maximal-weight SD representation of $-A$ is the same as that of A.

B   An Estimation of the Parameter h

In Algorithm 1, Let $k = k_0 k_1 \cdots k_r$, $2^r \le k \le 2^{r+1} - 1$, be given in its maximal-weight SD representation, then from Lemma 6, we have $\Pr\{k_i = 0, 0 < i < r\} = \frac{\frac{r(r-1)}{2}}{(r-1)\times 2^r} = \frac{r}{2^{r+1}}$. Thus, the average value of h can be given by $\overline{h} = r - (r-1)\frac{r}{2^{r+1}}$. The following table shows some values of $\overline{h}(r)$ as a function of r.

Table 1. Some values of $\overline{h}(r)$, $\text{Max}\{h(r)\}$, $\text{Min}\{h(r)\}$ and r.

r	2	3	4	5	6	7	8	9	10	12	$r > 15$
$\overline{h}(r)$	1.75	2.62	3.62	4.69	5.77	6.84	7.89	8.93	9.96	11.98	r
$\text{Max}\{h(r)\}$	2	3	4	5	6	7	8	9	10	12	r
$\text{Min}\{h(r)\}$	0	0	0	0	0	0	0	0	0	0	0

Although the value of h can be as small as 0, it can be seen that the average value $\overline{h}(r)$ is approximately equal to its maximal value r when $r \ge 10$.

XTR Extended to $\mathrm{GF}(p^{6m})$

Seongan Lim[1], Seungjoo Kim[1], Ikkwon Yie[2,*],
Jaemoon Kim[2,*], and Hongsub Lee[1]

[1] KISA (Korea Information Security Agency),
5th FL., Dong-A Tower, 1321-6, Seocho-Dong, Seocho-Gu, Seoul 137-070, Korea
{seongan, skim, hslee}@kisa.or.kr
[2] Department of Mathematics, Inha University,
YongHyun-Dong, Nam-Gu, Incheon, Korea
{ikyie, jmkim}@math.inha.ac.kr

Abstract. A. K. Lenstra and E. R. Verheul in [2] proposed a very efficient way called XTR in which certain subgroup of the Galois field $\mathrm{GF}(p^6)$ can be represented by elements in $\mathrm{GF}(p^2)$. At the end of their paper [2], they briefly mentioned on a method of generalizing their idea to the field $\mathrm{GF}(p^{6m})$. In this paper, we give a systematic design of this generalization and discuss about optimal choices for p and m with respect to performances. If we choose m large enough, we can reduce the size of p as small as the word size of common processors. In such a case, this extended XTR is well suited for the processors with optimized arithmetic on integers of word size.

1 Introduction

After Diffie-Hellman (DH) key agreement protocol was published, many related key agreement protocols have been proposed. Very recently, in [2] A. K. Lenstra and E. Verheul proposed an efficient computational tool called XTR (Efficient and Compact Subgroup Trace Representation) and showed that it can be adopted to various public key systems including key exchange protocols. Their scheme results in relatively efficient system with respect to the computational and communicational complexity compared to currently known public key schemes using subgroups. At the end of their paper, they mentioned very briefly that XTR can be generalized in a straightforward way using the extension field of the form $\mathrm{GF}(p^{6m})$ and made some general comments with the focus on the case $p = 2$.

In this paper, we carry out the generalization in detail and discuss about optimal choices of the parameters p and m. The idea is mostly straightforward, but we need to be more systematic to find out optimal choices of p and m among the possible cases. In more detail, the generalization is done in two steps. First, we propose a systematic design for XTR-like system in $\mathrm{GF}(p^{6m})$ using an irreducible cubic polynomial $F(c, X) = X^3 - cX^2 + c^{p^m} X - 1 \in \mathrm{GF}(p^{2m})[X]$ for

* Yie and Kim's work was supported by INHA Univ. Research Grant (INHA-21072).

S. Vaudenay and A. Youssef (Eds.): SAC 2001, LNCS 2259, pp. 301–312, 2001.
© Springer-Verlag Berlin Heidelberg 2001

any m. Then we determine the required properties of the parameters m, p, and c for $F(c, X)$ under the efficiency and security considerations. We are focusing on the case that is efficient in the limited applications such as smart cards. We use an optimal normal basis to represent elements of $\mathrm{GF}(p^{2m})$ over $\mathrm{GF}(p)$. Hence we consider the case where $2m + 1$ is a prime and p is a primitive element modulo $2m + 1$. We suggest to use m such that either $2m + 1$ is a Fermat prime or both m and $2m + 1$ are primes. With such a choice of m, a randomly chosen prime p has a better chance to be a primitive element in Z_{2m+1}.

We estimated the required computational complexity for XTR extended to $\mathrm{GF}(p^{6m})$ under considerations as the above, and compare the result with XTR in $\mathrm{GF}(P^6)$ where P and p^{6m} have the same bit sizes. The result shows us that the required number of bit operations for both are about the same.

Modern workstation microprocessors are designed to calculate in units of data known as words. For large prime p, multiple machine words are required to represent elements of prime field $\mathrm{GF}(p)$ on microprocessors, since typical word sizes are not large enough. This representation causes two possible computational difficulties: carries between words must be treated and reduction modulo p must be performed with operands of multiple span words.

Hence we see that using prime number p as small as the word size of common processors for XTR relieves the above computational difficulties for operation in $\mathrm{GF}(P)$ for large prime P. In this way, the proposed generalization maintains the communicational advantages as in XTR and it enhances the computational advantages over XTR since there is no need of multiprecision computation especially when the system is implemented under workstation processors with optimized arithmetic on integers of word size.

Unfortunately, there are some drawbacks in extending XTR to $\mathrm{GF}(p^{6m})$. As m gets larger there are fewer prime numbers p, q that can be used to establish an XTR-like system in $\mathrm{GF}(p^{6m})$. Also it takes longer to generate the primes p, q with the required properties. But generating p, q is a one-time task and it is not a serious disadvantage in many cases.

Our paper is organized as the following. In Section 2, we describe the generalized system in such a way that it can be formulated for any m, setting aside any security or complexity related concerns. In Section 3, we discuss about the security of the system and determine the choices of parameters. In Section 4, we estimate the computational complexity for the proposed XTR-like scheme in $\mathrm{GF}(p^{6m})$. Then we discuss about optimal choices of parameters. In Section 5, we conclude our paper with comparison of efficiency and security for cryptographic schemes using this generalization with those using the original XTR under the Galois fields with about the same sizes. We also discuss about an efficient way of parameter generation of XTR system extended to $\mathrm{GF}(p^{6m})$ and recommend good choices of m for current use in Appendix A.

2   A Description of XTR Extended to $\mathrm{GF}(p^{6m})$

In this section we describe the XTR system using elementary symmetric polynomials and give a systematic way to generalize XTR-like system to $\mathrm{GF}(p^{6m})$. Following [2], we start by setting up some notations. Given an element $c \in \mathrm{GF}(p^{2m})$, we define the cubic polynomial $F(c, X)$ as following:

$$F(c, X) = X^3 - cX^2 + c^{p^m} X - 1 = (X - h_0)(X - h_1)(X - h_2),$$

where the roots h_i's are taken from the splitting field of $F(c, X)$. We set $c_n = h_0^n + h_1^n + h_2^n$ for any integer n. Then from the root-coefficient relations of a cubic equation, we have

Lemma 1. *For any integers n and t we have*

1. *$c_1 = c$, $h_0 h_1 + h_1 h_2 + h_0 h_2 = c^{p^m}$, and $h_0 h_1 h_2 = 1$;*
2. *$c_{-n} = c_{np^m} = c_n^{p^m}$;*
3. *Either all h_i's are in $\mathrm{GF}(p^{2m})$ or $F(c, X)$ is irreducible over $\mathrm{GF}(p^{2m})$ and all h_i's have order dividing $p^{2m} - p^m + 1$;*
4. *$(c_n)_t = c_{tn} = (c_t)_n$.*

Proof. Item 1 is nothing but the root-coefficient relation. Items 2 and 3 can be proved exactly the same way as Lemma 2.3.2 of [2] is proved.

To prove item 4, note that

$$c_n = h_0^n + h_1^n + h_2^n, \quad c_n^{p^m} = c_{-n} = (h_1 h_2)^n + (h_0 h_2)^n + (h_0 h_1)^n.$$

This implies that

$$F(c_n, X) = X^3 - c_n X^2 + c_n^{p^m} X - 1 = (X - h_0^n)(X - h_1^n)(X - h_2^n).$$

And thus we see that

$$(c_n)_t = (h_0^n)^t + (h_1^n)^t + (h_2^n)^t = h_0^{nt} + h_1^{nt} + h_2^{nt}.$$

Hence we see that $(c_n)_t = c_{nt} = (c_t)_n$ for any integer n, t.

It can be easily checked that any irreducible cubic polynomial $f(x) = x^3 - ax^2 + bx - 1 \in \mathrm{GF}(p^{2m})$ is of the form $f(x) = x^3 - ax^2 + a^{p^m} x - 1$ if the order of the roots $h_0, h_1, h_2 \in \mathrm{GF}(p^{6m})$ of $f(x) = 0$ divides $p^{2m} - p^m + 1$.

Recall that the elementary symmetric polynomials σ_k of degree k in the indeterminates X_1, X_2, X_3 are given by

$$\sigma_1 = X_1 + X_2 + X_3, \quad \sigma_2 = X_1 X_2 + X_2 X_3 + X_1 X_3, \quad \sigma_3 = X_1 X_2 X_3.$$

Here is a theorem, due to Newton, so-called 'fundamental theorem on symmetric polynomials'.

Theorem 1. *(Theorem 1.75 in [7]) Let $\sigma_1, \sigma_2, \sigma_3$ be the elementary symmetric polynomials in X_1, X_2, X_3 over a commutative ring R, and let $s_0 = 3$, $s_n = X_1^n + X_2^n + X_3^n \in R[X_1, X_2, X_3]$ for $n \geq 1$. Then the following equality*

$$s_k - s_{k-1}\sigma_1 + s_{k-2}\sigma_2 - s_{k-3}\sigma_3 = 0$$

holds for $k \geq 3$.

As a direct application of Newton's Theorem the following lemma can be easily proved.

Lemma 2. *Then for any positive integer n we have*

1. $c_{n+2} = c_{n+1}c - c_n c^{p^m} + c_{n-1}$, $c_{n-1} = c_{n+2} - c_{n+1}c + c_n c^{p^m}$;
2. $c_{2n} = c_n^2 - 2c_n^{p^m}$;
3. $c_{2n+1} = c_n c_{n+1} - c c_n^{p^m} + c_{n-1}^{p^m}$;
4. $c_{2n-1} = c_n c_{n-1} - c^{p^m} c_n^{p^m} + c_{n+1}^{p^m}$.

As in [2], we denote $S_n(c) = (c_{n-1}, c_n, c_{n+1})$ for any integer n. Then by Lemma 2.2, we see that $S_{-1}(c) = (c^{2p^m} - 2c, c^{p^m}, 3)$, $S_0(c) = (c^{p^m}, 3, c)$, and $S_1(c) = (3, c, c^2 - 2c^{p^m})$. Thus, if we do not care about efficiency or security we can define XTR-like key exchange system as following for any prime p and positive integer m and for any $c \in GF(p^{2m})$.

XTR-DH key exchange system in $GF(p^{6m})$:

1. Alice chooses a random integer n and computes $S_n(c)$ then sends c_n to Bob.
2. Bob chooses a random integer t and computes $S_t(c)$ then sends c_t to Alice.
3. Alice and Bob share the key c_{nt} that can be obtained by computing either $S_n(c_t) = ((c_t)_{n-1}, (c_t)_n, (c_t)_{n+1})$ or $S_t(c_n) = ((c_n)_{t-1}, (c_n)_t, (c_n)_{t+1})$.

All the XTR-based schemes given in [2] can be extended to $GF(p^{6m})$ similarly. In the following sections, we will discuss about the parameter selections to meet various security levels and to boost up the efficiency.

3   Parameter Selection for Security Consideration

Various XTR-based public key systems or key exchange protocols rely their security on the Discrete Logarithm Problem(DLP) in the base $g \in GF(p^{6m})$, where g is a root of the cubic equation $F(c, X) = 0$. Therefore, in order to make XTR-based schemes secure, we need to use parameters p, m, c such that the DLP in the base g is difficult. In this section, we follow the method in [1] to determine the size of the subgroup generated by g to prevent known attacks on DLP in extension fields.

Up to this point one of the best attacks known for the DLP is the Index Calculus Method using Number Field Sieve. For a DLP in the base $g \in GF(p^{6m})$, the asymptotic complexity of the Index Calculus Method using Number Field Sieve is

$$L(p^s, 1/3, 1.923 + o(1)),$$

where s is the smallest divisor of $6m$ such that g is contained in a subfield of $GF(p^{6m})$ isomorphic to $GF(p^s)$. Thus, for the security reason, it is desirable to use $g \in GF(p^{6m})$, which is not contained in any proper subfield of $GF(p^{6m})$. Note that if a root g of the polynomial $F(c, X)$ is not contained any proper subfield of $GF(p^{6m})$ then $F(c, X)$ is irreducible over $GF(p^{2m})$. Hence we have $c = \mathrm{Tr}(g)$ and the roots of $F(c, X)$ are conjugates of g over $GF(p^{2m})$. Here, $\mathrm{Tr} : GF(p^{6m}) \rightarrow GF(p^{2m})$ is the trace projection defined by $\mathrm{Tr}(x) = x + x^{p^{2m}} + x^{p^{4m}}$ for $x \in GF(p^{6m})$.

The following Lemma, which follows directly from Lemma 2.4 of [1], gives a sufficient condition for a subgroup of $GF(p^{6m})^*$ not to be contained in any proper subfield of $GF(p^{6m})$.

Lemma 3. *Let q be a prime factor of $\Phi_{6m}(p)$. Then the subgroup of $GF(p^{6m})^*$ of order q is not contained in any proper subfield of $GF(p^{6m})$.*

Here $\Phi_n(X)$ is the n-th cyclotomic polynomial for a positive integer n not divisible by p. For computations with cyclotomic polynomials, we can use Theorem 3.27 in [7]:

$$\Phi_{6m} = \prod_{d|6m} (X^{6m/d} - 1)^{\mu(d)},$$

where $\mu(\cdot)$ is the Möbius function.

As we shall see in the next section, we will take m to be a prime or a power of 2 (so that $2m + 1$ is a Fermat prime) for easy selection of the prime p. Thus we have

$$\Phi_{6m}(X) = \frac{(X^{6m} - 1)(X^2 - 1)(X^3 - 1)(X^m - 1)}{(X^{2m} - 1)(X^{3m} - 1)(X^6 - 1)(X - 1)} = \frac{X^{2m} - X^m + 1}{X^2 - X + 1}$$

if m is a prime, or

$$\Phi_{6m}(X) = \frac{(X^{6m} - 1)(X^m - 1)}{(X^{2m} - 1)(X^{3m} - 1)} = X^{2m} - X^m + 1$$

if m is a power of 2.

When we use a system based on the DLP in a multiplicative subgroup of size q of the Galois field, the sizes of q and the underlying Galois field that guarantee the security required currently can be determined according to the table in [4]. The recommended size for q is much larger than $p^2 - p + 1$ for small or medium sized p. Thus in our case we need to take the size q of the subgroup of $GF(p^{6m})^*$ to be much larger than $p^2 - p + 1$. In addition, if we take q to be a prime factor of $p^{2m} - p^m + 1$ then the calculation in the previous paragraph tells us that q is a prime factor of $\Phi_{6m}(p)$. Then by Lemma 3 the subgroup of order q is not contained in any proper subfield of $GF(p^{6m})$.

Also it is well-known that birthday-type attacks can be applied to get x from the given g^x and the complexity of birthday attacks can be estimated by $O(\sqrt{q})$ where q is the order of g. Currently the recommended size for q is $q > 2^{160}$. Thus to make our system as secure as the DLP problem in $GF(p^{6m})$ against currently known attacks, it is enough to choose $g \in \mathrm{GF}(p^{6m})$ so that the order of g is a prime factor of $p^{2m} - p^m + 1$ and larger than $\max(p^2 - p + 1, 2^{160})$.

4   Parameter Selection for Efficiency Cconsiderations

The most basic computation required in XTR-like schemes is to compute $S_n(c)$ from any given $c \in \mathrm{GF}(p^{2m})$ and a positive integer n. ¿From Lemma 2 we have

Lemma 4. *Let c be any element of* $\mathrm{GF}(p^{2m})$ *and $A(c)$ be the 3×3 matrix given by*

$$A(c) = \begin{pmatrix} 0 & 0 & 1 \\ 1 & 0 & -c^{p^m} \\ 0 & 1 & c \end{pmatrix}.$$

For any integer $n \geq 1$, we have $S_{n+1}(c) = S_n(c)A(c)$.

By modifying the 'square and multiply' method, we can compute $S_n(c)$ as follows:

Algorithm 2 *Let $c \in GF(p^{2m})$ and a positive integer n be given. ¿From the binary expansion of $n = \sum_{i=0}^{k} m_i 2^i$ define a sequence $\{t_i\}$ by*

$$t_0 = m_0$$
$$t_i = 2t_{i-1} + m_i, \ i \geq 1.$$

To compute $S_n(c) = (c_{n-1}, c_n, c_{n+1})$, follow the steps:

Step 1. *Set $S_{t_0}(c) = (c^{p^m}, 3, c)$ if $t_0 = 0$, or $S_{t_0}(c) = (3, c, c^2 - 2c^{p^m})$ if $t_0 = 1$.*
Step 2. *Compute S_{t_i} from $S_{t_{i-1}}$ for $i = 1, 2, \cdots, k$.*
Step 3. *Output S_{t_k}.*

Step 2 of the above algorithm is performed as following. For a fixed i let us let $d = t_{i-1}$ for simplicity. If $m_i = 0$, then

$$S_{t_i} = S_{2d} = (c_d c_{d+1} - c^{p^m} c_d^{p^m} + c_{d+1}^{p^m}, c_d^2 - 2c_d^{p^m}, c_d c_{d+1} - cc_d^{p^m} + c_{d-1}^{p^m}).$$

If $m_i = 1$, then

$$S_{t_i} = S_{2d+1} = (c_d^2 - 2c_d^{p^m}, c_d c_{d+1} - cc_d^{p^m} + c_{d-1}^{p^m}, c_{d+1}^2 - 2c_{d+1}^{p^m}).$$

As we can see from Lemma 2 and Algorithm 2, the most frequently performed operations in our system are the following three types:

$$x^2, xy, xz - yz^{p^m} \quad \text{for } x, y, z \in \mathrm{GF}(p^{2m}).$$

In order to make the XTR system in GF(p^{6m}) efficient, it is enough to perform these operations efficiently. Thus, we consider the case when GF(p^{2m}) has an optimal normal basis (ONB) of type I, that is, the case when $2m + 1$ is a prime number and $p \pmod{2m + 1}$ is a primitive element in Z_{2m+1}. In this case, the cyclotomic polynomial

$$\Phi_{2m+1}(X) = \frac{X^{2m+1} - 1}{X - 1} = X^{2m} + X^{2m-1} + \cdots + X + 1 \in \text{GF}(p)[X]$$

is irreducible and the set $B_1 = \{\alpha, \alpha^p, \alpha^{p^2}, \ldots, \alpha^{p^{2m-1}}\}$ of roots becomes an Optimal Normal Basis for GF(p^{2m}). Since $p \pmod{2m + 1}$ is a primitive element in Z_{2m+1}, this basis B_1 is setwise equal to the (almost) polynomial basis $B_2 = \{\alpha, \alpha^2, \alpha^3, \ldots, \alpha^{2m}\}$.

The complexity for elementary operations in GF(p^{2m}) is well studied by Lenstra in [1]. Our focus is on the most frequently performed operations

$$x^2, xy, xz - yz^{p^m} \quad \text{for } x, y, z \in \text{GF}(p^{2m})$$

as was in [2]. We have similar result as in [2] on the complexity for these operations.

Lemma 5. *Let p and $2m + 1$ be prime numbers, where $p \pmod{2m + 1}$ is a primitive element in Z_{2m+1}. Then for $x, y, z \in \text{GF}(p^{2m})$, we have*

1. *Computing x^{p^m} is for free.*
2. *Computing x^2 takes 80 percent of the complexity taken for multiplications in GF(p^{2m}).*
3. *Computing xy takes $4m^2$ multiplications in GF(p).*
4. *Computing $xz - yz^{p^m}$ takes $4m^2$ multiplications in GF(p).*

Proof. Since we can use either of the two bases B_1 and B_2 at our convenience, all the items are straightforward. Thus we only prove item 4 in detail. Set $t = 2m$. First we represent x, y, z by using the normal basis B_1,

$$x = \sum_{i=0}^{t-1} a_i \alpha^{p^i}, y = \sum_{i=0}^{t-1} b_i \alpha^{p^i} z = \sum_{i=0}^{t-1} c_i \alpha^{p^i},$$

and we get

$$z^{p^m} = \sum_{i=0}^{m-1} c_i \alpha^{p^{i+m}} + \sum_{i=m}^{t-1} c_i \alpha^{p^{i-m}}.$$

Then we have

$$xz - yz^{p^m} = \sum_{i=0}^{t-1} \sum_{j=0}^{m-1} a_i c_j' \alpha^{p^i} \alpha^{p^{j+m}} + \sum_{i=0}^{t-1} \sum_{j=m}^{t-1} b_i c_j'' \alpha^{p^i} \alpha^{p^{j-m}},$$

where $c_j' = c_j + c_{m+j}$ and $c_j'' = c_j + c_{j-m}$. Getting c_j', c_j'' from c_j's is a free operation. Now we convert the basis into B_2 and then the computation for

$\alpha^{p^j} \alpha^{p^{i\pm m}}$ becomes free. Hence we need t^2 multiplications in $\mathrm{GF}(p)$ to compute $xy - yz^{p^m}$.

We note here that we haven't used any high speed multiplication algorithms in the proof. When we apply more efficient multiplication algorithms, then the required bit complexity will be reduced. Comparing Lemma 2.1.1 in [2] and the above Lemma 5, we see that the number of bit operations for computing

$$x^2, xy, xz - yz^p$$

are about the same in the following two cases when $|P| = m|p|$,

- $x, y, z \in \mathrm{GF}(P^2)$ with $P = 2 \pmod 3$;
- $x, y, z \in \mathrm{GF}(p^{2m})$ where $2m + 1$ is a prime and $\langle p \rangle = Z^*_{2m+1}$.

Now based on Lemma 5, we have the following estimate for the computational complexity of Algorithm 2.

Theorem 3. *Let $c \in \mathrm{GF}(p^{2m})$ and a positive integer n be given. Then it takes at most $11.2m^2 \log n$ multiplications in $\mathrm{GF}(p)$ to compute $S_n(c) = (c_{n-1}, c_n, c_{n+1})$.*

Also we estimate the required bit complexity to compute $Tr(g^a g^{bk})$ from given $Tr(g)$ and $S_k(Tr(g))$ for unknown k. This computation is required for XTR-Nyberg-Rueppel signature scheme as in [2].

Theorem 4. *Let $g \in \mathrm{GF}(p^{6m})$ be an element of prime order q and suppose $Tr(g)$ and $S_k(Tr(g))$ be given for some unknown positive integer k. Let a, b be positive integers with less than q. Then $Tr(g^a g^{bk})$ can be computed at a cost of $(11.2 \log(a/b \pmod q) + 11.2 \log b + 36)m^2$ multiplications in $\mathrm{GF}(p)$.*

We specify the steps as following :

- Compute $e = \frac{a}{b} \pmod q$.
- Compute $Tr(g^{k+e})$.
- Compute $Tr(g^{(k+e)b}) = Tr(g^a g^{bk})$

We focus on the second item here. We have

$$S_k(Tr(g))A(c)^e = [S_0(Tr(g))A(c)^k]A(c)^e = [S_0(Tr(g))A(c)^e]A(c)^k$$

In order to get $Tr(g^{k+e})$, what we need is $[S_0(Tr(g))A(c)^e]C(A(c)^k)$, where $C(A(c)^k)$ is the center column of the matrix $A(c)^k$. Since we have already given $S_k(Tr(g))$, and we know that

$$S_k(Tr(g))^T = \begin{pmatrix} S_{-1}(c) \\ S_0(c) \\ S_1(c) \end{pmatrix} C(A(c)^k),$$

where $S_k(Tr(g))^T$ is the transpose of $S_k(Tr(g))$. Hence we have the very same formula as in XTR,

$$C(A(c)^k) = \begin{pmatrix} S_{-1}(c) \\ S_0(c) \\ S_1(c) \end{pmatrix}^{-1} S_k(Tr(g))^T.$$

This implies that computing $C(A(c)^k)$ takes constant time for any k when $S_k(Tr(g))$ is given. In fact, it takes at most $9 \times 4m^2 = 36m^2$ multiplications in GF(p). Hence the complexity to compute $Tr(g^{k+e})$ can be estimated as $(11.2\log e + 36)m^2$ multiplications in GF(p). And the complexity to compute the third item is $11.2\log bm^2$ multiplications in GF(p). Thus we can estimate the complexity to get $Tr(g^a g^{bk})$ by $(11.2\log(a/b) \pmod{q}) + 11.2\log b + 36)m^2$ multiplications in GF(p).

Thus we see that the computational complexity is about the same for the original XTR and our extension to $GF(p^{6m})$. But the multiprecision problem that occurs in the operations in GF(P) for large P can be removed when we use GF(p) with p as small as the word size of the processor.

Now we pose another condition on m so that it is easy to generate the prime p. Since we are using ONB, p is necessarily a primitive element in Z_{2m+1}. But as a matter of parameter generation, we will decide m first and choose the prime p somewhat randomly. So it is desirable that Z_{2m+1} has more primitive elements.

The number of primitive elements in Z_{2m+1} is $\phi(\phi(2m+1)) = \phi(2m)$, where $\phi(\cdot)$ is the Euler totient function. Thus we want to make $\phi(2m)$ as big as possible compared to m. There are two possible directions we can take to this end. The first is to take m to be a power of 2. But then $2m+1$ must be a Fermat prime, which is very rare. The other is to take m to be a prime. There are reasonably many choice of primes m for which $2m+1$ is also a prime.

For appropriate selection of m for current recommendation is given in the Appendix A. After m and the sizes of p, q have been established, we generate p, q and $c = Tr(g)$. Algorithms for generating parameters are also given in Appendix A.

5   Conclusion

Thus most of the details in XTR can be generalized systematically to the finite field GF(p^{6m}) using the trace projection Tr : GF(p^{6m}) $\to$ GF(p^{2m}). Hence it is straightforward to see that the schemes as XTR-DH, XTR-ElGamal encryption, and XTR-Nyberg-Rueppel signatures can be extended to GF(p^{6m}).

For security concerns, all the details were given in Section 5 of [2]. They've discussed about the DLP in GF(p^t), and hence it can be applied to cases in GF(p^{6m}). The communicational and computational advantages of the XTR schemes can be obtained in the generalization as long as we choose m so that either $2m+1$ is a Fermat prime or both $m, 2m+1$ are primes and we don't have any multiprecision operations if we select the size of p as small as the word size of common processors in the generalization but we might need longer time to generate prime numbers p and q in such cases. But the prime numbers are needed to generate only once, the generalized version of XTR is more preferable for limited applications as smart cards.

Acknowledgment

We appreciate the anonymous referees of the SAC 2001 for their comments and suggestions.

References

1. Arjen K. Lenstra, *Using Cyclotomic Polynomials to Construct Efficient Discrete Logarithm Cryptosystems over Finite Fields*, ACISP'97 (1997), **LNCS 1270**, pp. 127–138.
2. Arjen K. Lenstra, Eric R. Verheul, *The XTR public key system*, Advances in Cryptology – CRYPTO'00 **LNCS 1880** (2000), pp. 1–19
3. Arjen K. Lenstra, Eric R. Verheul, *Key improvements to XTR*, Advances in Cryptology – Asiacrypt'00 **LNCS 1976** (2000), pp. 220–233
4. A.K. Lenstra, E.R. Verheul, *Selecting Cryptographic Key Sizes*, http://www.cryptosavvy.com (1999).
5. Arjen K. Lenstra, Eric R. Verheul, *Fast irreduciblility and subgroup membership testing in XTR*, Proceedings of the PKC'01 **LNCS 1992** (2001), pp. 73–86
6. A.E.Brouwer, R.Pellikaan, E.R. Verheul, *Doing More with Fewer Bits*, Advances in Cryptology – Asiacrypt'99, **LNCS 1716** (1999), pp. 321–332.
7. Rudolf Lidl, Harald Niederreiter, *Introduction to finite fields and their applications*, Cambridge, 1994.

A   An Efficient Way for Parameter Generation

In this appendix, we describe an efficient way for parameter generation one by one. At first, we decide the size of the field $GF(p^{6m})$ and which m to use. Then we select the primes p, q. And finally, we select $c \in GF(p^{2m}) \setminus GF(p^m)$.

As we discussed in Section 4, m will be chosen so that either $2m + 1$ is a Fermat prime or both m and $2m + 1$ are primes. In the Table 1 below, we give a list of m which meet this criterion. The numbers in the column titled 'ratio' are the proportion of primitive elements of Z_{2m+1}. The case $m = 1$ is the original XTR. Not much improvement is achieved in the cases $m = 2, 3$. Thus we do not recommend to use these cases. Note that the ratio tends to $1/2$ as m gets larger.

For given m as in the above table, we choose the appropriate size for the field characteristic p so that the size of p^{6m} is about the same as the recommended size for prime fields with respect to the security concerns in the DLP in prime fields. For example, see Table 2.

Now we consider the generation of the prime numbers p, q in our scheme. We follow the scheme of generating prime numbers as in [1] rather than using the

Table 1. Choices for m and the corresponding extension fields

m	ratio	field extensions
1	1/3	$\mathrm{GF}(p) \to \mathrm{GF}(p^2) \to \mathrm{GF}(p^6)$
2	2/5	$\mathrm{GF}(p) \to \mathrm{GF}(p^4) \to \mathrm{GF}(p^{12})$
3	2/7	$\mathrm{GF}(p) \to \mathrm{GF}(p^6) \to \mathrm{GF}(p^{18})$
5	4/11	$\mathrm{GF}(p) \to \mathrm{GF}(p^{10}) \to \mathrm{GF}(p^{30})$
8	8/17	$\mathrm{GF}(p) \to \mathrm{GF}(p^{16}) \to \mathrm{GF}(p^{48})$
11	10/23	$\mathrm{GF}(p) \to \mathrm{GF}(p^{22}) \to \mathrm{GF}(p^{66})$
23	22/47	$\mathrm{GF}(p) \to \mathrm{GF}(p^{46}) \to \mathrm{GF}(p^{138})$
29	28/59	$\mathrm{GF}(p) \to \mathrm{GF}(p^{58}) \to \mathrm{GF}(p^{174})$
41	40/83	$\mathrm{GF}(p) \to \mathrm{GF}(p^{80}) \to \mathrm{GF}(p^{240})$

Table 2. Choices for m and the corresponding size of a finite field

field size	1024 bit	2048 bit	2700 bit	5100 bit
recommended m for p of 16 bits	11	23	29	41
recommended m for p of 32 bits	8	11	23	29
recommended m for p of 64 bits	5	8	11	23

method given in [2]. In general setting, we are interested in the cases using small or medium sized prime number p covered in [1].

Here we describe the method of generating prime numbers p and q that we need. First we determine the $|p^{6m}|$ and $|q|$ (the bit sizes of p^{6m} and q) for general security considerations according to [4]. And then we decide m so that p is of the word size of the processor to be used.

We repeat selecting p until $p^{2m} - p^m + 1$ has a prime factor of size $|q|$. We refer the result of [1] that points it works sufficiently quickly in practice. As m grows, hence p gets smaller, there are fewer appropriate p, m's than XTR case (assuming comparable levels of security). The exact distribution of such primes p, q is not known until now. Since this is one-time cost, it's not a serious disadvantage.

Now consider the generation of $c = Tr(g)$. The most elementary way to generate $c = Tr(g)$ where $g \in GF(p^{6m})$ of the order q with q divides $p^{2m} - p^m + 1$. As usual, first we randomly generate $h \in GF(p^{6m})$ and check if $h^{\frac{p^{2m}-p^m+1}{q}} \neq 1$ and set $g = h^{\frac{p^{2m}-p^m+1}{q}}$. Then such g has the order q. We compute $Tr(g)$. But here we have lemmas which will come in handy when we construct a suitable generator g of a subgroup. And the proofs of these lemmas are similar to the XTR in $GF(P^6)$.

Lemma 6. *Let m be a positive integer such that either $2m+1$ is a Fermat prime or both m and $2m+1$ are primes. Suppose $F(c, X)$ is irreducible over $\mathrm{GF}(p^{2m})$ and $g \in \mathrm{GF}(p^{6m})$ is a root of $F(c, X)$. Then we have $c = \mathrm{Tr}(g)$ and $c_n = \mathrm{Tr}(g^n)$ and the multiplicative order q of g divides $p^{2m} - p^m + 1$.*

Lemma 7. *$F(c, X)$ is reducible over $GF(p^{2m})$ if and only if $c_{p^m+1} \in GF(p^m)$*

By using this lemma, we have a similar algorithm as in XTR to generate $c = Tr(g)$ where $g \in GF(p^{6m})$ of prime order q that is not contained any subfield of $GF(p^{6m})$.

Algorithm to generate $c = Tr(g)$ for our purpose

1. Choose $\tilde{c} \in GF(p^{2m}) \setminus GF(p^m)$.
2. Check if $\tilde{c}_{p^m+1} \in GF(p^m)$. If it is, go to 1.
3. Compute $\tilde{c}_{\frac{p^{2m}-p^m+1}{q}}$ and check if it is 3. If it is not 3, then set $c = \tilde{c}_{\frac{p^{2m}-p^m+1}{q}}$.

   This c is what we wanted.

The Two Faces of Lattices in Cryptology

Phong Q. Nguyen

École Normale Supérieure, Département d'Informatique,
45 rue d'Ulm, 75005 Paris, France
pnguyen@ens.fr and http://www.di.ens.fr/~pnguyen/

Abstract. Lattices are regular arrangements of points in n-dimensional space, whose study appeared in the 19th century in both number theory and crystallography. Since the appearance of the celebrated Lenstra-Lenstra-Lovász lattice basis reduction algorithm twenty years ago, lattices have had surprising applications in cryptology. Until recently, the applications of lattices to cryptology were only negative, as lattices were used to break various cryptographic schemes. Paradoxically, several positive cryptographic applications of lattices have emerged in the past five years: there now exist public-key cryptosystems based on the hardness of lattice problems, and lattices play a crucial rôle in a few security proofs. In this talk, we will try to survey the main examples of the two faces of lattices in cryptology. The full material of this talk appeared in [2]. A preliminary version can be found in [1].

References

1. P. Q. Nguyen and J. Stern. Lattice reduction in cryptology: An update. In *Algorithmic Number Theory, Proc. of ANTS-IV*, volume 1838 of *LNCS*. Springer-Verlag, 2000.
2. P. Q. Nguyen and J. Stern. The two faces of lattices in cryptology. In *Cryptography and Lattices, Proc. of CALC '01*, volume 2146 of *LNCS*. Springer-Verlag, 2001.

S. Vaudenay and A. Youssef (Eds.): SAC 2001, LNCS 2259, p. 313, 2001.
© Springer-Verlag Berlin Heidelberg 2001

New (Two-Track-)MAC
Based on the Two Trails of RIPEMD
Efficient, Especially on Short Messages
and for Frequent Key-Changes

Bert den Boer[1,*], Bart Van Rompay[2,**], Bart Preneel[2], and Joos Vandewalle[2]

[1] TNO/TPD, Stieltjesweg 1, 2628 CK Delft, The Netherlands
bdenboer@tpd.tno.nl
[2] Katholieke Universiteit Leuven, ESAT-COSIC,
Kasteelpark Arenberg 10, 3001 Leuven-Heverlee, Belgium
bart.vanrompay@esat.kuleuven.ac.be

Abstract. We present a new message authentication code. It is based on a two trail construction, which underlies the unkeyed hash function RIPEMD-160. It is in comparison with the MDx-MAC based on RIPEMD-160, much more efficient on short messages (that is on messages of 512 or 1024 bits) and percentage-wise a little bit more efficient on long messages. Moreover, it handles key-changes very efficiently. This positive fact remains if we compare our Two-Track-MAC with HMAC based on RIPEMD-160.

1 Introduction

Message Authentication Codes (MACs) are symmetric-key cryptographic primitives used to provide data integrity and symmetric data origin authentication. Given a message M to be authenticated and a secret key K (shared between two parties), the MAC algorithm computes an authentication tag $A = MAC(K, M)$ for the message. The pair (M, A) is passed from sender to receiver who can verify the authentication tag by computing the MAC of the message himself (as he knows the key).

The goal of an adversary (who does not know the key) is to forge a MAC for a message of his choice (selective forgery), or for an arbitrary message (existential forgery). Here it is assumed that the adversary has knowledge of a number of messages M^i and their corresponding authentication tags $A^i = MAC(K, M^i)$. In the case of a chosen-text attack the opponent is even able to request the MAC for a number of messages of his choice (before forging a MAC on a new, and different, message).

---

[*] Algorithm invented while working at debis Information Security Services – Bonn, Germany.
[**] The work described in this paper has been supported in part by the Commission of the European Communities through the IST Programme under Contract IST-1999-12324 and by the Concerted Research Action (GOA) Mefisto-666.

S. Vaudenay and A. Youssef (Eds.): SAC 2001, LNCS 2259, pp. 314–324, 2001.

It is a common approach to construct MAC algorithms from existing cryptographic hash functions, as such schemes require little additional implementation effort. They are also generally faster than MACs which are based on block ciphers. A cryptographic hash function is a function which compresses an input of arbitrary length into a hash value of fixed length, while also satisfying some additional cryptographic properties (preimage resistance and collision resistance). A hash function usually works by iteration of a compression function, which has a fixed-length message input operating on an internal state variable. The final value of this internal state then serves as hash value.

To build a MAC algorithm from a hash function it is necessary to include a secondary input, the secret key, in the computation. Early proposals such as the envelope method [6], where the key material is simply prepended and appended to the message input to the hash function, were shown to have significant weaknesses [4,5]. MDx-MAC and HMAC have emerged as the most secure alternatives. MDx-MAC [4], which can be based on MD5, RIPEMD, SHA or similar algorithms, makes some small changes to the hash function used, while HMAC [1] is a black box construction that can be based on any hash function.

In this paper we will present a new MAC algorithm, called Two-Track-MAC (or TTMAC in short). It has been submitted as a candidate algorithm for the NESSIE project [3]. The algorithm is based on the RIPEMD-160 hash function [2] (making only small changes to the hash function). We will show that the structure of RIPEMD, which consists of two parallel trails, has been exploited to double the size of the internal state, and that this allows to significantly reduce the overhead in the computation of the MAC for short messages, compared to the other MAC constructions. Another advantage of our proposal is better efficiency in the case of frequent key changes. These properties are very useful in applications, e.g., banking applications, where many short messages need to be authenticated (with frequent key changes). Although there is no formal proof of security for our construction, based on the heuristic arguments presented in Section 3, we believe it is very unlikely that an attack can be found on Two-Track-MAC, which would not also breach the security of RIPEMD-160.

The remainder of this paper is organized as follows. In Section 2 we present our new MAC. Section 3 discusses the security, and Section 4 the efficiency of our proposal. In Section 5 we suggest a more general construction method that can be used to construct new schemes. Section 6 concludes the paper. Pseudo-code for our algorithm is given in the Appendix.

2   Presentation of Two-Track-MAC

The unkeyed hash function RIPEMD-160 (for a description we refer to [2]) uses two trails in its compression function. If we separate those two trails then each trail can be seen as a transformation of a 160-bit input I, controlled by a message M, consisting of sixteen words of 32 bits. Those 160 bits of the input I (and of the output) consist of five words of 32 bits. Call the output of the different trails $L(I, M)$ and $R(I, M)$ (left respectively right trail output for an input I and a message M), then our proposal for a MAC on a relative short message M (of 512 bits) and a key K of 160 bits is (in short notation)

$$R(K, M) - L(K, M).$$

Or as $R(K, M)$ can be viewed as five words A_i of 32 bits : $(A_0, A_1, A_2, A_3, A_4)$, and similarly the value $L(K, M)$ as $(B_0, B_1, B_2, B_3, B_4)$, we get an output $E = (E_0, E_1, E_2, E_3, E_4)$ of five 32-bit words. Here

$$E_i = A_i - B_i \text{ (subtraction modulo } 2^{32}) \text{ for } i = 0, 1, 2, 3, 4.$$

Then the 160-bit string E is the MAC of the 512-bit message M. Figure 1 gives a schematic view of this computation.

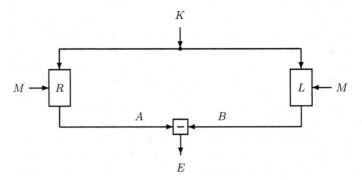

Fig. 1. High level view of TTMAC for a message of a single block.

If the message is longer, i.e. $M = M_1 M_2 M_3 \cdots M_n$ where each M_i is of length 512 bits, we define, using a new operation L^* and a new operation R^*, the 160-bit quantity A, respectively the 160-bit quantity B. $A = (A_0, A_1, A_2, A_3, A_4) = L^*(K, M_1)$, where each A_i is a 32 bit word. And $B = (B_0, B_1, B_2, B_3, B_4) = R^*(K, M_1)$ as the result of the right trail. The operation L^* is based on the operation L, which had a straightforward inverse operation on the first (160 bits long) argument. This new operation L^* has a simple feedback with the first argument, i.e.

$$L^*(I, M) = L(I, M) - I$$

(this is five times a subtraction modulo 2^{32}). Similarly the operation R^* is defined in shorthand as

$$R^*(I, M) = R(I, M) - I$$

(this is five times a subtraction modulo 2^{32}). Now we introduce two 160-bit blocks C and D of five 32-bit words, $C = (C_0, C_1, C_2, C_3, C_4)$ and $D = (D_0, D_1, D_2, D_3, D_4)$, which are defined as follows:

$$C_2 = A_3 - B_0,$$

$$C_3 = A_4 - B_1,$$
$$C_4 = A_0 - B_2,$$
$$C_0 = (A_1 + A_4) - B_3,$$
$$C_1 = A_2 - B_4,$$
$$D_1 = (A_4 + A_2) - B_0,$$
$$D_2 = A_0 - B_1,$$
$$D_3 = A_1 - B_2,$$
$$D_4 = A_2 - B_3,$$
$$D_0 = A_3 - B_4.$$

All subtractions and additions are modulo 2^{32}. These 160-bit blocks C and D are the starting values for the left, respectively, right trail to incorporate the next 512-bit message block M_2. If there are more message blocks M_i the iteration is the same. So we have

$$HL(1) = A = L^*(K, M_1),$$
$$HR(1) = B = R^*(K, M_1),$$

and then iteratively (for $i = 2, ..., n - 1$) the three operations

$$(A, B) \to (C, D),$$
$$HL(i) = A = L^*(C, M_i),$$
$$HR(i) = B = R^*(D, M_i).$$

For the last message block M_n however, the role of the left and right trails is interchanged:

$$(A, B) \to (C, D),$$
$$HL(n) = A = R^*(C, M_n),$$
$$HR(n) = B = L^*(D, M_n).$$

Once we have $HL(n)$ and $HR(n)$ we define our MAC as $TTMAC(K, M)$ by $HL(n) - HR(n)$ (five times a subtraction modulo 2^{32}). In Figure 2 a schematic view of the computation is given for a message consisting of two blocks.

The same preprocessing rules as in the RIPEMD-160 hash function are used to format the message input to the algorithm (the message is padded to a bitlength which is a multiple of 512). An additional output transformation can be used to reduce the length of the MAC result. This transformation calculates the necessary number of output words, in such a manner that all of the normal output words are used. Let the normal 160-bit result be $E = (E_0, E_1, E_2, E_3, E_4)$, and denote the final (shortened) MAC result with F, consisting of t 32-bit words F_i ($t = 1, 2, 3,$ or 4). For a 32-bit MAC we compute (using addition modulo 2^{32})

$$F_0 = E_0 + E_1 + E_2 + E_3 + E_4.$$

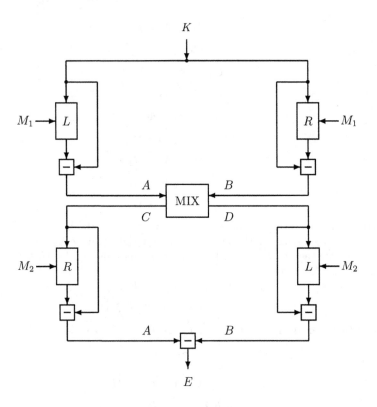

Fig. 2. High level view of TTMAC for a message of two blocks.

For a MAC result of 64, 96 or 128 bits we compute respectively the first two, the first three or all four of the following values (all additions are modulo 2^{32}):

$$F_0 = E_0 + E_1 + E_3,$$

$$F_1 = E_1 + E_2 + E_4,$$

$$F_2 = E_2 + E_3 + E_0,$$

$$F_3 = E_3 + E_4 + E_1.$$

3   Security of Our Proposal

3.1   Philosophy on the Security

The idea for the security is simple: Now we have an internal state variable $(HL(i), HR(i))$ of 320 bits. This is twice as long as for other MAC constructions (e.g., MDx-MAC and HMAC) based on RIPEMD-160 (or on SHA-1). Only in the case of very weak transformations a cryptanalyst is allowed to hope on so-called internal collisions. In almost all attacks, which do not attack the very heart of the MAC (in our case the two trails of RIPEMD-160), forgery is based on internal collisions.

Another attack is possible if the MAC on a message contains all the information (or lacks "only" 32 bits of information) of the internal variable on a longer message, containing the first message as a prefix. In our case we have an internal state of 320 bits, so we can use 160 bits of information (the difference between the left and the right state variable, this depends on both trails) as the MAC output without compromising the internal state. Furthermore, we have used the idea of interchanging the left and right trails for the last message block as a free extra defence against such extension attacks. Note that other MAC constructions need to apply the compression function with some secret key material at the end of the computation in order to prevent these attacks. In our case, the secret key is only used as initial value for the two trails.

So now the worry for the cryptographer are the two trails of RIPEMD-160 itself. A single trail has one important weakness: it is a bijective operation, where the attacker can choose the bijection, which is parameterized by the 512-bit quantity M_i. But as long as two trails are used, parametrized by the same 512-bit quantity M_i, and only a sum will come out in the open, there is no danger that an attacker can invert the operation. Moreover, we have used feedback to counter a straightforward inverse operation. (We do not use feedback on messages of 512 bits, because there the feedback from the left trail would cancel out the feedback from the right trail, in other words we do not need feedback there). All this makes the transformation of a new 512-bit message block on the 320-bit internal variable a one-way operation.

Suppose a cryptanalist discovers a message N, such that the function $L^*(\cdot, N)$, from a 160-bit leftside argument to a 160-bit output, has only a few short cycles (and many relative short tails ending in those cycles). Such a discovery is useless because we have chosen to mix the outputs of the two trails, as soon as the functions L^* and R^* have outputted their results. (Otherwise it might be possible for the cryptanalyst to generate collisions for the left trail by appending blocks to the message, and seperately trying to find collisions for the right trail.) This mixing is thorough in the following sense: Denote the outcome of the left trail by A, the outcome of the right trail by B (as we did before), denote the MAC, in case we are done (in case this was the last message block), with E. Now denote C and D as the starting values for the new trails (in case it was not the last block). Then each pair out of the five values A, B, C, D and E has (just) enough information to determine the other three values. This ensures all kinds of injectivity properties.

The information theoretic uncertainty about the pair (C, D) can, of course, not be larger than the uncertainty about the key. That is 160 bits. But the goal of the design is that RIPEMD-160 is so "complicated", that the "virtual" uncertainty about (C, D) is 320 bits. The construction is such that the information in any pair of the five tuple (A, B, C, D, E) is 320 bits. So if the uncertainty about (C, D) is 320 bits then the uncertainty about for example the pair (A, E) is also 320 bits. This means that given all information about E, the uncertainty about A is still 160 bits. In other words the attacker has "virtually" no information about the starting value for the new left trail. Of course he has "virtual" information about the pair (A, B), but that is more "complicated" than having information about a single trail.

3.2   Resistance against General Attacks

The resistance of Two-Track-MAC against forgery attacks which are generally applicable, depends on the following parameters: the keylength k which is 160 bits, the output length m which can be between 32 and 160 bits (in 32 bit steps), and the length l of the internal state which is 320 bits.

A first possible approach for an adversary is trying all possible keys (once he recovers the key he is able to forge the MAC for any message he chooses). For a key length k and output length m, such an attack requires about 2^k trials and k/m known text-MAC pairs (for verification of the attack).

Alternatively, the adversary can just guess the MAC corresponding to a chosen message. His success probability will be $1/2^m$ although this attack is not verifiable. The parameter m should be chosen long enough according to the needs of the application.

The forgery attack based on internal collisions requires about $2^{l/2}$ known text-MAC pairs to find an internal collision (with a birthday attack), and 2^{l-m} chosen texts to distinguish the internal collision from the external ones (this is shown in [4]). Table 1 below summarizes the difficulty of these general attacks applied to Two-Track-MAC.

Table 1. Resistance of Two-Track-MAC against general attacks. The output length m can take the following values: 32, 64, 96, 128 or 160 bits.

attack	trials	success prob.	known pairs	chosen texts
key search	2^{160}		$160/m$	
guessing the MAC		$1/2^m$		
internal collision			2^{160}	2^{320-m}

4   Short Comparison on the Efficiency of Two-Track-MAC

Our MAC uses only a few percent more operations on a message as RIPEMD-160 would do to get an unkeyed hash of the message (about 97% of the speed of RIPEMD-160 is achieved). This is already the case for the shortest possible message of 512 bits. In contrast, both MDx-MAC and HMAC require an extra computation of the underlying compression function (using a secret key) at the end of the MAC computation. So that is relatively costly on messages of just one (512-bit) block (less than 50% of the speed of unkeyed hashing is achieved).

Also, since the secret key only serves as an initial value for the computation, a key-change will not slowdown the speed of the computation of Two-Track-MAC. In the case of HMAC or MDx-MAC a keychange costs respectively two or six extra computations with the underlying compression function.

5   General Construction

Our new MAC construction is not dependent on RIPEMD-160 alone. It just needs two operations L and $R : (T1 \times T2) \mapsto T1$. The set $T1$ should be big enough to make collissions improbable, for example $GF(2^{160})$. The size of the set $T2$ should be chosen big if messages are expected to be long. The operations L and R are allowed to be invertible if the second argument is fixed, but the operations L^* and R^* (including feedback from the first input) should be infeasible to invert. The operations L and R might be bijective in the first argument, but they should behave unpredictable on changes in the second argument, if the first argument is unknown (but perhaps fixed). It would even be better if the change in the output of the function L, say, is unpredictable with known first argument, i.e., the only way to know the effect of a change is to compute the new function value. Based on the experience that a first version of RIPEMD was partially broken, it is recommended that L and R should be as different as possible. In the case that $T1$ contains all 160-bit strings, one can use the same transitions $(A, B) \rightarrow (C, D)$ as we use for Two-Track-MAC based on RIPEMD-160. One of course needs to define also a padding rule as the message-length needs to be a multiple of some fixed quantity. With those transitions and a padding rule one can define the MAC on a message of any length.

6   Conclusion

We have presented a new message authentication code based on the two trail construction which underlies RIPEMD-160. The main advantage of the scheme is that it is more efficient than other schemes based on RIPEMD-160, especially in the case of short messages and frequent key-changes. We also suggested a more general construction method which can be used to construct new schemes.

A   Pseudo-code for Two-Track-MAC

The TTMAC algorithm computes a 160-bit MAC value for an arbitrary message, under a 160-bit key. The result can be transformed into a shorter value by use of the output transformation (not reflected in the pseudo-code below). First we define all the constants and functions.

TTMAC: definitions

nonlinear functions at bit level: exor, mux, -, mux, -

$$f(j, x, y, z) = x \oplus y \oplus z \qquad\qquad (0 \le j \le 15)$$
$$f(j, x, y, z) = (x \wedge y) \vee (\neg x \wedge z) \qquad (16 \le j \le 31)$$
$$f(j, x, y, z) = (x \vee \neg y) \oplus z \qquad\qquad (32 \le j \le 47)$$
$$f(j, x, y, z) = (x \wedge z) \vee (y \wedge \neg z) \qquad (48 \le j \le 63)$$
$$f(j, x, y, z) = x \oplus (y \vee \neg z) \qquad\qquad (64 \le j \le 79)$$

added constants (hexadecimal)

$$c(j) = \mathtt{00000000_x} \qquad (0 \le j \le 15)$$
$$c(j) = \mathtt{5A827999_x} \qquad (16 \le j \le 31) \qquad \lfloor 2^{30} \cdot \sqrt{2} \rfloor$$
$$c(j) = \mathtt{6ED9EBA1_x} \qquad (32 \le j \le 47) \qquad \lfloor 2^{30} \cdot \sqrt{3} \rfloor$$
$$c(j) = \mathtt{8F1BBCDC_x} \qquad (48 \le j \le 63) \qquad \lfloor 2^{30} \cdot \sqrt{5} \rfloor$$
$$c(j) = \mathtt{A953FD4E_x} \qquad (64 \le j \le 79) \qquad \lfloor 2^{30} \cdot \sqrt{7} \rfloor$$
$$c'(j) = \mathtt{50A28BE6_x} \qquad (0 \le j \le 15) \qquad \lfloor 2^{30} \cdot \sqrt[3]{2} \rfloor$$
$$c'(j) = \mathtt{5C4DD124_x} \qquad (16 \le j \le 31) \qquad \lfloor 2^{30} \cdot \sqrt[3]{3} \rfloor$$
$$c'(j) = \mathtt{6D703EF3_x} \qquad (32 \le j \le 47) \qquad \lfloor 2^{30} \cdot \sqrt[3]{5} \rfloor$$
$$c'(j) = \mathtt{7A6D76E9_x} \qquad (48 \le j \le 63) \qquad \lfloor 2^{30} \cdot \sqrt[3]{7} \rfloor$$
$$c'(j) = \mathtt{00000000_x} \qquad (64 \le j \le 79)$$

selection of message word

$$r(j) \qquad = j \qquad\qquad\qquad (0 \le j \le 15)$$
$$r(16..31) = 7, 4, 13, 1, 10, 6, 15, 3, 12, 0, 9, 5, 2, 14, 11, 8$$
$$r(32..47) = 3, 10, 14, 4, 9, 15, 8, 1, 2, 7, 0, 6, 13, 11, 5, 12$$
$$r(48..63) = 1, 9, 11, 10, 0, 8, 12, 4, 13, 3, 7, 15, 14, 5, 6, 2$$
$$r(64..79) = 4, 0, 5, 9, 7, 12, 2, 10, 14, 1, 3, 8, 11, 6, 15, 13$$
$$r'(0..15) = 5, 14, 7, 0, 9, 2, 11, 4, 13, 6, 15, 8, 1, 10, 3, 12$$
$$r'(16..31) = 6, 11, 3, 7, 0, 13, 5, 10, 14, 15, 8, 12, 4, 9, 1, 2$$
$$r'(32..47) = 15, 5, 1, 3, 7, 14, 6, 9, 11, 8, 12, 2, 10, 0, 4, 13$$
$$r'(48..63) = 8, 6, 4, 1, 3, 11, 15, 0, 5, 12, 2, 13, 9, 7, 10, 14$$
$$r'(64..79) = 12, 15, 10, 4, 1, 5, 8, 7, 6, 2, 13, 14, 0, 3, 9, 11$$

amount for rotate left (rol)

$$s(0..15) = 11, 14, 15, 12, 5, 8, 7, 9, 11, 13, 14, 15, 6, 7, 9, 8$$
$$s(16..31) = 7, 6, 8, 13, 11, 9, 7, 15, 7, 12, 15, 9, 11, 7, 13, 12$$
$$s(32..47) = 11, 13, 6, 7, 14, 9, 13, 15, 14, 8, 13, 6, 5, 12, 7, 5$$
$$s(48..63) = 11, 12, 14, 15, 14, 15, 9, 8, 9, 14, 5, 6, 8, 6, 5, 12$$
$$s(64..79) = 9, 15, 5, 11, 6, 8, 13, 12, 5, 12, 13, 14, 11, 8, 5, 6$$
$$s'(0..15) = 8, 9, 9, 11, 13, 15, 15, 5, 7, 7, 8, 11, 14, 14, 12, 6$$
$$s'(16..31) = 9, 13, 15, 7, 12, 8, 9, 11, 7, 7, 12, 7, 6, 15, 13, 11$$
$$s'(32..47) = 9, 7, 15, 11, 8, 6, 6, 14, 12, 13, 5, 14, 13, 13, 7, 5$$
$$s'(48..63) = 15, 5, 8, 11, 14, 14, 6, 14, 6, 9, 12, 9, 12, 5, 15, 8$$
$$s'(64..79) = 8, 5, 12, 9, 12, 5, 14, 6, 8, 13, 6, 5, 15, 13, 11, 11$$

It is assumed that the message after padding consists of n 16-word blocks that will be denoted with $M_i[j]$, with $1 \leq i \leq n$ and $0 \leq j \leq 15$. The key used and the MAC value obtained consist of five words each, respectively $(K_0, K_1, K_2, K_3, K_4)$ and $(E_0, E_1, E_2, E_3, E_4)$. The symbols $\boxplus$ and $\boxminus$ denote respectively addition and subtraction modulo 2^{32}; rol_s denotes cyclic left shift (rotate) over s positions. The pseudo-code for TTMAC is then given below.

TTMAC: pseudo-code

$C_0 := K_0;\ C_1 := K_1;\ C_2 := K_2;\ C_3 := K_3;\ C_4 := K_4;$

$D_0 := K_0;\ D_1 := K_1;\ D_2 := K_2;\ D_3 := K_3;\ D_4 := K_4;$

for $i := 1$ *to* n {

$\quad A_0 := C_0;\ A_1 := C_1;\ A_2 := C_2;\ A_3 := C_3;\ A_4 := C_4;$

$\quad B_0 := D_0;\ B_1 := D_1;\ B_2 := D_2;\ B_3 := D_3;\ B_4 := D_4;$

$\quad$ *if* $(i\ !=n)$ *for* $j := 0$ *to* 79 {

$\quad\quad T := \text{rol}_{s(j)}\,(A_0 \boxplus f(j, A_1, A_2, A_3) \boxplus M_i[r(j)] \boxplus c(j)) \boxplus A_4;$

$\quad\quad A_0 := A_4;\ A_4 := A_3;\ A_3 := \text{rol}_{10}(A_2);\ A_2 := A_1;\ A_1 := T;$

$\quad\quad T := \text{rol}_{s'(j)}\,(B_0 \boxplus f(79 - j, B_1, B_2, B_3) \boxplus M_i[r'(j)] \boxplus c'(j)) \boxplus B_4;$

$\quad\quad B_0 := B_4;\ B_4 := B_3;\ B_3 := \text{rol}_{10}(B_2);\ B_2 := B_1;\ B_1 := T;$

$\quad$ }

$\quad$ *else for* $j := 0$ *to* 79 {

$\quad\quad T := \text{rol}_{s'(j)}\,(A_0 \boxplus f(79 - j, A_1, A_2, A_3) \boxplus M_i[r'(j)] \boxplus c'(j)) \boxplus A_4;$

$\quad\quad A_0 := A_4;\ A_4 := A_3;\ A_3 := \text{rol}_{10}(A_2);\ A_2 := A_1;\ A_1 := T;$

$\quad\quad T := \text{rol}_{s(j)}\,(B_0 \boxplus f(j, B_1, B_2, B_3) \boxplus M_i[r(j)] \boxplus c(j)) \boxplus B_4;$

$\quad\quad B_0 := B_4;\ B_4 := B_3;\ B_3 := \text{rol}_{10}(B_2);\ B_2 := B_1;\ B_1 := T;$

$\quad$ }

$\quad A_0 := A_0 \boxminus C_0;\ A_1 := A_1 \boxminus C_1;\ A_2 := A_2 \boxminus C_2;\ A_3 := A_3 \boxminus C_3;$

$\quad A_4 := A_4 \boxminus C_4;$

$\quad B_0 := B_0 \boxminus D_0;\ B_1 := B_1 \boxminus D_1;\ B_2 := B_2 \boxminus D_2;\ B_3 := B_3 \boxminus D_3;$

$\quad B_4 := B_4 \boxminus D_4;$

$\quad$ *if* $(i\ !=n)$ {

$\quad\quad C_2 := A_3 \boxminus B_0;\ C_3 := A_4 \boxminus B_1;\ C_4 := A_0 \boxminus B_2;\ C_0 := (A_1 \boxplus A_4) \boxminus B_3;$

$\quad\quad C_1 := A_2 \boxminus B_4;$

$\quad\quad D_1 := (A_4 \boxplus A_2) \boxminus B_0;\ D_2 := A_0 \boxminus B_1;\ D_3 := A_1 \boxminus B_2;$

$\quad\quad D_4 := A_2 \boxminus B_3;\ D_0 := A_3 \boxminus B_4;$

$\quad$ }

}

$E_0 := A_0 \boxminus B_0;\ E_1 := A_1 \boxminus B_1;\ E_2 := A_2 \boxminus B_2;\ E_3 := A_3 \boxminus B_3;$

$E_4 := A_4 \boxminus B_4.$

For a short message of up to 512 bits (one 16-word block after padding), no feedback or mixing is required and the following simplified pseudo-code can be used.

TTMAC (one message block): pseudo-code

$A_0 := K_0;\ A_1 := K_1;\ A_2 := K_2;\ A_3 := K_3;\ A_4 := K_4;$

$B_0 := K_0;\ B_1 := K_1;\ B_2 := K_2;\ B_3 := K_3;\ B_4 := K_4;$

for $j := 0$ *to* 79 {

$\quad T := \text{rol}_{s'(j)}\,(A_0 \boxplus f(79 - j, A_1, A_2, A_3) \boxplus M[r'(j)] \boxplus c'(j)) \boxplus A_4;$

$\quad A_0 := A_4;\ A_4 := A_3;\ A_3 := \text{rol}_{10}(A_2);\ A_2 := A_1;\ A_1 := T;$

$\quad T := \text{rol}_{s(j)}\,(B_0 \boxplus f(j, B_1, B_2, B_3) \boxplus M[r(j)] \boxplus c(j)) \boxplus B_4;$

$\quad B_0 := B_4;\ B_4 := B_3;\ B_3 := \text{rol}_{10}(B_2);\ B_2 := B_1;\ B_1 := T;$

}

$E_0 := A_0 \boxminus B_0;\ E_1 := A_1 \boxminus B_1;\ E_2 := A_2 \boxminus B_2;\ E_3 := A_3 \boxminus B_3;$

$E_4 := A_4 \boxminus B_4.$

References

1. Mihir Bellare, Ran Canetti and Hugo Krawczyk, "Keying hash functions for message authentication," *Proc. Crypto'96, LNCS 1109*, Springer-Verlag, 1996, pp 1-15.
2. Hans Dobbertin, Antoon Bosselaers and Bart Preneel, "RIPEMD-160: a strengthened version of RIPEMD," *Fast Software Encryption, LNCS 1039*, Springer-Verlag, 1996, pp 71-82.
3. NESSIE Project – New European Schemes for Signatures, Integrity and Encryption – home page: http://cryptonessie.org.
4. Bart Preneel and Paul van Oorschot, "MDx-MAC and building fast MAC's from hash functions," *Proc. Crypto'95, LNCS 963*, Springer-Verlag, 1995, pp 1-14.
5. Bart Preneel and Paul van Oorschot, "On the security of two MAC algorithms," *Proc. Eurocrypt'96, LNCS 1070*, Springer-Verlag, 1996, pp 19-32.
6. Gene Tsudik, "Message authentication with one-way hash functions," *ACM Computer Communications Review*, Vol. 22, No. 5, 1992, pp. 29-38.

Key Revocation with Interval Cover Families

Johannes Blömer and Alexander May

Department of Mathematics and Computer Science,
University of Paderborn, 33095 Paderborn, Germany
{bloemer,alexx}@uni-paderborn.de

Abstract. We present data structures for complement covering with intervals and their application for digital identity revocation. We give lower bounds showing the structures to be nearly optimal. Our method improves upon the schemes proposed by S. Micali [5,6] and Aiello, Lodha, Ostrovsky [1] by reducing the communication between a Certificate Authority and public directories while keeping the number of tokens per user in the public key certificate small.

1   Introduction

Digital identities play an essential role in many cryptographic applications. Infrastructures for digital identities are built by means of public-key cryptography and Certification Authorities. The schemes differ in how digital identities can checked to be valid and how the identities can be revoked.

A digital identity is validated by a certificate issued by a Certification Authority (CA). The CA initially uses a public key generation process to create a public key/secret key pair. The public key together with a fingerprint is published. A user u who wants to establish his own digital identity creates a new public key/secret key pair and sends the public key together with identifying information to the CA. The Certificate Authority checks u's identity to ensure that the user is really the person he/she claims to be. After that, the CA signs with its secret key a certificate containing u's public key, the identifying information and an expiration date of this certification. Hence, anyone is able to check the certificate issued by the CA with the CA's public key. For accepting u's public key, one must not trust the user u himself but the CA. To establish higher levels of trust, one can use a hierarchy of CAs.

A digital identity is valid as long as its certificate has not expired. In contrast to this, we must also have a mean for revoking users. Assume u's identity is stolen or compromised before the certificate expiration date. The thief can sign arbitrary messages with u's secret key. Hence, as in the case of credit cards, one must establish an immediate identity revocation.

There are many solutions proposed in literature how to revoke digital identities. The first one is a centralized online solution where a trusted database holds the status of each public key. The database answers queries about public keys. However, these answers must be authenticated by the database to avoid man-in-the-middle attacks. In many cases the method is impractical because an online access is required.

Another solution, the Certificate Revocation List (CRL), is widely used in practice. In this offline approach, the CA makes a list of all users revoked thus far and signs it. This list is distributed at regular intervals – for example during a daily update period – to many public directories. A public directory is untrusted but one insists that it cannot cheat and must return a user's revocation status when queried. The main drawback of this scheme is the time it takes to check a key's validity. One must first check the CA signature and then look at the whole list of revoked users. Consider a fixed update time and let r be the total number of revoked users up to this point. The CA has to communicate a CRL of size $O(r)$ to each public directory in order to update the status of the keys. The time to check a key's validity using the CRL is also $O(r)$.

There are two other offline schemes proposed by Kocher [3] and Naor, Nissim [7]. They make use of authenticated hash trees. For a fixed update time and r as defined above the communication from the CA to the public directories is reduced to $O(\log r)$. In order to check a user's identity, one receives $O(\log r)$ hash values from the directory and computes another $O(\log r)$ hash values. These values are compared with the public directory data in a specified way. Additionally, the root signature of the authenticated hash tree is checked.

The main drawbacks of the offline solutions mentioned so far are:

- The information send by the CA must be authenticated. Therefore, signing the data is necessary. In order to prove the status, the signature must be checked.
- The proof length – the amount of data that has to be checked for validation – is a function of r. Since normally one must prove a key's validity very often, the proof length is the main bottleneck of digital identity revocation.

S. Micali [5,6] proposed an elegant solution for these two problems based on an idea of offline/online signatures [2], which in turn builds on a work of Lamport [4]. He suggests to add an additional number y – called the user's 0-token – into the certificate. In order to create the 0-token, the CA picks a random number x and a one-way hash function f and computes $y = f^{(l)}(x) := f(f(\ldots f(x)))$, that is, the function f is applied l times to x in order to compute the 0-token y. The parameter l corresponds to the number of update periods, e.g. the days till expiration. On day 1, if user u is not revoked, the CA publishes the 1-token $f^{(l-1)}(x)$ of u. Since f is a one-way function, y can easily be computed from $f^{(l-1)}(x)$ by applying f once, but it is infeasible to find a valid 1-token $\tilde{x}$ with $f(\tilde{x}) = y$. Hence, u can take the 1-token as a proof that his key is valid on day 1. In general, the CA publishes the i-token $f^{(l-i)}$ on day i. This i-token serves as a day-i proof for the validity of u's key. Applying f i times and comparing the result with the 0-token proves the key's validity. In the sequel, we will use the terms token and proof synonymously. Notice that in contrast to the schemes of Kocher [3] and Naor, Nissim [7] this scheme needs only one proof for key validation and no signature of the CA in the daily update period.

Let $U = \{1, 2, \ldots, n\}$ be the set of users and let 2^U denote the power set of U. For a fixed update time let $R \subseteq U$ be the set of all users revoked so far. We set $r = |R|$. The complement $\bar{R} = U - R$ of R is the set of non-revoked users. The problem with Micali's scheme is that each of the $n - r$ non-revoked users in $\bar{R}$ obtains his own proof during an update period. Hence in each update period the CA has to communicate $n - r$ tokens to a public directory. We denote this as the CA-to-directory communication.

ALO (Aiello, Lodha, Ostrovsky) [1] proposed two schemes that reduce the CA-to-directory communication. These schemes are called Hierarchical and Generalized Scheme. The main building block of the ALO schemes is a set $F \subseteq 2^U$. The set F has the property that each set $\bar{R}$ of non-revoked users can be written as the union of the elements in a subset $S(\bar{R})$ of F. Each element $S_j \in F$ has its own 0-token. For each set $S_j \in F$, each user $u \in S_j$ stores the 0-token of S_j in his certificate. That is, the certificate of user u contains $|\{S_j \in F : u \in S_j\}|$ different tokens. We denote the maximal number $\max_{u \in U} |\{S_j \in F : u \in S_j\}|$ of tokens per certificate by $\mathcal{T}$.

In order to issue day-i proofs for the non-revoked users $u \in \bar{R}$, the CA computes a cover $S(\bar{R}) = \{S_{j_1}, S_{j_2}, \ldots, S_{j_m}\}$, $\bigcup_{1 \leq k \leq m} S_{j_k} = \bar{R}$ of the set $\bar{R}$. Next, it publishes the m i-tokens of the sets $S_{j_1}, S_{j_2}, \ldots, S_{j_m}$. Since these sets cover the set $\bar{R}$, each non-revoked user u is contained in at least one set S_j. Recall that u stores the 0-token of S_j in his certificate. Hence, the i-token of the set S_j is a day-i proof for user u.

There may be different ways to cover $\bar{R}$ by elements $S_j \in F$. In ALO's schemes, the CA always takes the minimal number of subsets for the cover in order to minimize the number m of proofs. Let

$$\max_{\bar{R} \subseteq U : |\bar{R}| = n - r} \{m : \text{CA needs } m \text{ sets to cover } \bar{R}\}$$

be the maximal number of proofs that the CA has to publish for a set $\bar{R}$ of size $n - r$. We denote this maximal number of proofs by $\mathcal{P}$.

Note that Micali's revocation scheme fits this description. To obtain Micali's revocation scheme, define F as $F = \{\{1\}, \{2\}, \ldots \{n\}\}$. Hence, the users only have to store one 0-token in their certificate.

Let us define three demands on our key revocation scenario in order of decreasing priority:

Proof of key's validity: In our scenario, a user must prove a key's validity very often. Therefore, we insist on only one proof for key validation as in the revocation schemes of Micali and ALO. The schemes of Kocher and Naor, Nissim do not meet this requirement.

CA-to-directory communication ($\mathcal{P}$): The CA-to-directory communication corresponds to the maximal number of proofs the CA has to send to a public directory. The maximal number of proofs is denoted by $\mathcal{P}$. We have to keep the CA-to-directory communication small to allow frequent update periods. Thus, we want to minimize $\mathcal{P}$.

Tokens per certificate ($\mathcal{T}$): We denote the number of tokens per certificate by $\mathcal{T}$. To make the scheme practical (especially for smart card applications),

$\mathcal{T}$ must be kept small, since checking long certificates is inefficient. But assuming that checked keys are stored, certificates normally have to be checked only once.

As mentioned above, Micali's scheme has $\mathcal{T} = 1$ token per certificate. However, the CA-to-directory communication is $\mathcal{P} = n - r$ if r is the number of revoked users. ALO's Hierarchical Scheme improves upon Micali's scheme by reducing the CA-to-directory communication $\mathcal{P}$ to $r \log_2(n/r)$ while increasing the number of 0-tokens $\mathcal{T}$ per certificate to $\log_2 n$. The Generalized Scheme of ALO needs at most $r(\log_c(n/r)+1)$ proofs per update period and $\mathcal{T} \leq (2^{c-1}-1) \log_c n$ tokens per certificate. Due to the $(2^{c-1} - 1)$ factor, this scheme is only practical for $c = 2$ or $c = 3$, otherwise the certificates become too large.

Our results build on the work of ALO. We propose a new method for covering the set $\bar{R}$ of non-revoked users by intervals. Therefore, we define a new class for covering problems called interval cover family (ICF). Our ICFs are constructed using interval trees.

The set R of revoked users partitions U in subintervals of non-revoked users, which can be represented by the nodes of an interval tree. Our task is to find a scheme which covers any interval with sets of an ICF. Furthermore, we want that each user $u \in U$ is in a small number of sets. This property is important because as in ALO's schemes, user u must include in his certificate all the 0-tokens of sets that contain u.

Micali's [5,6] and ALO's Hierarchical scheme [1] also belong to the class of algorithms using interval cover families. The ICF in [5,6] is the simplest one. It covers intervals by single elements. Thus, the length of the covering intervals is always 1. In ALO's Hierarchical scheme, the set of non-revoked users is covered by intervals with interval lengths that are powers of 2. In this paper, we propose two new methods for covering intervals that might be interesting for other areas of covering problems as well.

In Section 2, we introduce the class ICF of interval cover families. Revocation Scheme 1 (RS1) is presented in Section 3. It is a generalization of ALO's Hierarchical scheme. The length of the covering intervals is a power of $c \geq 2$. For RS1, we obtain the upper bounds $\mathcal{P} \leq (r+1)(2 \log_c n - 1)$ and $\mathcal{T} \leq \frac{(c+1)^2}{4} \log_c n$ for some constant parameter c.

Our second Revocation Scheme (RS2) presented in Section 4 leads to a CA-to-directory communication of $\mathcal{P} \leq (r+1)(\log_c n + 1)$, while keeping the number of 0-tokens per certificate upper bounded by $\mathcal{T} \leq \frac{(c+1)^2}{2} \log_c n(1 + o(1))$.

Since the new bounds for $\mathcal{T}$ are polynomial in c our systems are practical for larger parameters c than ALO's schemes. Thus, we can reduce the CA-to-directory communication $\mathcal{P}$ by choosing a large c. Since this communication is done during each update period, the system becomes more efficient.

In Section 5, we study the relations of the class ICF to the task of key revocation. Using a more refined analysis, we show that RS1 has a maximal CA-to-directory communication of $\mathcal{P} \leq 2r(\log_c n - \lfloor \log_c r \rfloor)$. Assuming the revoked users to be uniformly distributed, we can further reduce the bound of RS2 to an expected upper bound of $\mathcal{P} \leq (r + 1 - \frac{r(r-1)}{n}))(\log_c n + 1)$.

Table 1. Comparison of our schemes with Micali's and ALO's schemes

Scheme	$\mathcal{P}$ proofs from CA to directories	$\mathcal{T}$ tokens per certificate
Micali	$n - r$	1
ALO's Hierarchical	$r \log_2(\frac{n}{r})$	$\log_2 n$
ALO's Generalized	$r(\log_c(\frac{n}{r}) + 1)$	$(2^{c-1} - 1) \log_c n$
RS1	$2r(\log_c n - \lfloor \log_c r \rfloor)$	$\frac{(c+1)^2}{4} \log_c n$
RS2	$(r + 1)(\log_c n + 1)$	$\frac{(c+1)^2}{2} \log_c n(1 + o(1))$

If we only want to minimize the CA-to-directory communication, an optimal solution can be obtained from Yao's range query data structures [8]. However, Yao's construction results in prohibitively many tokens per certificate. For the first time in this area, we also prove lower bounds for the number $\mathcal{T}$ of 0-tokens (see Section 6). For example, Corollary 22 provides a lower bound of $\mathcal{T} \geq (\frac{c}{e} - 1) \cdot \log_c n$, where e is the Euler number. This shows that if c is constant, the trade-off in our revocation schemes between CA-to-directory communication $\mathcal{P}$ and the number of 0-tokens $\mathcal{T}$ is optimal up to a constant.

2   Definitions

Consider the universe $U = \{1, 2, \ldots, n\}$ of users with personal identification numbers 1 to n. Let 2^U denote the power set of U. Let $R \subseteq U$ be the subset of revoked users, $\bar{R} = U - R$ the complement of R. In our schemes, the CA has to find a family of sets that covers the subset $\bar{R}$ of all non-revoked users. Then the CA issues the i-tokens for all the sets in the cover. A day-i proof for a non-revoked u is a set that contains u.

Definition 1 (interval set) *The* interval set $V = [a, b]$, $V \subseteq U$ *is defined as* $[a, b] := \{x \in \mathbb{N} \mid a \leq x \leq b\}$. *The interval set* $[a, a]$ *is briefly written as* $[a]$. *The* length *of an interval set* $[a, b]$ *is defined as* $b - a + 1$.

Definition 2 (interval cover) *We call a family of subsets* $S \subseteq 2^U$ *an* interval cover (IC) *of the interval set* I *iff* $\bigcup_{V \in S} V = I$ *and all subsets* V *are interval sets. If* $|S| \leq k$, S *is called a* k-IC.

Definition 3 (interval cover family) $F \subseteq 2^U$ *is an* interval cover family (ICF) *of* U *iff for every interval set* $I \subseteq U$, *there is a subset* S *of* F *such that* S *is an IC of* I. F *is a* k-ICF *of* U *iff there is at least one* k-IC $S \subseteq F$ *of* I *for every* $I \subseteq U$.

Lemma 4 *Assume we have a k-ICF F of the universe $U = \{1, \ldots, n\}$ and an arbitrary $R \subseteq U$ with $|R| = r$. Then F covers $\bar{R}$ with at most $(r+1)k$ interval sets.*

Proof: Notice that a subset R of size $|R| = r$ partitions the interval $[1, n]$ in at most $r + 1$ subintervals $\bar{R}_1 \cup \cdots \cup \bar{R}_{r+1} = \bar{R}$. Thus, it suffices to cover these subintervals for covering $\bar{R}$. Since each $\bar{R}_i$ is coverable by F with at most k interval sets, the claim follows. ⊡

The number of interval sets needed to cover a set $\bar{R}$ in Lemma 4 corresponds to the maximal number of proofs – denoted $\mathcal{P}_F$ – the CA must send to the public directories during an update period. Hence, for a k-ICF F the size of $\mathcal{P}_F$ is always upper-bounded by $(r+1)k$.

It is also important for the practicality of a revocation scheme that the size of F is polynomial in n and that for every subset $\bar{R}$ of non-revoked users the corresponding ICs can be computed in time polynomial in $\log(n)$.

For an ICF F and every $u \in U$, we define $h_F(u)$ as the *multiplicity* of u in F, that is the number of sets in F containing the element u. Because every set in F that contains u can be part of an interval cover, user u's certificate must include all $h_F(u)$ 0-tokens that contain u. Thus, the maximal number of 0-tokens, denoted $\mathcal{T}_F := \max_u\{h_F(u)\}$, corresponds to the maximal length of a user's public key certificate. This length should be polynomial in $\log(n)$. There is a trade-off between the number of proofs $\mathcal{P}_F$ the CA must send to a public directory and the number of 0-tokens $\mathcal{T}_F$ in a revocation scheme. For instance in the Micali scheme [5,6], we have $\mathcal{P} = n - r$ and $\mathcal{T} = 1$. In their Generalized scheme, ALO [1] had $\mathcal{P} \leq r(\log_c(n/r) + 1)$ and $\mathcal{T} \leq (2^{c-1} - 1)\log_c n$.

In the next section, we present a $(2k - 1)$-ICF F with $\mathcal{T}_F = O(kn^{2/k})$ for some system parameter k. Taking $k = \log_c n$ leads to $\mathcal{T} = O(c^2 \log_c n)$.

3   A Revocation Scheme Using ICFs

First, we introduce a notation on intervals.

Definition 5 (combinational sum) *The* combinational sum *of an interval $[a_1, b_1]$ with an interval $[a_2, b_2]$ is defined as the interval $[\min\{a_1, a_2\}, \max\{b_1, b_2\}]$. We also say, we combine interval $[a_1, b_1]$ with $[a_2, b_2]$. Let $W = \{[a_1, b_1], [a_2, b_2], \ldots, [a_m, b_m]\}$ be a set of disjoint intervals. We define the* maximal combinational sum *of W that is contained in an interval $[a, b]$ as the interval $[\min_{a_i} : a_i \geq a, \max_{b_j} : b_j \leq b]$.*

Note that the combinational sum of two intervals $[a_1, b_1]$ and $[a_2, b_2]$ may contain elements which are neither in $[a_1, b_1]$ nor in $[a_2, b_2]$.

Next, we define an interval tree T for the interval $[1, n]$ and a parameter k, that might depend on n. The construction is recursive.

Construction of the interval tree T

- The root is labelled with the interval $[1, n]$.
- Each node labelled with an interval $[a, b]$ of length greater than 1 has $n^{1/k}$ children. The children partition the interval $[a, b]$ into equally long pieces. That is, the children are roots of the interval trees for the intervals $[a + i \cdot \frac{b-a+1}{n^{1/k}}, a + (i + 1) \cdot \frac{b-a+1}{n^{1/k}} - 1]$, $0 \leq i < n^{1/k}$ (for simplicity, we assume that $n^{1/k}$ is an integer to avoid rounding).

We store the following contents in each node of the interval tree.

- Each node stores the interval of its label.
- Moreover, each node stores the combinational sums of its label with the labels of its right siblings.

Let combinational sums of nodes be defined as the combinational sums of their labels.

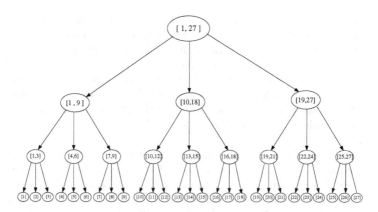

Fig. 1. The interval tree T for $n = 3^3$, $k = 3$

Example: In Figure 1, the node with label $[10, 12]$ stores the interval sets $[10, 12]$, $[10, 15]$ and $[10, 18]$. Its father $[10, 18]$ stores the intervals $[10, 18]$ and $[10, 27]$.

Since in level i the nodes are labelled with intervals of length $n^{(k-i)/k}$, in level k we have interval length 1 and the recursive construction stops. Thus, the interval tree has depth k.

We define the ICF F as the union of all the sets of intervals stored in the nodes of the interval tree T. However, we exclude the root label interval $[1, n]$.

Next, we want to show that F is a $(2k - 1)$-ICF, that is, we want to show that we can cover every interval set $I \subseteq [1, n]$ by at most $2k - 1$ sets in F. In order to prove this, we present an algorithm that needs a maximum of $2k - 1$ combinational sums for covering any interval I.

Algorithm Cover Scheme 1 (CS1)

INPUT: interval $I = [a, b]$

> FOR level $i = 1$ TO k in the interval tree T DO
> Take the maximal combinational sums of the label intervals in
> level i of T that is contained in the yet uncovered parts of $[a, b]$.
> IF $[a, b]$ is covered completely, EXIT.

OUTPUT: interval sets $I_1, I_2, \ldots, I_m$ with $\bigcup_{j=1\ldots m} I_j = [a, b]$ and $m \leq 2k - 1$.

Example: In Figure 1 on input $I = [2, 21]$, the Algorithm CS1 covers I by taking the intervals $[10, 18]$ (level 1, stored in node $[10, 18]$), $[4, 9]$ and $[19, 21]$ (level 2, stored in nodes $[4, 6]$ and $[19, 21]$) and $[2, 3]$ (level 3, stored in $[2]$).

Lemma 6 *The union F of all intervals stored in the interval tree T is a $(2k-1)$-ICF.*

Proof: We have to show that CS1 needs at most $2k - 1$ interval sets to cover $[a, b]$. Notice that CS1 covers the whole interval $[a, b]$ successively from the middle to the borders. In level 1 of the interval tree T, one gets at most one combinational sum $[a_1, b_1]$. The uncovered parts $[a, a_1 - 1]$ and $[b_1 + 1, b]$ both yield at most one additional interval in level 2. This holds because the maximal combinational sums in level 2 are always of the form $[a_2, a_1 - 1]$ respectively $[b_1 + 1, b_2]$. Analogously, we get at most two additional intervals in the subsequent levels. This leads to the upper bound of $2k - 1$. ⊟

We define the memory requirement $|F|$ of an ICF F to be the number of interval sets in F. The running time of a k-ICF F on input $I = [a, b]$ is the time to find a k-IC S for I. Further, we define the running time of a k-ICF to be the maximal running time taken over all choices of input intervals I. The following lemma shows that our $(2k - 1)$-ICF F can be efficiently implemented.

Lemma 7 *The ICF F has memory requirement $O(n^{1+1/k})$ and running time $O(kn^{1/k})$.*

Proof: In every set of siblings at most $\sum_{i=1}^{n^{1/k}} i = O(n^{2/k})$ intervals are stored. This follows from the fact that each node contains its label and all combinational sums with its right siblings. Hence, F contains at most $O(n^{2/k}) \sum_{i=0}^{k-1} (n^{1/k})^i = O(n^{1+1/k})$ interval sets.

The operations in each level can easily be implemented to run in time $O(n^{1/k})$, that is in the number of children. Thus, the total running time is $O(kn^{1/k})$. ⊟

Definition 8 (RS1) Revocation Scheme 1 (RS1) *uses the $(2k - 1)$-ICF F and Algorithm CS1 in order to cover all interval sets $\bar{R} = U - R$ of non-revoked users.*

Theorem 9 *RS1 is a revocation scheme with $\mathcal{P}_F \leq (r+1)(2k-1)$ and $\mathcal{T}_F \leq \frac{1}{4}k(n^{1/k}+1)^2$.*

Proof: The number of proofs $\mathcal{P}_F \leq (r+1)(2k-1)$ follows from Lemma 4.

It remains to show the upper bound for $\mathcal{T}_F$. Because the node labels in each level partition the interval $[1, n]$, every element $u \in U$ is stored in exactly one label per level. We want to determine the number of interval sets of a single level in which a user u is contained. Therefore, consider the node with the label interval containing u. Combinational sums are only taken among the $n^{1/k}$ siblings of this node. When enumerating the siblings from left to right, it is easy to see that the ith sibling is in exactly $i \cdot (n^{1/k} - i + 1)$ interval sets. This function in i takes its maximum for $i = (n^{1/k}+1)/2$, leading to $\max_i\{i \cdot (n^{1/k}-i+1)\} = (\frac{n^{1/k}+1}{2})^2$. Hence, user u can be in at most $(\frac{n^{1/k}+1}{2})^2$ intervals sets per level. Summing over the k levels, we obtain $\mathcal{T}_F \leq \frac{1}{4}k(n^{1/k}+1)^2$. $\qquad\square$

Corollary 10 *Choosing $k = \log_c n$ for some constant c, we obtain a $(2 \log_c n - 1)$-ICF F that needs $O(n)$ memory and $O(\log_c n)$ running time. RS1 is a key revocation scheme with $\mathcal{P}_F \leq (r+1)(2\log_c n - 1)$ and $\mathcal{T}_F \leq \frac{(c+1)^2}{4}\log_c n$.*

Note that our result improves upon the generalized scheme of ALO, who had $\mathcal{T} \leq (2^{c-1}-1)\log_c n$. A refined analysis of $\mathcal{P}_F$ is given in Section 5.

Even with refined analysis, there remain two problems with the $(2k-1)$-ICF presented above. First, we always assume an upper bound of $2k-1$ for the number of intervals taken by algorithm CS1. Consider a small interval $[a, b]$ with length much shorter than n. Algorithm CS1 will take its first combinational sum in a level i that is close to the leaves in level k. It is easy to see that in this case, CS1 outputs at most $2(k-i)+1$ interval sets. Therefore, Lemma 6 gives a pessimistic bound. Second, after using the first combinational sum in level i, we just need combinational sums of the rightmost or leftmost sibling nodes in the subsequent levels. But we store combinational sums of all sibling nodes. In the next section, we show how to avoid these problems.

4 Another Revocation Scheme Based on ICFs

We take an interval tree T' similar to the interval tree T in Section 3. The nodes and their labels remain the same as in T, only their content is changed.

Definition 11 (partial sums) *For each set of sibling nodes in a tree, we call the combinational sums of the leftmost sibling v with all other siblings to the right the* right partial sums. *The combinational sums of v's father's leftmost sibling with v and all of v's siblings are called the* upper right partial sums *(for an example, see below). The combinational sums of the rightmost sibling w with all other siblings to the left – except the leftmost sibling – are called the* left partial sums. *Analogously, the combinational sums of w's father's rightmost sibling with w and all of w's siblings are called the* upper left partial sums.

Let $W = \{[a_1, b_1], [a_2, b_2], \ldots, [a_m, b_m]\}$ *be a set of partial sums. We define the* maximal partial sum *of W that is contained in an interval $[a, b]$ as the interval* $[\min_{a_i} : a_i \geq a, \max_{b_j} : b_j \leq b]$.

Example: In Figure 1, the right partial sums of the set $\{[13], [14], [15]\}$ of sibling nodes are $[13]$, $[13, 14]$ and $[13, 15]$. The left partial sums are $[15]$ and $[14, 15]$. The upper right partial sums are the interval sets $[10, 13]$, $[10, 14]$, $[10, 15]$ and the upper left partial sums are $[15, 18]$, $[14, 18]$ and $[13, 18]$.

Notice that we should omit those upper right partial sums where the father of the leftmost sibling v is itself a leftmost sibling, since these combinational sums yield always the label of the father. This holds analogous for the upper left partial sums.

Node contents of the interval tree T'

- Each node stores its label.
- For any sibling nodes in level $k - 2j$, $0 \leq j < \frac{k-1}{2}$, the leftmost sibling v stores the right partial sums. Additionally, v stores the upper right partial sums.
- For any sibling nodes in level $k - 2j$, $0 \leq j < \frac{k-1}{2}$, the rightmost sibling w stores the left partial sums. In addition, w stores the upper left partial sums.
- Any sibling nodes in level i of the recursion tree are divided in i equally large parts. In any part, each node stores the combinational sums of its label with the labels of its right siblings in this part.

Again, the ICF F' is defined as the union of all intervals stored in the nodes.

Algorithm Cover Scheme 2 (CS2)

INPUT: interval $I = [a, b]$

    level $i := 1$
    UNTIL a combinational sum is taken DO
        Take the maximal combinational sum of the label intervals
        in level i of T' that is contained in $[a, b]$. (That combinational
        sum may consist of up to i intervals.) $i := i + 1$
    FOR level $j = i + (k - i \bmod 2)$ TO k STEP 2 DO
        Take the maximal partial sums of the yet uncovered parts of $[a, b]$.
        IF $[a, b]$ is covered completely, EXIT.

OUTPUT: interval sets $I_1, I_2, \ldots, I_m$ with $\bigcup_{j=1\ldots m} I_j = [a, b]$ and $m \leq k + 1$.

Example: We cover the interval $[2, 21]$ using Algorithm CS2 and the interval tree T' of Figure 1. CS2 outputs the combinational sum $[10, 18]$ in level 1 and the upper partial sums $[2, 9]$ and $[19, 21]$ in level 3.

Lemma 12 *The union F' of all intervals stored in the nodes of the interval tree T' is a $(k+1)$-ICF.*

Proof: Let Algorithm CS2 take a combinational sum in level i. Since in level i, maximal combinational sums of label intervals can be divided into i parts, we take at most i intervals. In the remaining $k-i$ levels, CS2 can take at most 2 intervals in each of the levels $k-2j$, $0 \le j < \frac{k-i}{2}$. These are at most $\lceil \frac{k-i}{2} \rceil$ levels. Thus, we obtain the upper bound $i + 2 \cdot \lceil \frac{k-i}{2} \rceil \le i + 2 \cdot \frac{k-i+1}{2} = k+1$. $\square$

Lemma 13 *The $(k+1)$-ICF F' needs $O(n^{1+1/k})$ memory and $O(kn^{1/k})$ running time.*

Proof: The memory requirement of F' is the amount of partial sums and combinational sums. We have at most $2n^{1/k}$ right and left partial sums per set of siblings. The upper partial sums sum up to another $2n^{1/k}$ intervals. Ignoring that these intervals are only taken in each second level we get an upper bound of $4n^{1/k} \cdot \sum_{i=0}^{k-1}(n^{1/k})^i = O(n)$ for the partial sums. Further, each set of siblings stores $O(n^{2/k})$ combinational sums. Summing over the levels gives an upper bound of $O(n^{2/k}) \cdot \sum_{i=0}^{k-1}(n^{1/k})^i = O(n^{1+1/k})$ which is also an upper bound for the total memory requirement.

Since the operations in each level can be implemented to run in time $O(n^{1/k})$, the running time is $O(kn^{1/k})$. $\square$

Definition 14 (RS2) *The Revocation Scheme 2 (RS2) uses the $(k+1)$-ICF F' and Algorithm CS2 in order to cover all interval sets $\bar{R} = U - R$ of non-revoked users.*

Theorem 15 *The $(k+1)$-ICF F' yields a revocation scheme with $\mathcal{P}_{F'} \le (r+1)(k+1)$ and $\mathcal{T}_{F'} \le \frac{k+1}{2} \cdot n^{2/k} + \frac{k}{4} \cdot (2n^{1/k} + 1) + \frac{1}{2}(\log k + 1)n^{1/k}$.*

Proof: The upper bound for the number of proofs $\mathcal{P}_{F'}$ follows from Lemma 4.

To complete the proof, we must show that $\mathcal{T}_{F'} \le \frac{k+1}{2} \cdot n^{2/k} + \frac{k}{4} \cdot (2n^{1/k} + 1) + \frac{1}{2}(\log k + 1)n^{1/k}$. Let us start with the partial sums and consider a set of sibling nodes as enumerated from left to right. It is easy to see that the ith sibling is in $n^{1/k} - i + 1$ right partial sums and in $i - 1$ left partial sums. This gives a total of $n^{1/k}$ partial sums for each element $u \in U$. Analogously, one can show that each element u is in $n^{1/k}$ upper partial sums. Since each upper partial sum consists of $n^{1/k}$ combinational sums, we get another $n^{2/k}$ intervals. Thus, we have a total of $n^{2/k} + n^{1/k}$ partial sums for every element $u \in U$ in the levels $k - 2j$, $0 \le j < \frac{k-1}{2}$. Summing over these levels gives us an upper bound of $\lceil \frac{k-1}{2} \rceil \cdot (n^{2/k} + n^{1/k}) \le \frac{k}{2} \cdot (n^{2/k} + n^{1/k})$.

In addition to the partial sums, we divide all sibling nodes in level i of T' in parts of size $\frac{n^{1/k}}{i}$ and compute the combinational sums with their right siblings in that part. Analogous to the proof of Theorem 9 every element in level i is in no more than $\frac{1}{4}(\frac{n^{1/k}}{i} + 1)^2$ of these intervals. Summing over the levels gives

$$\frac{1}{4}\sum_{i=1}^{k}\left(\frac{n^{1/k} + i}{i}\right)^2 = \frac{1}{4}\left(\sum_{i=1}^{k}\frac{n^{2/k}}{i^2} + \sum_{i=1}^{k}\frac{2n^{1/k}}{i} + \sum_{i=1}^{k}1\right)$$

$$\leq \frac{1}{4}\left(\frac{\pi^2}{6}\cdot n^{2/k} + 2(\log k + 1)\cdot n^{1/k} + k\right)$$

Together with the partial sums computed before we get the desired upper bound for the number of 0-tokens

$$T_{F'} \leq \frac{k+1}{2}\cdot n^{2/k} + \frac{k}{4}\cdot(2n^{1/k} + 1) + \frac{1}{2}(\log k + 1)n^{1/k}.$$

$\square$

Corollary 16 *Taking $k = \log_c n$, RS2 is a revocation scheme with $P_{F'} \leq (r + 1)(\log_c n + 1)$ and $T_{F'} \leq \frac{(c+1)^2}{2}\cdot\log_c n(1 + o(1))$.*

If we compare this result with the revocation scheme RS1 of Section 3, we roughly halve the number of proofs P by doubling T. Since we have update periods frequently, it is preferable to make P small by slightly enlarging T.

5   ICFs and Key Revocation

In the previous sections, we studied the covering of *arbitrary intervals* $[a, b]$ by ICFs and connected this to the task of key revocation by Lemma 4. But Lemma 4 yields a pessimistic bound:

- We always expect that r revoked users yield $r + 1$ intervals of non-revoked users. This is no longer true if r becomes large.
- The intervals representing non-revoked users are not arbitrary but disjoint, that is the intervals do not overlap. Further, the average length of the intervals that have to be covered depends on the parameter r.

5.1   The Expected Number of Intervals

In the following, we assume that the revoked users R are uniformly distributed over the interval $[1, n]$ and look for the expected number of intervals of non-revoked users. Let $i_1, i_2, \ldots, i_r$ be the revoked users in sorted order, that is $i_1 < i_2 < \cdots < i_r$. We call i_j and i_{j+1} a pair iff $i_{j+1} = i_j + 1$. Note that pairs of revoked users do not introduce a new interval that must be covered, since they enclose an interval of non-revoked users of size 0. Let X be the random variable for the number of intervals. Then

$$E_R(X) \leq r + 1 - E_R(\text{number of pairs}).$$

We obtain an upper bound since revoked users at the interval borders 1 and n never yield an additional interval. Thus, the borders 1 and n always pair. The expected number of pairs is

$$E_R(\text{number of pairs}) = \frac{\binom{n-2}{r-2}}{\binom{n}{r}} \cdot (n-1) = \frac{r(r-1)}{n}.$$

We summarize this in the following lemma.

Lemma 17 *Let the revoked users be distributed uniformly over* $[1, n]$ *and let* F *be a* k-*ICF. Then* F *yields a key revocation system with an expected upper bound of* $\mathcal{P}_F \leq (r + 1 - \frac{r(r-1)}{n})k$ *for the CA-to-directory communication.*

5.2   Key Revocation with RS1 for Growing r

Lemmas 4 and 17 still give pessimistic bounds, since they assume that arbitrary intervals are covered. But CS1 does not always use $2k - 1$ intervals to cover an interval $[a, b]$. If the interval length of $[a, b]$ is small, Algorithm CS1 will not use any intervals in the upper levels of the interval tree T. However, if the number r of revoked users increases then the average interval lengths of intervals that must be covered decreases. Hence, we expect some amortization of costs with growing r.

This fact was studied in ALO [1]. They proved an upper bound of $T \leq r \log_2(\frac{n}{r})$. Note, that the logarithmic term decreases with increasing r. We show our algorithm to be a generalization of [1] by showing a bound of $2r(\log_c n - \lfloor \log_c r \rfloor) - m$ for arbitrary $c > 2$ and $m = r - c^{\lfloor \log_c r \rfloor}$. Therefore, we adapt the proof techniques of [1]. For $c = 2$, our scheme reduces to the Hierarchical Scheme proposed in ALO[1].

Definition 18 *Let* $P(n, R)$ *be the number of proofs using the revocation scheme RS1 for covering* $U - R$, *where* $U = \{1, 2, \ldots, n\}$. *We define* $P(n, r) = \max_{R:|R|=r}\{P(n, R)\}$ *to be the worst case number of proofs for a revocation set* R *of* r *users.*

Assume $n = c^k$.

Theorem 19 *For* $r = c^l$, $l \geq 0$, *RS1 has* $P(n, r) \leq 2r \log_c(\frac{n}{r})$ *for* $c > 2$, $c \in \mathbb{N}$.

Proof: Similar to the proof in ALO [1]. The proof is given in the full version of the paper. ⌑

In the following theorem, we prove the upper bound for $P(n, r)$ for arbitrary r.

Theorem 20 *For* $r = c^l + m$, *RS1 yields* $P(n, r) \leq 2r(\log_c n - \lfloor \log_c r \rfloor) - m$.

Proof: The proof is given in the full version of the paper. ⌑

6   Lower Bounds for $\mathcal{T}_F$ in a k-ICF F

In this section, we show lower bounds for the number of tokens $\mathcal{T}_F$. Let $U = \{1, 2, \ldots, n\}$ be the set of users. We cover arbitrary interval sets $[a, b] \subseteq U$ of non-revoked users by k-ICFs. Comparing the lower bounds to our results in Section 3 and 4 will prove that our revocation schemes are up to constants optimal.

Theorem 21 *Let F be a k-ICF of U, then $\mathcal{T}_F \geq \sqrt[k]{k!} \cdot (n+1)^{1/k} - k$.*

Proof: We prove a lower bound for covering the interval sets $[1, 1], [1, 2], \ldots$, $[1, n]$ with the k-ICF F. This yields a lower bound for covering all interval sets $I \subseteq U$. The bound is proven by induction. For $k = 1$ an optimal family F_1 covering these sets must contain all of the n interval sets. But each of these sets contains the element 1. Thus, $\mathcal{T}_{F_1} = h_{F_1}(1) = n$. The identity

$$\mathcal{T}_{F_k} = h_{F_k}(1), \tag{1}$$

is an invariant of the proof, where F_k denotes an optimal covering scheme with at most k interval sets. The inductive step is done from $k - 1$ to k.

Assume, there's an optimal family F_k covering the sets $[1, 1], [1, 2], \ldots , [1, n]$ with at most k interval sets and minimal $\mathcal{T}_{F_k}$. We show in Lemma 24 that we can assume wlog $\mathcal{T}_{F_k} = h_{F_k}(1)$. Hence, invariant (1) holds.

Now, consider the interval sets of F_k containing 1. Let these be the sets $[1, a_1], [1, a_2], \ldots , [1, a_t]$, where $a_1 = 1$ because F_k must cover the single user 1. By construction, $t = \mathcal{T}_{F_k}$. An auxiliary set is defined by $[1, a_{t+1}]$ with $a_{t+1} = n + 1$. The intervals sets $[1, a_{i+1} - 1], 1 \leq i \leq t$ are covered by taking the interval set $[1, a_i]$ and an optimal covering in F_k of the remaining interval $[a_i + 1, a_{i+1} - 1]$ with at most $k - 1$ sets.

The element $a_i + 1$ is the first element in the interval $[a_i + 1, a_{i+1} - 1]$. Hence, it plays the role of the element 1 when covering the sets $[a_i + 1, a_i + 1], [a_i + 1, a_i + 2], \ldots , [a_i + 1, a_{i+1} - 1]$. By equation (1), the element $a_i + 1$ is critical, because it has maximal multiplicity of all the elements in $[a_i + 1, a_{i+1} - 1]$. Thus, element $a_i + 1$ must be contained in $h_{F_{k-1}}(1)$ interval sets and the induction hypothesis applies with $k - 1$ and interval length $a_{i+1} - a_i - 1$. Additionally, $a_i + 1$ is contained in the $t - i$ interval sets $[1, a_{i+1}], \ldots , [1, a_t]$. Since $h_{F_k}(1) = t$ and 1 is the element of maximal multiplicity, we obtain for $1 \leq i \leq t$

$$\sqrt[k-1]{(k-1)!} \cdot (a_{i+1} - a_i)^{\frac{1}{k-1}} - (k-1) + t - i \leq t \tag{2}$$

$$\Rightarrow \quad a_{i+1} \leq a_i + \frac{(i+k-1)^{k-1}}{(k-1)!}. \tag{3}$$

Solving the recurrence in (3) for $a_1 = 1$ yields

$$a_{i+1} \leq a_{i-1} + \frac{(i+k-2)^{k-1}}{(k-1)!} + \frac{(i+k-1)^{k-1}}{(k-1)!} \leq \cdots \leq \frac{1}{(k-1)!} \sum_{j=1}^{i+k-1} j^{k-1}$$

$$< \frac{1}{(k-1)!} \int_{j=0}^{i+k} j^{k-1} dj = \frac{(i+k)^k}{k!}$$

But we know $a_{t+1} = n + 1$, which leads to

$$\frac{(t + k)^k}{k!} \geq n + 1$$

$$\Rightarrow \quad t \geq \sqrt[k]{k!} \cdot (n + 1)^{1/k} - k.$$

$\square$

Using Stirling's formula $k! \sim \sqrt{2\pi k} \left(\frac{k}{e}\right)^k > \left(\frac{k}{e}\right)^k$, we conclude

Corollary 22

$$\mathcal{T}_F > \frac{k}{e} \cdot (n + 1)^{1/k} - k > \left(\frac{n^{1/k}}{e} - 1\right) k$$

Taking $k = \log_c n$ yields

Corollary 23

$$\mathcal{T}_F \geq \left(\frac{c}{e} - 1\right) \cdot \log_c n.$$

Lemma 24 *Let F_k be a k-ICF covering $[1, 1], [1, 2], \ldots, [1, n]$ with minimal $\mathcal{T}_{F_k}$. F_k can be turned into a k-ICF with $h_{F_k}(1) = \mathcal{T}_{F_k}$.*

Proof: The proof is given in the full version of the paper. $\square$

Theorem 25 *Let F be a k-ICF of U, then $\mathcal{T}_F > \left(\frac{1}{6}\right)^{\frac{1}{k-1}} n^{\frac{2}{k}}$.*

Proof: Let $F = \{S_1, S_2, \ldots, S_m\}$. Since F is a k-ICF, for every interval set $I \in U$ there exist at most k interval sets $S_{i_1}, S_{i_2}, \ldots, S_{i_k}$ such that $S(I) := \{S_{i_1}, S_{i_2}, \ldots, S_{i_k}\}$ is a cover of I. Note, that some S_{i_j} might be empty and there might be several $S(I)$ in F that cover I. For each interval set I we consider an arbitrary but fixed $S(I)$.

Assume $\mathcal{T}_F \leq \left(\frac{1}{6}\right)^{\frac{1}{k-1}} n^{\frac{2}{k}}$. Fix some interval set $S_i = [a, b]$ of F and consider the number of times this set can be contained in a cover $S(I)$ of some interval set $I = [s, t]$, where $s < a$ or $b < t$. Consider the case $s < a$. By assumption, user $a - 1$ is contained in at most $\left(\frac{1}{6}\right)^{\frac{1}{k-1}} n^{\frac{2}{k}}$ sets of F. On the other hand, every cover $S(I)$ containing $S_i = [a, b]$ must contain a set S_j which includes the element $a - 1$. Next consider the interval $S_j \cup S_i = [c, d]$. Assuming $s < c$ and arguing as above, $S(I)$ must contain one of the $\left(\frac{1}{6}\right)^{\frac{1}{k-1}} n^{\frac{2}{k}}$ intervals containing $c - 1$. Continuing in this way and using the fact that F is a k-ICF, we conclude that there are at most $\frac{1}{6} n^{\frac{2(k-1)}{k}}$ covers $S(I)$ in which a set S_i can participate. This holds for any S_i:

$$|\{I \subseteq U : S_i \in S(I)\}| \leq \frac{1}{6} n^{\frac{2(k-1)}{k}}. \tag{4}$$

Next, we count the number of elements with multiplicities in all the $\binom{n}{2}$ interval sets I in U. Since there is one interval of length n, two intervals of length $n-1$, etc., we get $\sum_{I \in U} |I| = \sum_{i=1}^{n} i \cdot (n - i + 1) = \frac{1}{6}n^3 + \frac{1}{2}n^2 + \frac{1}{3} > \frac{1}{6}n^3$. Since the interval sets S_i that cover I can overlap, the following inequality holds

$$\sum_{I \in U} \sum_{S_i \in S(I)} |S_i| \geq \sum_{I \in U} |I| > \frac{1}{6}n^3. \tag{5}$$

Using inequality (4), we also obtain

$$\sum_{I \in U} \sum_{S_i \in S(I)} |S_i| = \sum_{S_i \in F} |S_i| \cdot |\{I \subseteq U : S_i \in S(I)\}|$$

$$\leq \frac{1}{6}n^{\frac{2(k-1)}{k}} \sum_{S_i \in F} |S_i| \tag{6}$$

Combining (5) and (6) leads to

$$\sum_{S_i \in F} |S_i| > n^{3 - \frac{2(k-1)}{k}} = n^{1 + \frac{2}{k}}.$$

Note that $\sum_{S_i \in F} |S_i| = \sum_{u \in U} h_F(u)$. Taking the average number of the multiplicities $h_F(u)$ yields that there must be an element u with $h_F(u) > n^{\frac{2}{k}}$, contradicting the assumption that each element is in at most $(\frac{1}{6})^{\frac{1}{k-1}} n^{\frac{1}{k}}$ sets. $\boxdot$

Definition 26 *We call a k-ICF F δ-optimal, if for all k-ICFs $\bar{F}$: $\frac{T_F}{T_{\bar{F}}} = O(\delta)$.*

Theorem 27 *RS1 uses a $\min_k\{n^{3/k}, kn^{2/k}\}$-optimal k-ICF. The $(k+1)$-ICF F' constructed for RS2 is $\min_k\{n^{1/k}, k\}$-optimal.*

Proof: RS1 uses a $(2k-1)$-ICF F with $T_F = O(kn^{2/k})$. This can be turned into a k-ICF with $T_F = O(kn^{4/k})$. Dividing by the lower bounds of Corollary 22 and Theorem 25 gives the $\min_k\{n^{3/k}, kn^{2/k}\}$-optimality. RS2 uses a $(k+1)$-ICF F' with $T_{F'} = O(kn^{2/k})$. Applying Corollary 22 and Theorem 25 proves the claim. $\boxdot$

Corollary 28 *For $k = \log_c n$, the $(2k-1)$-ICF F used in RS1 and the $(k+1)$-ICF F' used in RS2 are 1-optimal.*

Note, that we obtain the lower bound in Theorem 21 by covering the intervals $[1,1], [1,2], \ldots, [1,n]$. This is only a small subset of all the intervals in $[1,n]$ and the left border is fixed by the element 1. It seems that making both borders variable introduces a factor of $n^{2/k}$, but we can not prove this yet. Thus, we expect a lower bound of $T_F = \Omega(kn^{2/k})$ for any k-ICF F. This would yield 1-optimality for the ICF F' in RS2 independent of the choice of k.

7   Conclusion

We introduced a new class called ICF for key revocation. Micali's scheme [5,6] and ALO's Hierarchical scheme [1] belong to this class. We improved upon the former results by reducing the critical update cost for CA-to-directory communication. In practice, the performances of our revocation schemes depend on the expected number r of revoked users. If one expects r to be a small fraction of n, then RS2 is preferable. It avoids a factor of 2 in the communication. RS1 should perform better for large r. We have shown the first lower bounds in this area, proving our schemes to be optimal up to constants.

References

1. W. Aiello, S. Lodha, R. Ostrovsky, "Fast Digital Identity Revocation", Proc. Crypto '98, Lecture Notes in Computer Science Vol. 1462, pages 137-152, 1998
2. S. Even, O. Goldreich, S. Micali, "On-line/Off-line Digital Signing", Proc. Crypto '89, Lecture Notes in Computer Science Vol. 435, pages 263-275, 1989
3. P. Kocher, "A quick introduction to Certificate Revocation Trees", unpublished manuscript, 1998
4. L. Lamport, "Password authentication with insecure communication", Communications of ACM, 24, pages 770-771, 1981
5. S. Micali, "Enhanced Certificate Revocation System", Technical memo MIT/ LCS/TM-542, ftp://ftp-pubs.lcs.mit.edu/pub/lcs-pubs/tm.outbox, 1995
6. S. Micali, "Certificate Revocation System", U.S. Patent number 5666416, 1997
7. M. Naor and K. Nissim, "Certificate Revocation and Certificate Update", Proceedings of 7th USENIX Security Symposium, pages 217-228, 1998
8. Andrew C. Yao, "Space-time trade-off for answering range queries (extended abstract)", Proceedings of the Fourteenth Annual ACM Symposium on Theory of Computing, pages 128-136, 1982

Timed-Release Cryptography

Wenbo Mao

Hewlett-Packard Laboratories,
Filton Road, Stoke Gifford, Bristol BS34 8QZ, United Kingdom
wm@hplb.hpl.hp.com

Abstract. Let n be a large composite number. Without factoring n, the computation of a^{2^t} (mod n) given a, t with $\gcd(a, n) = 1$ and $t < n$ can be done in t squarings modulo n. For $t \ll n$ (e.g., $n > 2^{1024}$ and $t < 2^{100}$), no lower complexity than t squarings is known to fulfill this task. Rivest et al suggested to use such constructions as good candidates for realising timed-release crypto problems.

We argue the necessity for a zero-knowledge proof of the correctness of such constructions and propose the first practically efficient protocol for a realisation. Our protocol proves, in $\log_2 t$ standard crypto operations, the correctness of $(a^e)^{2^t}$ (mod n) with respect to a^e where e is an RSA encryption exponent. With such a proof, a Timed-release Encryption of a message M can be given as $a^{2^t} M$ (mod n) with the assertion that the correct decryption of the RSA ciphertext M^e (mod n) can be obtained by performing t squarings modulo n starting from a. Timed-release RSA signatures can be constructed analogously.

Keywords Timed-release cryptography, Time-lock puzzles, Non-parallelisability, Efficient zero-knowledge protocols.

1 Introduction

Let n be a large composite natural number. Given $t < n$ and $\gcd(a, n) = 1$, without factoring n, the validation of

$$X \equiv a^{2^t} \pmod{n} \tag{1}$$

can be done in t squarings mod n. However if $\phi(n)$ (Euler's phi function of n) is known, then the job can be completed in $O(\log n)$ multiplications via the following two steps:

$$u \stackrel{\text{def}}{=} 2^t \pmod{\phi(n)}, \tag{2}$$

$$X \stackrel{\text{def}}{=} a^u \pmod{n}. \tag{3}$$

For $t \ll n$ (e.g., $n > 2^{1024}$ and $t < 2^{100}$), it can be anticipated that factoring of n (and hence computing $\phi(n)$ for performing the above steps) will be much more difficult than performing t squarings. Under this condition we do not know any other method which, without using the factorisation of n, can compute

S. Vaudenay and A. Youssef (Eds.): SAC 2001, LNCS 2259, pp. 342–357, 2001.

a^{2^t} (mod n) in time less than t squarings. Moreover, because each squaring can only be performed on the result of the previous squaring, it is not known how to speedup the t squarings via parallelisation of multiple processors. Parallelisation of each squaring step cannot achieve a great deal of speedup since a squaring step only needs a trivial computational resource and so any non-trivial scale of parallelisation of a squaring step is likely to be penalised by communication delays among the processors.

These properties suggest that the following language (notice that each element in the language associates a non-secret natural number t)

$$L(a, n) = \{ a^{2^t} \ (\text{mod } n) \mid \gcd(a, n) = 1, \ t = 1, 2, ..., \} \tag{4}$$

forms a good candidate for the realisation of timed-release crypto problems. Rivest, Shamir and Wagner pioneered the use of this language in a time-lock puzzle scheme [11]. In their scheme a puzzle is a triple (t, a, n) and the instruction for finding its solution is to perform t squarings mod n starting from a which leads to a^{2^t} (mod n). A puzzle maker, with the factorisation knowledge of n, can construct a puzzle efficiently using the steps in (2) and (3) and can fine tune the difficulty for finding the solution by choosing t in a vast range. For instance, the MIT Laboratory for Computer Science (LCS) has implemented the time-lock puzzle of Rivest et al into "The LCS35 Time Capsule Crypto-Puzzle" and started its solving routine on 4th April 1999. It is estimated that the solution to the LCS35 Time Capsule Crypto-Puzzle will be found in 35 years from 1999, or on the 70 years from the inception of the MIT-LCS [10]. (Though we will discuss a problem of this puzzle in §1.2.)

1.1   Applications

Boneh and Naor used a sub-language of $L(a, n)$ (details to be discussed in §1.2) and constructed a timed-release crypto primitive which they called "timed commitments" [3]. Besides several suggested applications they suggested an interesting use of their primitive for solving a long-standing problem in fair contract signing. A previous solution (due to Damgård [6]) for fair contract signing between two remote and mutually distrusted parties is to let them exchange signatures of a contract via gradual release of a secret. A major drawback with that solution is that it only provides a *weak fairness*. Let us describe this weakness by using, for example, a discrete-logarithm based signature scheme. A signature being gradually released relates to a series of discrete logarithm problems with the discrete logarithm values having gradually decreasing magnitudes. Sooner or later before the two parties completes their exchange, one of them may find himself in a position of extracting a discrete logarithm which is sufficiently small with respect to his computational resource. It is well-known (e.g., the work of van Oorschot and Wiener on the parallelised rho method [13]) that parallelisation is effective for extracting small discrete logarithms. So the resourceful party (one who is able to afford vast parallelisation) can abort the exchange at that point and wins an advanced position unfairly. Boneh and Naor suggested to seal signatures under exchange using elements in $L(a, n)$. Recall the aforementioned non-parallelisable property for re-constructing the elements in $L(a, n)$, a roughly equal time can be imposed for both parties to open the sealed signatures regardless of their difference (maybe vast) in computing resources. In this way, they argued that *strong fairness* for contract signing can be achieved.

Rivest et al suggested several other applications of timed-release cryptography [11]:

- A bidder in an auction wants to seal his bid so that it can only be opened after the bidding period is closed.
- A homeowner wants to give his mortgage holder a series of encrypted mortgage payments. These might be encrypted digital cash with different decryption dates, so that one payment becomes decryptable (and thus usable by the bank) at the beginning of each successive month.
- A key-escrow scheme can be based on timed-release crypto, so that the government can get the message keys, but only after a fixed, pre-determined period.
- An individual wants to encrypt his diaries so that they are only decryptable after fifty years (when the individual may have forgotten the decryption key).

1.2   Previous Work and Unsolved Problem

With the nice properties of $L(a, n)$ we are only half way to the realisation of timed-release cryptography. In most imaginable applications where timed-release crypto may play a role, it is necessary for a problem constructor to prove (ideally in zero-knowledge) the correct construction of the problem. For example, without a correctness proof, the strong fairness property of the fair-exchange application is actually absent.

From the problem's membership in NP we know that there exists a zero-knowledge proof for a membership assertion regarding language $L(a, n)$. Such a proof can be constructed via a general method (e.g., the work of Goldreich et al [9]). However, the performance of a zero-knowledge proof in a general construction is not suitable for practical use. By the performance for practical use we mean an efficiency measured by a *small* polynomial in some typical parameters (e.g., the bit length of n). To our knowledge, there exists no practically efficient zero-knowledge protocols for proving the general case of the membership in $L(a, n)$.

Boneh and Naor constructed a practically efficient protocol for proving membership in a sub-language of $L(a, n)$ where $t = 2^k$ with k being any natural number. The time control that the elements in this sub-language can offer has the granularity 2. We know that the time complexity in bit operation for performing one squaring modulo n can be expressed by the lowest known result of $c \cdot \log n \cdot \log \log n$ (where $c > 1$ is a machine dependent value, a faster machine has a smaller c) if FFT (fast Fourier transform) is used for the implementation of squaring. Thus, the time complexity for computing elements in this sub-language is the step function

$$2^k \cdot c \cdot \log n \cdot \log \log n$$

which has a fast increasing step when k gets large. Boneh and Naor envisioned $k \in [30, ..., 50]$ for typical cases in applications. While it is evident that k decreasing from 30 downwards will quickly trivialise a timed-release crypto problem as 2^{30} is already at the level of a small polynomial in the secure bit length of n (usually 2^{10}), a k increasing from 30 upwards will harden the problem in such increasingly giant steps that imaginable services (e.g., the strong fairness for gradual disclosure of secret proposed in [3]) will quickly become unattractive or unusable. Taking the LCS35 Time Capsule for example, let the 35-year-opening-time capsule be in that sub-language (so the correctness can be efficiently proved with the protocol in [3]), then the only other elements in that sub-language with opening times close to 35 years will be 17.5 years and 70 years. We should notice that there is no hope to try to tune the size of n as a means of tuning the time complexity since changing $c \cdot \log n \cdot \log \log n$ will have little impact on the above giant step function.

Boneh and Naor expressed a desire for a finer time-control ratio than 2 and sketched a method to obtain a finer ratio with $t_0 = 1$ and $t_i = t_{i-1} + t_{i-2}$ for $i = 1, ..., k$. This method of reducing the ratio renders the ratio being bounded below by $\alpha = \frac{1+\sqrt{5}}{2}$ (≈ 1.618) while increasing the number of proof rounds from k to $\log_\alpha k$. They further mentioned that smaller values can be obtained by other such recurrences. It seems to us that if some recurrence method similar to above is used, then with ratio $\to 1$ (1 is the ideal ratio and will be that for our case), the number of proof rounds $\log_{\text{ratio}} k \to \infty$. So their suggested methods for reducing the time-control ratio are not practical for obtaining a desirable ratio.

The Time-Lock-Puzzle work of Rivest et al [11] did not provide a method for proving the correct construction of a timed-release crypto problem.

1.3   Our Work

We construct the first practically efficient zero-knowledge proof protocol for demonstrating the membership in $L(a, n)$ which runs in $\log_2 t$ steps, each an exponentiation modulo n, or $O(\log_2 t \cdot (\log_2 n)^3)$ bit operations in total (without using FFT). This efficiency suits practical uses. The membership demonstration can be conducted in terms of $(a^e)^{2^t} \pmod{n} \in L(a^e, n)$ on given a, a^e and t, where e is an RSA encryption exponent. Then we are able to provide two timed-release crypto primitives, one for timed release of a message, and the other for timed release of an RSA signature. In the former, a message M can be sealed in $a^{2^t} M \pmod{n}$, and the established membership asserts that the correct decryption of the RSA ciphertext $M^e \pmod{n}$ can be obtained by performing t squarings modulo n starting from a. The latter primitive can be constructed analogously.

Our schemes provide general methods for the use of timed-release cryptography.

1.4   Organisation

In the next section we agree on the notation to be used in the paper. In Section 3 we construct general methods for timed-release cryptography based on proven membership in $L(a, n)$. In Section 4 we construct our membership proof protocol working with an RSA modulus of a safe-prime structure. In Section 5 we will discuss how to generalise our result to working with a general form of composite modulus.

2   Notation

Throughout the paper we use the following notation. $\mathbb{Z}_n$ denotes the ring of integers modulo n. $\mathbb{Z}_n^*$ denotes the multiplicative group of integers modulo n. $\phi(n)$ denotes Euler's phi function of n, which is the order, i.e., the number of elements, of the group $\mathbb{Z}_n^*$. For an element $a \in \mathbb{Z}_n^*$, $\mathrm{ord}_n(a)$ denotes the multiplicative order modulo n of a, which is the least index i satisfying $a^i \equiv 1 \pmod n$; $\langle a \rangle$ denotes the subgroup generated by a; $\left(\frac{x}{n}\right)$ denotes the Jacobi symbol of x mod n. We denote by $J_+(n)$ the subset of $\mathbb{Z}_n^*$ containing the elements of the positive Jacobi symbol. For integers a, b, we denote by $\gcd(a, b)$ the greatest common divisor of a and b. For a real number r, we denote by $\lfloor r \rfloor$ the floor of r, i.e., r rounded down to the nearest integer.

3   Timed-Release Crypto with Proven Membership in $L(a, n)$

Let Alice be the constructor of a timed-release crypto problem. She begins with constructing a composite natural number $n = pq$ where p and q are two distinct odd prime numbers. Define

$$a(t) \stackrel{\text{def}}{=} a^{2^t} \pmod n, \tag{5}$$

$$a^e(t) \stackrel{\text{def}}{=} (a(t))^e \pmod n, \tag{6}$$

where e is a fixed natural number relatively prime to $\phi(n)$ (in the position of an RSA public exponent), and $a \not\equiv \pm 1 \pmod n$ is a random element in $\mathbb{Z}_n^*$. Alice can construct $a(t)$ using the steps in (2) and (3).

The following security requirements should be in place: n should be so constructed that $\mathrm{ord}_{\phi(n)}(2)$ is sufficiently large, and a should be so chosen that $\mathrm{ord}_n(a)$ is sufficiently large. Here, "sufficiently large" means "much larger than t" for the largest possible t that the system should accommodate.

In the remainder of this section, we assume that Alice has proven to Bob, the verifier, the following membership status (using the protocol in §4):

$$a^e(t) \in L(a^e, n). \tag{7}$$

Clearly, with e co-prime to $\phi(n)$, this is equivalent to another membership status:

$$a(t) \in L(a, n).$$

However in the latter case $a(t)$ is (temporarily) unavailable to Bob due to the difficulty of extracting the e-th root (of $a^e(t)$) in the RSA group.

3.1   Timed-Release of an Encrypted Message

For message $M < n$, to make it decryptable in time t, Alice can construct a "timed encryption":

$$TE(M, t) \stackrel{\text{def}}{=} a(t)M \ (\text{mod } n). \tag{8}$$

Let Bob be given the tuple $(TE(M, t), a^e(t), e, a, t, n)$ where $a^e(t)$ is constructed in (5) and (6) and has the membership status in (7) proven by Alice. Then from the relation

$$TE(M, t)^e \equiv a^e(t)M^e \ (\text{mod } n), \tag{9}$$

Bob is assured that the plaintext corresponding to the RSA ciphertext $M^e \ (\text{mod } n)$ can be obtained from $TE(M, t)$ by performing t squarings modulo n starting from a. We should note that in this encryption scheme, Alice is the sender and Bob, the recipient; so if Alice wants the message to be timed-release to Bob exclusively then she should send a to Bob exclusively, e.g., via a confidential channel.

Remark 1. As in the case of any practical public-key encryption scheme, M in (8) should be randomised using a proper plaintext randomisation scheme designed for providing the semantic security (e.g., the OAEP scheme for RSA [7]).

3.2   Timed-Release of an RSA Signature

Let e, n be as above and d satisfy $ed \equiv 1 \ (\text{mod } \phi(n))$ (so d is in the position of an RSA signing exponent). For message $M < n$ (see Remark 2 below), to make its RSA signature $M^d \ (\text{mod } n)$ releasable in time t, Alice can construct a "timed signature":

$$TS(M, t) \stackrel{\text{def}}{=} a(t)M^d \ (\text{mod } n). \tag{10}$$

Let Bob be given the tuple $(M, TS(M, t), a^e(t), e, a, t, n)$ where $a^e(t)$ is constructed in (5) and (6) and has the membership status in (7) proven by Alice. Then from the relation

$$TS(M, t)^e \equiv a^e(t)M \ (\text{mod } n), \tag{11}$$

Bob is assured that the RSA signature on M can be obtained from $TS(M, t)$ by performing t squarings modulo n starting from a.

Remark 2. As in the case of a practical digital signature scheme, in order to prevent existential forgery of a signature, M in (10) should denote an output from a cryptographically secure one-way hash function. If we further require the signature to an indistinguishability property (see §3.3), then the hashed result should be in $J_+(n)$. Padding M with a random string and then hashing, the probability for the hashed result in $J_+(n)$ is 0.5.

3.3   Security Analysis

Confidentiality of M in $TE(M, t)$ We assume that Alice has implemented properly our security requirements on the large magnitudes of $\mathrm{ord}_{\phi(n)}(2)$ and $\mathrm{ord}_n(a)$. Then we observe that $L(a, n)$ is a large subset of the quadratic residues modulo n, and the mapping $a \mapsto a(t)$ is one-way under the appropriate intractability assumption (here, integer factorisation). Consequently, our scheme for encrypting $M \in \mathbb{Z}_n^*$ in $TE(M, t)$ is a trapdoor one-way permutation since it is the multiplication, modulo n, of the message M to the trapdoor secret $a(t)$. In fact, from (9) we see that the availability of $TE(M, t)^e$ and $a^e(t)$ makes M^e available, and so without considering to go through t squarings, the underlying intractability of $TE(M, t)$ is reduced to that of RSA. Therefore, well-known plaintext randomisation schemes for RSA encryption (e.g., OAEP [7]), which have been proposed for achieving the semantic security (against adaptive chosen ciphertext attacks) can be applied to our plaintext message before the application of the permutation. The message confidentiality properties (i.e., the indistinguishability and non-malleability on the message M) of our timed-release encryption scheme should follow directly those of RSA-OAEP.

Thus, given the difficulty of extracting the e-th root of a random element modulo n, a successful extraction of $a(t)$ from $a^e(t)$, or of some information regarding M from $TE(M, t)$, will constitute a grand breakthrough in the area if they are done at a cost less than t squarings modulo n.

Unforgeability of M^d in $TS(M, t)$ First, recall that M here denotes an output from a cryptographically secure one-way hash function before signing in the RSA fashion. The unforgeability of M^d in $TS(M, t)$ follows directly that of $M^d \pmod{n}$ given in clear.

Secondly, the randomness of $a^e(t)$ ensures that of $TS(M, t)^e$. Thus the availability of the pair $(TS(M, t), TS(M, t)^e)$ does not constitute a valid signature of Alice on anything (such as on an adaptively chosen message). The availability of the pair $(TS(M, t), TS(M, t)^e)$ is equivalent to that of (x, x^e) which can be constructed by anybody using a random x.

Indistinguishability of M^d in $TS(M, t)$ The indistinguishability is the following property: with the timed-release signature $TS(M, t)$ on M and with the proven membership $a^e(t) \in L(a^e, n)$ but without going through t squarings mod n, one should not be able to tell whether $TS(M, t)$ has any verifiable relationship with a signature on M. This property should hold even if the signature pair (M, M^d) becomes available; namely, even if Bob has recovered the signature pair (M, M^d) (e.g., after having performed t squarings), he is still not able to convince a third party that $TS(M, t)$ is a timed-release signature of Alice on M. This property is shown below.

Let $\tilde{M} \in J_+(n)$ be any message of Bob's choice (e.g., Bob may have chosen it because $\tilde{M}^d$ may be available to him from a different context). We have

$$TS(M, t) \equiv a(t)M^d \equiv a(t) \left(\frac{M}{\tilde{M}} \right)^d \tilde{M}^d \equiv \tilde{a}\tilde{M}^d \pmod{n}.$$

So upon seeing Bob's allegation on a "verifiable relationship" between $TS(M, t)$ and M^d, the third party faces a problem of deciding which of M^d or $\tilde{M}^d$ is sealed in $TS(M, t)$. This boils down to deciding if $a(t) \in L(a, n)$ or if $\tilde{a} \in L(a, n)$ (both are in $J_+(n)$), which is still a problem of going through t squarings. Thus, even though the availability of M^d and $\tilde{M}^d$ does allow one to recognise that the both are in fact Alice's valid signatures, without verifying the membership status, one is unable to tell if any of the two has any connection with $TS(M, t)$ at all.

4   Membership Proof with Modulus of a Safe-Prime Structure

Let Alice have constructed her RSA modulus n with a safe-prime structure. This requires $n = pq$, $p' = (p - 1)/2$, $q' = (q - 1)/2$ where p, q, p' and q' are all distinct primes of roughly equal size. We assume that Alice has proven to Bob in zero-knowledge such a structure of n. This can be achieved via using, e.g., the protocol of Camenisch and Michels [4].[1]

Let $a \in \mathbb{Z}_n^*$ satisfy

$$\gcd(a \pm 1, n) = 1, \tag{12}$$

$$\left(\frac{a}{n} \right) = -1. \tag{13}$$

It is elementary to show that a satisfying (12) and (13) has the full order $2p'q'$. The following lemma observes a property of a.

Lemma 1. *Let n be an RSA modulus of a safe-prime structure and $a \in \mathbb{Z}_n^*$ of the full order. Then for any $x \in \mathbb{Z}_n^*$, either $x \in \langle a \rangle$ or $-x \in \langle a \rangle$.*

Proof It's easy to check $-1 \notin \langle a \rangle$. So $\langle a \rangle$ and the coset $(-1)\langle a \rangle$ both have the half the size of $\mathbb{Z}_n^*$, yielding $\mathbb{Z}_n^* = \langle a \rangle \cup (-1)\langle a \rangle$. Any $x \in \mathbb{Z}_n^*$ is either in $\langle a \rangle$ or in $(-1)\langle a \rangle$. The latter case means $-x \in \langle a \rangle$. $\square$

---

[1] Due to the current difficulty of zero-knowledge proof for a safe-prime-structured RSA modulus, we recommend to use the method in Section 5 which works with a general form of composite modulus. The role of Section 4 is to serve a clear exposition on how we solve the current problem in timed-release cryptography.

4.1   A Building Block Protocol

Let Alice and Bob have agreed on n (this is based on Bob's satisfaction on Alice's proof that n has a safe-prime structure).

Figure 1 specifies a perfect zero-knowledge protocol (SQ) for Alice to prove that for $a, x, y \in \mathbb{Z}_n^*$ with n of a safe-prime structure, a of the full order, and $x, y \in J_+(n)$, they satisfy (note, $\pm$ below means either $+$ or $-$, but not both)

$$\exists z : \quad x \equiv \pm a^z \pmod{n}, \qquad y \equiv \pm a^{z^2} \pmod{n}. \tag{14}$$

Alice should of course have constructed a, x, y to satisfy (14). She sends a, x, y to Bob.

Bob (has checked n of a safe-prime structure) should first check (12) and (13) on a for its full-order property (the check guarantees $a \not\equiv \pm 1 \pmod{n}$); he should also check $x, y \in J_+(n)$.

$SQ(a, x, y, n)$

Input Common: n: an RSA modulus with a safe-prime structure;
$\qquad\qquad\qquad a \in \mathbb{Z}_n^*$: an element of the full-order $2p'q' = \phi(n)/2$
$\qquad\qquad\qquad$ (so $a \not\equiv \pm 1 \pmod{n}$);
$\qquad\qquad\qquad x, y \in J_+(n)$: $x \not\equiv \pm y \pmod{n}$;
$\qquad$ Alice: z: $x \equiv \pm a^z \pmod{n}$, $y \equiv \pm a^{z^2} \pmod{n}$;

1. Bob chooses at random $r < n$, $s < n$ and sends to Alice: $C \stackrel{\text{def}}{=} a^r x^s \pmod{n}$;
2. Alice sends to Bob: $R \stackrel{\text{def}}{=} C^z \pmod{n}$;
3. Bob accepts if $R \equiv \pm x^r y^s \pmod{n}$, or rejects otherwise.

Fig. 1. Building Block Protocol

Remark 3. For ease of exposition this protocol appears in a non zero-knowledge format. However, the zero-knowledge property can be added to it using the notion of a commitment function: Instead of Alice sending R in Step 2, she sends a commitment $commit(R)$, after which Bob reveals r and s; this allows Alice to check the correct formation of C; the correct formation means that Bob has already known Alice's response.

Theorem 1. *Let a, x, y, n be as specified in the common input in Protocol SQ. The protocol has the following properties:*
Completeness *If the common input satisfies (14) then Bob will always accept Alice's proof;*

Soundness *If (14) does not hold for the common input, then Alice, even computationally unbounded, cannot convince Bob to accept her proof with probability greater than* $\frac{2p'+2q'-1}{2p'q'}$.[2]

Zero-knowledge *Bob gains no information about Alice's private input.*

Proof

Completeness Evident from inspection of the protocol.

Soundness Suppose that (14) does not hold for the common input (a, x, y, n) (here $x, y \in J_+(n)$) whereas Bob has accepted Alice's proof. By Lemma 1, the first congruence of (14) always holds for some $z = \log_a \pm x$. So it is the second congruence of (14) that does not hold for the same z. Let $\xi \in \mathbb{Z}_n^*$ satisfy

$$y \equiv \xi a^{z^2} \pmod{n} \text{ with } \xi \neq \pm 1. \tag{15}$$

Since Bob accepts the proof, he sees the following two congruences

$$C \equiv a^r x^s \pmod{n}, \tag{16}$$

$$R \equiv \pm x^r y^s \pmod{n}. \tag{17}$$

Since (16) implies

$$C^2 \equiv a^{2r} x^{2s} \pmod{n},$$

and by Lemma 1, both $\log_a C^2$ and $\log_a x^2$ $(= \log_a(\pm x)^2 = 2z)$ exist, we can write the following linear congruence with r and s as unknowns

$$\log_a C^2 \equiv 2r + 2zs \pmod{2p'q'}.$$

For $s = 1, 2, \cdots, 2p'q'$, this linear congruence yields $r = \frac{\log_a C^2 - 2zs}{2} \pmod{2p'q'}$. Therefore there exists exactly $2p'q'$ pairs of (r, s) to satisfy (16) for any fixed C (and the fixed a, x). Each of these pairs and the fixed x, y will yield an R from (17). Below we argue that for any two such pairs, denoted by (r, s) and (r', s'), if $\gcd(s - s', 2p'q') \leq 2$ then they must yield $R \not\equiv \pm R' \pmod{n}$. Suppose on the contrary for

$$a^r x^s \equiv C \equiv a^{r'} x^{s'} \pmod{n}, \quad \text{i.e.,} \quad a^{r-r'} \equiv x^{s'-s} \pmod{n}, \tag{18}$$

it also holds

$$x^r y^s \equiv R \equiv \pm R' \equiv \pm x^{r'} y^{s'} \pmod{n}, \quad \text{i.e.,} \quad x^{r-r'} \equiv \pm y^{s'-s} \pmod{n}. \tag{19}$$

Using the second congruence in (18), noticing $x \equiv \pm a^z$ and (15), we can transform the second congruence in (19) to

$$(\pm 1)^{[r-r'+z(s'-s)]} a^{[z^2(s'-s)]} \equiv x^{r-r'} \equiv \pm y^{s'-s} \equiv \pm \xi^{(s'-s)} a^{[z^2(s'-s)]} \pmod{n},$$

---

[2] The safe-prime structure of n implies $p' \approx q' \approx \sqrt{n}$ and hence this probability value is approximately $2/\sqrt{n}$.

which yields

$$\pm \xi^{(s'-s)} \equiv (\pm 1)^{[r-r'+z(s'-s)]} \equiv \pm 1, \quad \text{i.e.,} \quad \xi^{2(s'-s)} \equiv 1 \pmod{n}. \qquad (20)$$

Recall that $\xi \neq \pm 1$ and $y \equiv \xi a^{z^2} \equiv \pm \xi x^z \pmod{n}$ with $x, y \in J_+(n)$, we know $\text{ord}_n(\xi) \neq 2$ (i.e., ξ cannot be any square root of 1, since the two roots $\neq \pm 1$ will render $y \notin J_+(n)$). Thus, $\text{ord}_n(\xi)$ must be a multiple of p' or q' or both. However, we have assumed $\gcd(s' - s, 2p'q') \leq 2$, i.e., $\gcd(2(s' - s), 2p'q') = 2$, so $2(s' - s)$ cannot be such a multiple. Consequently (20) cannot hold and we reach a contradiction.

For any $s \leq 2p'q'$, it's routine to check that there are $2p' + 2q' - 2$ cases of s' satisfying $\gcd(2(s' - s), 2p'q') > 2$. Thus, if (14) does not hold, amongst $2p'q'$ possible R's matching the challenge C, there are in total $2p' + 2q' - 1$ of them (matching the s itself and the $2p' + 2q' - 2$ other s's) that may collide to Bob's fixing of R. Even computationally unbounded, Alice will have at best $\frac{2p'+2q'-1}{2p'q'}$ probability to have responded with a correct R.

Zero-Knowledge Immediate (see Remark 3). $\qquad \qquad \qquad \Box$

4.2 Proof of Membership in $L(a, n)$

For $t \geq 1$, we can express 2^t as

$$2^t = \begin{cases} 2^{[2 \cdot (t/2)]} = [2^{(t/2)}]^2 & \text{if } t \text{ is even} \\ 2^{[2 \cdot (t-1)/2+1]} = [2^{(t-1)/2}]^2 \cdot 2 & \text{if } t \text{ is odd} \end{cases}$$

Copying this expression to the exponent position of $a^{2^t} \pmod{n}$, we can express

$$a^{2^t} \pmod{n} \equiv \begin{cases} a^{[2^{(t/2)}]^2} & \text{if } t \text{ is even} \\ (a^{[2^{(t-1)/2}]^2})^2 & \text{if } t \text{ is odd} \end{cases} \qquad (21)$$

In (21) we see that the exponent 2^t can be expressed as the square of another power of 2 with t being halved in the latter. This observation suggests that repeatedly using SQ, we can demonstrate, in $\lfloor \log_2 t \rfloor$ steps, that the discrete logarithm of an element is of the form 2^t. This observation translates precisely into the protocol specified in Figure 2 which will terminate within $\lfloor \log_2 t \rfloor$ steps and prove the correct structure of $a(t)$. The protocol is presented in three columns: the actions in the left column are performed by Alice, those in the right column, by Bob, and those in the middle, by the both parties.

A run of $Membership(a, t, a(t), n)$ will terminate within $\lfloor \log_2 t \rfloor$ loops, and this is the completeness property. The zero-knowledge property follows that of SQ (also note Remark 4(ii) below). We only have to show the soundness property.

Theorem 2. *Let $n = (2p' + 1)(2q' + 1)$ be an RSA modulus of a safe-prime structure, $a \in \mathbb{Z}_n^*$ be of the full order $2p'q'$, and $t > 1$. Upon acceptance termination of $Membership(a, t, a(t), n)$, relation $a(t) \equiv \pm a^{2^t} \pmod{n}$ holds with probability greater than*

$$1 - \frac{\lfloor \log_2 t \rfloor (2p' + 2q' - 1)}{2p'q'}.$$

$Membership(a, t, a(t), n)$

Abort and reject if any checking by Bob fails, or accept upon termination.

Alice	both	Bob
	$u \overset{\text{def}}{=} a(t);$	$u \overset{?}{\in} J_+(n);\ a \overset{?}{\not\equiv} \pm u \pmod{n}$

While $t > 1$ do

$\left\{\begin{array}{l}
y \overset{\text{def}}{=} u; \\
\text{if } t \text{ is odd: } y \overset{\text{def}}{=} a(t-1); \\
x \overset{\text{def}}{=} a(\lfloor t/2 \rfloor); \\
\text{Sends } x, y \text{ to Bob;}
\end{array}\right.$

Bob side:

Receives x, y from Alice;

$x, y \overset{?}{\in} J_+(n);$

if t is odd: $y^2 \overset{?}{\equiv} u \pmod{n};$

if t is even: $y \overset{?}{\equiv} u \pmod{n};$

both:

$SQ(a, x, y, n);$

$u \overset{\text{def}}{=} x;$

$t \overset{\text{def}}{=} \lfloor t/2 \rfloor;$

When $t = 1$:

$u \overset{?}{\equiv} a^2 \pmod{n};$

Fig. 2. Membership Proof Protocol

Proof Denote by $SQ(a, x_1, y_1, n)$ and by $SQ(a, x_2, y_2, n)$ any two consecutive acceptance calls of SQ in $Membership$ (so in the first call, $y_1 = a(t)$ if t is even, or $y_1 = a(t-1)$ if t is odd; and in the last call, $x_2 = a^2$). When $t > 1$, such two calls prove that there exists z:

$$x_2 \equiv \pm a^z \pmod{n}, \quad y_2 \equiv \pm a^{z^2} \pmod{n}, \tag{22}$$

and either

$$x_1 = y_2 \equiv \pm a^{z^2} \pmod{n}, \quad y_1 \equiv \pm a^{z^4} \pmod{n}, \tag{23}$$

or

$$x_1 = y_2^2 \equiv a^{2z^2} \pmod{n}, \quad y_1 \equiv \pm a^{4z^4} \pmod{n}. \tag{24}$$

Upon $t = 1$, Bob further sees that $x_2 = a^2$. By induction, the exponents z (resp. z^2, z^4, $2z^2$, $4z^4$) in all cases of $\pm a^z$ (resp. $\pm a^{z^2}$, $\cdots$) in (22), (23) or (24) contain a single factor: 2. So we can write $a(t) = \pm a^{2^u} \pmod{n}$ for some natural number u.

Further note that each call of SQ causes an effect of having 2^u square-rooted in the integers which is equivalent to having u halved in the integers. Thus, exactly $\lfloor \log_2 u \rfloor$ calls (and no more) of SQ can be made. But Bob has counted $\lfloor \log_2 t \rfloor$ calls of SQ, therefore $u = t$.

Each acceptance call of SQ has the correctness probability of $1 - \frac{2p'+2q'-1}{2p'q'}$. So after $\lfloor \log_2 t \rfloor$ acceptance calls of SQ, the probability for $Membership$ to be correct is

$$(1 - \frac{2p' + 2q' - 1}{2p'q'})^{\lfloor \log_2 t \rfloor} > 1 - \frac{\lfloor \log_2 t \rfloor (2p' + 2q' - 1)}{2p'q'}. \qquad \square$$

Remark 4.

i) An acceptance run of $Membership(a, t, a(t), n)$ proves $\pm a(t) \in L(a, n)$, or $a^2(t) = a(t+1) \in L(a, n)$.

ii) It is obvious that by preparing all the intermediate values in advance, Protocol $Membership$ can be run in parallel to save the $\lfloor \log_2 t \rfloor$ rounds of interactions. This way of parallelisation should not be confused with another common method for parallelising a proof of knowledge protocol using a hash function to create challenge bits (which turns the proof publicly verifiable). Our parallelisation does not damage the zero-knowledge property.

iii) In most applications, $a(t)$ is the very number (solution to a puzzle) that should not be disclosed to Bob during the proof time. In such a situation, Alice should choose t to be even and render $a(t-1)$ to be the solution to a puzzle. Then a proof of $Membership(a, t, a(t), n)$ will not disclose $a(t-1)$. Note that such a proof does disclose to Bob $a(\lfloor t/2 \rfloor)$ which provides Bob with a complexity of $\lfloor t/2 \rfloor - 1$ squarings to reach $a(t-1)$. To compensate the loss of computation, proof of $Membership(a, 2t, a(2t), n)$ is necessary. Consequently, the proof runs one loop more than $Membership(a, t, a(t), n)$ does. Note that the above precautions are unnecessary for our applications in §3 where it is the e-th root of $a^e(t)$ that is the puzzle's solution; the disclosures of $a^e(t)$ or $a^e(\lfloor t/2 \rfloor)$ do not seem to reduce the time complexity for finding $a(t)$.

4.3   Performance

In each run of SQ, Alice (resp. Bob) performs one (resp. four) exponentiation(s) mod n. So in $Membership(a, t, a(t), n)$ Alice (resp. Bob) will perform $\lfloor \log_2 t \rfloor$ (resp. $4\lfloor \log_2 t \rfloor$) exponentiations mod n. These translate to $O(\lfloor \log_2 t \rfloor (\log_2 n)^3)$ bit operations.

In the LCS35 Time Capsule Crypto-Puzzle [10], $t = 79685186856218$ is a 47-bit binary number. Thus the verification for that puzzle can be completed within $4 \times 47 = 188$ exponentiations mod n.

The number of bits to be exchanged is measured by $O((\lfloor \log_2 t \rfloor)(\log_2 n))$.

5   Use of Modulus of a General Form

When n does not have a safe-prime structure, the error probability of SQ can be much larger than what we have measured in Theorem 1. The general method for Alice to introduce an error in her proof (i.e., to cheat) is to fix y in (15) with some $\xi \neq \pm 1$. For y so fixed before Bob's challenge C, Bob is actually awaiting for $R \equiv \pm C^z \xi^s \pmod{n}$ in which $\xi^s \pmod{n}$ is the only value that

Alice does not know (she does not know it because Bob's random choice of s is perfectly hidden in C). Therefore in order to respond with the correct R, it is both necessary and sufficient for Alice to guess $\xi^s \pmod{n}$ correctly. Notice that while it is unnecessary and can be too difficult for Alice to guess s, guessing $\xi^s \pmod{n}$ need not be very difficult and the probability of a correct guess is bounded by $\frac{1}{\text{ord}_n(\xi)}$. Thus, in order for Alice to achieve a large error probability (meaning, to ease her cheating), she should use ξ of a small order.

The above cheating scenario provides the easiest method for Alice to cheat and yet is general enough for covering the cases that the soundness of SQ should consider. Multiplying both x and y with some small-order elements will only make the cheating job more difficult. Therefore it suffices for us to anticipate the above general cheating method.

To this end it becomes apparent that in order to limit Alice's cheating probability we should prevent her from constructing y in (15) using ξ of a small order. Using a safe-prime-structured modulus $n = (2p' + 1)(2q' + 1)$ achieves this purpose exactly because then the least order available to Alice is $\min(\frac{1}{p'}, \frac{1}{q'})$ which is satisfiably small (using ξ of order 2 either does not constitute an attack, or will cause detection of $y \notin J_+(n)$).

While a zero-knowledge proof of n being in a safe-prime structure is computationally inefficient to date, it is rather easy to construct a zero-knowledge proof protocol for proving that $\phi(n)$ is free of small odd prime factors up to a bound B. Boyar et al [2] constructed a practically efficient zero-knowledge proof protocol for proving that $\phi(n)$ is relatively prime to n. As in [8], we can apply the same idea to prove that $\phi(n)$ is relatively prime to Δ (i.e., using Δ in place of n) where

$$\Delta = \prod_{\substack{\text{primes } \ell : \\ 2 < \ell < B}} \ell. \tag{25}$$

Supposing that n is a Blum integer (which can be efficiently proved using, e.g., the protocol of van de Graaf and Peralta [12]), then after applying the protocol of Boyar et al using Δ in (25) in place of n, we can be sure that the error probability of SQ is bounded by B^{-1}. Notice that the multiplication attack using the square roots of 1 with the negative Jacobi symbol (in place of ξ in (15)) is not possible since that will be detected by the Jacobi symbol checking conducted on the input values. Thus, if Alice is required to repeat running SQ $\frac{k}{\log_2 B}$ times, then Bob is sure that her cheating probability (i.e., for (14) not to hold) is bounded by 2^{-k}.

5.1   Performance of Membership Proof Using General Form of Modulus

With the soundness probability of SQ bounded by B^{-1}, for each case of x, y, $SQ(a, x, y, n)$ need to be run $\frac{k}{\log_2 B}$ times to achieve an acceptable soundness probability 2^{-k}. Thus in *Membership*, SQ is run $\frac{\lfloor \log_2 t \rfloor k}{\log_2 B}$ times. Since in each run of SQ, Alice (resp. Bob) performs one (resp. four) exponentiation(s) mod n. So in $Membership(a, t, a(t), n)$ Alice (resp. Bob) will perform $\frac{\lfloor \log_2 t \rfloor k}{\log_2 B}$ (resp. $\frac{4\lfloor \log_2 t \rfloor k}{\log_2 B}$ exponentiations mod n. Adding to this is the cost for running k times the protocol of Boyar et al, each run of that protocol costs one modulo exponentiation for both parties. Thus, the total cost in number of exponentiations mod n of the membership proof for Alice is

$$\frac{\lfloor \log_2 t \rfloor k}{\log_2 B} + k,$$

and that for Bob is

$$\frac{4\lfloor \log_2 t \rfloor k}{\log_2 B} + k.$$

In the LCS35 Time Capsule Crypto-Puzzle [10] where $\lfloor \log_2 t \rfloor = 47$, if we consider $B = 2^{10}$ and $k = 100$, then the quantity for Alice is 570 and that for Bob is 1980. Therefore, the LCS35 Time Capsule Crypto-Puzzle using a general-form modulus (Blum integer) can be verified with 1980 modulo exponentiations.

Zero-knowledge proof of a Blum integer using the protocol in [12] has a performance similar to one modulo exponentiation for Alice; the workload of that protocol for Bob is trivial since it only involves multiplications and evaluations of Jacobi symbols. Thus, considering the same low soundness probability of 2^{-100}, we should add 100 modulo exponentiations to Alice's workload to reach 670 modulo exponentiations.

6   Conclusion

We have constructed an efficient zero-knowledge protocol for providing general solutions to timed-release cryptographic problems (encryption and signature). These schemes have proven correctness on time control which can be fine tuned to the granularity in number of multiplications.

Successful timed-release cryptographic problems have been constructed upon the integer-factoring based intractability. An important feature that such intractability offers is non-parallelisability. An open question is that can other intractability offer this feature? (We know that the problem of extraction of discrete logarithm can be parallelised [13].)

Acknowledgments

I would like to thank Steven Galbraith, Kenny Paterson, David Soldera and the anonymous referees of SAC'01 for their helpful comments on a draft of this paper.

References

1. Blum, M. Coin Flipping by Telephone: A Protocol for Solving Impossible Problems, Proceedings of the 24th IEEE Computer Conference, pages 133–137, 1981.
2. Boyar, J., Friedl, K. and Lund, C. Practical zero-knowledge proofs: Giving hints and using deficiencies, Advances in Cryptology — Proceedings of EUROCRYPT 89 (J.-J. Quisquater and J. Vandewalle, eds.), Lecture Notes in Computer Science 434, Springer-Verlag 1990, pages 155–172.
3. Boneh, D. and Naor, M. Timed commitments (extended abstract), Advances in Cryptology: Proceedings of CRYPTO'00, Lecture Notes in Computer Science 1880, Springer-Verlag 2000, pages 236–254.
4. Camenisch J. and Michels, M. Proving in zero-knowledge that a number is the product of two safe primes, In Advances in Cryptology — EUROCRYPT 99 (J. Stern ed.), Lecture Notes in Computer Science 1592, Springer-Verlag 1999, pages 106–121.
5. Chaum, D. Zero-knowledge undeniable signatures, Advances in Cryptology: Proceedings of CRYPTO 90 (I.B. Damgaard, ed.) Lecture Notes in Computer Science 473, Springer-Verlag 1991, pages 458-464.
6. Damgård, I. Practical and probably secure release of a secret and exchange of signatures, Advances in Cryptology — Proceedings of EUROCRYPT 93 (T. Helleseth ed.), Lecture Notes in Computer Science 765, Springer-Verlag 1994, pages 200–217.
7. Fujisaki, E., Okamoto, T. Pointcheval, D. and Stern, J. RSA-OAEP is Secure under the RSA Assumption, To appear in Advances in Cryptology: Proceedings of CRYPTO 01, Springer-Verlag 2001.
8. Galbraith, S., Mao, W. and Paterson, K. RSA-based undeniable signatures for general moduli, to appear in the 2002 RSA Conference, Cryptographers' Track, February 2002.
9. Goldreich, O., Micali, S. and Wigderson, A. How to prove all NP statements in zero-knowledge and a methodology of cryptographic protocol design, Advances in Cryptology — Proceedings of CRYPTO 86 (A.M. Odlyzko ed.), Lecture Notes in Computer Science, Springer-Verlag 263 (1987), pages 171–185.
10. Rivest, R.L. Description of the LCS35 Time Capsule Crypto-Puzzle, http://www.lcs.mit.edu/about/tcapintro041299, April 4th, 1999.
11. Rivest, R.L., Shamir, A. and Wagner, D.A. Time-lock puzzles and timed-release crypto, Manuscript. Available at ⟨http://theory.lcs.mit.edu/~rivest/RivestShamirWagner-timelock.ps⟩.
12. van de Graaf, J. and Peralta, R. A simple and secure way to show that validity of your public key, (C. Pomerance ed.), *CRYPTO '87*, Springer LNCS 293, (1988) 128–134.
13. van Oorschot, P.C. and Wiener, M.J. Parallel collision search with cryptanalytic applications, *J. of Cryptology*, Vol.12, No.1 (1999), pages 1–28.

Author Index

Lecture Notes in Computer Science

For information about Vols. 1–2175
please contact your bookseller or Springer-Verlag

Vol. 2213: M.J. van Sinderen, L.J.M. Nieuwenhuis (Eds.), Protocols for Multimedia Systems. Proceedings, 2001. XII, 239 pages. 2001.

Vol. 2214: O. Boldt, H. Jürgensen (Eds.), Automata Implementation. Proceedings, 1999. VIII, 183 pages. 2001.

Vol. 2215: N. Kobayashi, B.C. Pierce (Eds.), Theoretical Aspects of Computer Software. Proceedings, 2001. XV, 561 pages. 2001.

Vol. 2216: E.S. Al-Shaer, G. Pacifici (Eds.), Management of Multimedia on the Internet. Proceedings, 2001. XIV, 373 pages. 2001.

Vol. 2217: T. Gomi (Ed.), Evolutionary Robotics. Proceedings, 2001. XI, 139 pages. 2001.

Vol. 2218: R. Guerraoui (Ed.), Middleware 2001. Proceedings, 2001. XIII, 395 pages. 2001.

Vol. 2219: S.T. Taft, R.A. Duff, R.L. Brukardt, E. Ploedereder (Eds.), Consolidated Ada Reference Manual. XXV, 560 pages. 2001.

Vol. 2220: C. Johnson (Ed.), Interactive Systems. Proceedings, 2001. XII, 219 pages. 2001.

Vol. 2221: D.G. Feitelson, L. Rudolph (Eds.), Job Scheduling Strategies for Parallel Processing. Proceedings, 2001. VII, 207 pages. 2001.

Vol. 2223: P. Eades, T. Takaoka (Eds.), Algorithms and Computation. Proceedings, 2001. XIV, 780 pages. 2001.

Vol. 2224: H.S. Kunii, S. Jajodia, A. Sølvberg (Eds.), Conceptual Modeling – ER 2001. Proceedings, 2001. XIX, 614 pages. 2001.

Vol. 2225: N. Abe, R. Khardon, T. Zeugmann (Eds.), Algorithmic Learning Theory. Proceedings, 2001. XI, 379 pages. 2001. (Subseries LNAI).

Vol. 2226: K.P. Jantke, A. Shinohara (Eds.), Discovery Science. Proceedings, 2001. XII, 494 pages. 2001. (Subseries LNAI).

Vol. 2227: S. Boztaş, I.E. Shparlinski (Eds.), Applied Algebra, Algebraic Algorithms and Error-Correcting Codes. Proceedings, 2001. XII, 398 pages. 2001.

Vol. 2228: B. Monien, V.K. Prasanna, S. Vajapeyam (Eds.), High Performance Computing – HiPC 2001. Proceedings, 2001. XVIII, 438 pages. 2001.

Vol. 2229: S. Qing, T. Okamoto, J. Zhou (Eds.), Information and Communications Security. Proceedings, 2001. XIV, 504 pages. 2001.

Vol. 2230: T. Katila, I.E. Magnin, P. Clarysse, J. Montagnat, J. Nenonen (Eds.), Functional Imaging and Modeling of the Heart. Proceedings, 2001. XI, 158 pages. 2001.

Vol. 2232: L. Fiege, G. Mühl, U. Wilhelm (Eds.), Electronic Commerce. Proceedings, 2001. X, 233 pages. 2001.

Vol. 2233: J. Crowcroft, M. Hofmann (Eds.), Networked Group Communication. Proceedings, 2001. X, 205 pages. 2001.

Vol. 2234: L. Pacholski, P. Ružička (Eds.), SOFSEM 2001: Theory and Practice of Informatics. Proceedings, 2001. XI, 347 pages. 2001.

Vol. 2235: C.S. Calude, G. Păun, G. Rozenberg, A. Salomaa (Eds.), Multiset Processing. VIII, 359 pages. 2001.

Vol. 2237: P. Codognet (Ed.), Logic Programming. Proceedings, 2001. XI, 365 pages. 2001.

Vol. 2239: T. Walsh (Ed.), Principles and Practice of Constraint Programming – CP 2001. Proceedings, 2001. XIV, 788 pages. 2001.

Vol. 2240: G.P. Picco (Ed.), Mobile Agents. Proceedings, 2001. XIII, 277 pages. 2001.

Vol. 2241: M. Jünger, D. Naddef (Eds.), Computational Combinatorial Optimization. IX, 305 pages. 2001.

Vol. 2242: C.A. Lee (Ed.), Grid Computing – GRID 2001. Proceedings, 2001. XII, 185 pages. 2001.

Vol. 2244: D. Bjørner, M. Broy, A.V. Zamulin (Eds.), Perspectives of System Informatics. Proceedings, 2001. XIII, 548 pages. 2001.

Vol. 2245: R. Hariharan, M. Mukund, V. Vinay (Eds.), FST TCS 2001: Foundations of Software Technology and Theoretical Computer Science. Proceedings, 2001. XI, 347 pages. 2001.

Vol. 2246: R. Falcone, M. Singh, Y.-H. Tan (Eds.), Trust in Cyber-societies. VIII, 195 pages. 2001. (Subseries LNAI).

Vol. 2247: C. P. Rangan, C. Ding (Eds.), Progress in Cryptology – INDOCRYPT 2001. Proceedings, 2001. XIII, 351 pages. 2001.

Vol. 2248: C. Boyd (Ed.), Advances in Cryptology – ASIACRYPT 2001. Proceedings, 2001. XI, 603 pages. 2001.

Vol. 2249: K. Nagi, Transactional Agents. XVI, 205 pages. 2001.

Vol. 2250: R. Nieuwenhuis, A. Voronkov (Eds.), Logic for Programming, Artificial Intelligence, and Reasoning. Proceedings, 2001. XV, 738 pages. 2001. (Subseries LNAI).

Vol. 2251: Y.Y. Tang, V. Wickerhauser, P.C. Yuen, C.Li (Eds.), Wavelet Analysis and Its Applications. Proceedings, 2001. XIII, 450 pages. 2001.

Vol. 2252: J. Liu, P.C. Yuen, C. Li, J. Ng, T. Ishida (Eds.), Active Media Technology. Proceedings, 2001. XII, 402 pages. 2001.

Vol. 2253: T. Terano, T. Nishida, A. Namatame, S. Tsumoto, Y. Ohsawa, T. Washio (Eds.), New Frontiers in Artificial Intelligence. Proceedings, 2001. XXVII, 553 pages. 2001. (Subseries LNAI).

Vol. 2254: M.R. Little, L. Nigay (Eds.), Engineering for Human-Computer Interaction. Proceedings, 2001. XI, 359 pages. 2001.

Vol. 2256: M. Stumptner, D. Corbett, M. Brooks (Eds.), AI 2001: Advances in Artificial Intelligence. Proceedings, 2001. XII, 666 pages. 2001. (Subseries LNAI).

Vol. 2258: P. Brazdil, A. Jorge (Eds.), Progress in Artificial Intelligence. Proceedings, 2001. XII, 418 pages. 2001. (Subseries LNAI).

Vol. 2259: S. Vaudenay, A.M. Youssef (Eds.), Selected Areas in Cryptography. Proceedings, 2001. XI, 359 pages. 2001.

Vol. 2260: B. Honary (Ed.), Cryptography and Coding. Proceedings, 2001. IX, 416 pages. 2001.

Vol. 2264: K. Steinhöfel (Ed.), Stochastic Algorithms: Foundations and Applications. Proceedings, 2001. VIII, 203 pages. 2001.